Research, Fabrication and Applications of Bi-2223 HTS Wires

World Scientific Series in Applications of Superconductivity and Related Phenomena

ISSN 2424-8533

Series Editor: Guy Deutscher *(Tel-Aviv University, Israel)*

Published

Vol. 1 Research, Fabrication and Applications of Bi-2223 HTS Wires
edited by Kenichi Sato

Forthcoming

MgB_2 Superconducting Wires: Basics and Applications
edited by René Flükiger

Editor-in-Chief
Guy Deutscher

World Scientific Series in Applications of Superconductivity and Related Phenomena
VOL **1**

Research, Fabrication and Applications of Bi-2223 HTS Wires

Edited by
Kenichi Sato
Sumitomo Electric Industries Ltd, Japan

World Scientific

NEW JERSEY • LONDON • SINGAPORE • BEIJING • SHANGHAI • HONG KONG • TAIPEI • CHENNAI • TOKYO

Published by

World Scientific Publishing Co. Pte. Ltd.
5 Toh Tuck Link, Singapore 596224
USA office: 27 Warren Street, Suite 401-402, Hackensack, NJ 07601
UK office: 57 Shelton Street, Covent Garden, London WC2H 9HE

British Library Cataloguing-in-Publication Data
A catalogue record for this book is available from the British Library.

World Scientific Series in Applications of Superconductivity and Related Phenomena — Vol. 1
RESEARCH, FABRICATION AND APPLICATIONS OF BI-2223 HTS WIRES

ISBN 978-981-4749-25-1

In-house Editor: Song Yu

Typeset by Stallion Press
Email: enquiries@stallionpress.com

Printed in Singapore

This book is dedicated to my wife Motoko, daughter Yuri, and son Takaaki.

Preface

Prof. Guy Deutscher serves as scientific Editor-in-Chief of a new series of "Applications of Superconductivity". He wishes to acknowledge the help of the Executive Committee of the IEA Agreement on Superconductivity in preparation of this Series. This book is focusing on Bi-2223. Bi–Sr–Ca–Cu–O oxide superconducting material (BSCCO) was discovered on Christmas Eve, 24th December, 1987 by Dr. Hitoshi Maeda, *et al.* There are three compounds in the BSCCO system, and Bi-2223 ($Bi_2Sr_2Ca_2Cu_3O_{10}$) has the highest critical temperature of around 110 K. Bi-2223 has many features; not only high critical temperature but also non-rare earth elements, and well aligned crystals through mechanical deformation.

Around 28 years have passed since the discovery of BSCCO. There are so many research and development works on Bi-2223 superconducting wires from fundamental aspects and the fabrication process to applications such as current leads, power cables, magnets, and motors. Especially, there are many daily operating apparatus incorporated with Bi-2223 superconducting wires due to their electro-magnetic, mechanical and thermal performance, and industrial productivity for long length wires with an affordable economic point of view.

The purpose of this book is to cover all aspects of Bi-2223 superconducting wires from fundamental research, and the fabrication process to applications. This book contains about 40 chapters written by distinguished experts in the world. Bi-2223 superconducting wires have possibilities to realize much higher performance than those of today. I really hope that this book could contribute to the future progress of oxide superconducting wires, including Bi-2223.

Finally, I invited Dr. Hitoshi Maeda to write an invited preface for this book, and he accepted my invitation. Unfortunately, he passed away on May 24th, 2014. My invitation could not be realized. For his memory, I would like to show his handwritten memorandum on his discovery of BSCCO which he wrote on November 20th, 2003 on his visit to Tsinghua University (Courtesy of Professor Zhenghe Han).

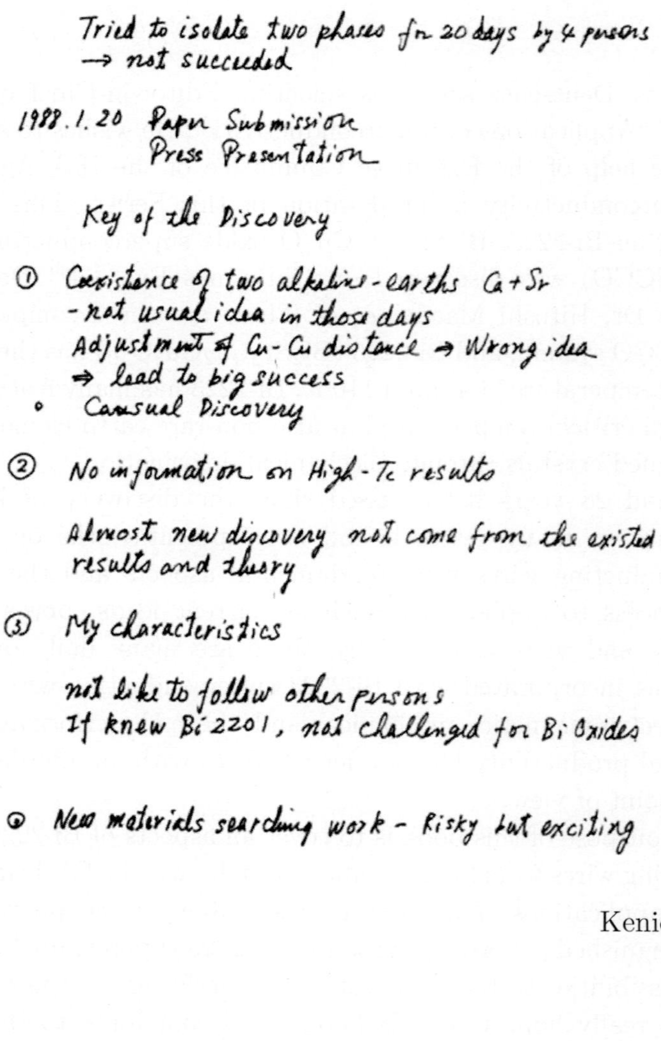

Tried to isolate two phases for 20 days by 4 persons
→ not succeeded

1988.1.20 Paper Submission
Press Presentation

Key of the Discovery

① Coexistance of two alkaline-earths Ca+Sr
- not usual idea in those days
Adjustment of Cu-Cu distance ⇒ Wrong idea
⇒ lead to big success
∘ Casual Discovery

② No information on High-Tc results

Almost new discovery not come from the existed results and theory

③ My characteristics

not like to follow other persons
If knew Bi 2201, not challenged for Bi Oxides

∘ New materials searching work - Risky but exciting

Kenichi Sato
Editor

Table of Contents

Preface vii

Part 1 Research 1

1.1. Materials Aspects of Bi-2223 3

1.2. Oxygen Doping of Bi-2223 Wires 17

1.3. Magnetic Field and Temperature Dependence of Critical Currents of Bi-2223 Wires 29

1.4. Electro-Magnetic Properties of Bi-2223 Wires 39

1.5. AC Loss of Bi-2223 Wires 49

1.6. Mechanical Properties of Bi-2223 Wires 61

1.7. Thermal Properties of Bi-2223 Wires 73

1.8. Thermal Stability of Bi-2223 Wires 105

1.9. Thin Films of Bi-2223 123

Part 2 Fabrication 135

2.1. Development and Manufacture of Bi-2223 Wires 137

2.2. Microstructure Evolution of Bi-2223 Wires 151

2.3. Lamination of Metals onto Bi-2223/Ag Wires 165

2.4. Quality Assurance of Bi-2223 Wires 181

2.5. Silver Contribution to Bi-2223 Wires in Terms of Cost 199

Part 3 Applications 211

3.1. Current Leads for ITER 213

3.2. Current Leads for Conduction-Cooled Magnets 225

3.3. Saturated Iron Core Type Fault Current Limiter 237

3.4. Resistive-Type Fault Current Limiter 249

3.5. AmpaCity Project — World's First Superconducting Cable and Fault Current Limiter Installation in a German City Center 263

3.6. AC Cable: Yokohama Project 279

3.7. Research and Development of Bi-2223-Based AC Power Cables in Russia 289

3.8. DC Cable for Data Center 301

3.9. DC Cables for Metropolitan Networks: St. Petersburg Project 315

3.10. The Development and Demonstration of a 360 m/10 kA HTS DC Power Cable 331

3.11. DC cable for railway 339

3.12. Economical Aspects of Superconducting Cable 347

3.13. MRI Magnet for Human Brain 357

3.14. LTS/Bi-2223 NMR Magnets Operated Beyond 23.5 T (1 GHz) 367

3.15. Compact HTS Beamline and Magnetic Resonance Magnets 379

3.16. Conduction-Cooled Magnets for Industrial Applications 403

3.17. Dipole Magnet for Beam Line Switching 415

3.18. Ship Propulsion Motor Employing Bi-2223 and MgB_2 Superconductors 427

3.19. Axial-Gap Superconducting Synchronous Motors Cooled by Liquid Nitrogen 451

3.20. Radial-Gap Motor for Ship Propulsion 463

3.21. Induction/Synchronous Motor for Automobile 473

3.22. Development of High-Temperature Superconducting DC Motor for Automobiles 485

PART 1
Research

Chapter 1.1

Materials Aspects of Bi-2223

Jun-ichi Shimoyama

Department of Physics and Mathematics, Aoyama Gakuin University, 5-10-1 Fuchinobe, Chuo-ku, Sagamihara 252-5258, Japan

shimo@phys.aoyama.ac.jp

Among a large number of cuprate superconductors, Bi-2223 is one of the well-developed one as superconducting tapes because of its high T_c, chemical stability and many other reasons. In this chapter, characteristic features of Bi-2223 are summarized and its superconducting properties as practical materials are compared with other superconductors. Furthermore, potentials of Bi-2223 materials are discussed from a viewpoint of controlling chemical composition.

1. Introduction

The successive discoveries of cuprate superconductors with high critical temperature T_c up to 135 K in 1986–1993 had opened possibility of applications of superconducting technologies at high temperatures, such as the boiling point of liquid nitrogen 77 K. Although metallic superconductors, such as Nb–Ti, Nb_3Sn, and Nb, have been extensively used as wires, tapes, films, and other materials thus far, they are applicable only at very low temperatures, liquid helium temperature 4.2 K or lower, due to their low T_c, which limited application fields of superconducting technologies. In particular, robust cryostats equipped with high performance heat-insulating layers have been always indispensable. Cuprate superconductors, however, have layered crystal structure composed of superconducting and blocking (=non-superconducting) layers, resulting in various

Table 1. High-T_c cuprate superconductors, which can be candidate materials applicable at 77 K.

System	Chemical Formula	Abbr.	$T_{c(max)}$ [K]	γ	Crystal Shape
RE-based	$REBa_2Cu_3O_y$	RE123	96	~7	block
	$RE_2Ba_4Cu_7O_y$	RE247	95		block, plate
	$(RE,Ca)_2Ba_4Cu_8O_{16}$	RE124	90		block
Bi-based	$Bi_2Sr_2Ca_{n-1}Cu_nO_y$	Bi22$(n-1)n$	110	~100	thin plate
	$(Bi,Pb)_2Sr_2Ca_{n-1}Cu_nO_y$		116	~50	thin plate
Tl-based	$TlBa_2Ca_{n-1}Cu_nO_y$	Tl12$(n-1)n$	132		plate
	$Tl_2Ba_2Ca_{n-1}Cu_nO_y$	Tl22$(n-1)n$	127	~80	plate
Hg-based	$HgBa_2Ca_{n-1}Cu_nO_y$	Hg12$(n-1)n$	135	~80	block
	$(Hg,Re)Ba_2Ca_{n-1}Cu_nO_y$		135	~30	block

*γ values are typical values at carrier optimally-doped state.

anisotropic properties. Characteristic features of superconducting cuprates, which can be synthesized as sintered bulks by solid-state reaction under ambient pressure and/or below 1 MPa, with higher T_c than 90 K are listed in Table 1.

For practical applications at higher temperature, Hg- and Tl-based superconductors are attractive because of their high T_cs. In fact, developments of superconducting materials had been attempted for these compounds for a decade after their discoveries. However, studies for developing these materials are almost stopped at the present stage, because any advantageous points were not found in critical current properties of their polycrystalline materials compared to those of RE- and Bi-based superconductors. This is partly due to containing highly volatile components at synthesis temperatures, Hg- or Tl-based superconductor, which is considered to deteriorate grain coupling, and poor cleavability, that will be mentioned later. Although, the Bi-based superconductors also contain volatile components, Bi and/or Pb, their equilibrium vapor pressures at synthesis temperature are much lower than those of Hg- and Tl-based superconductors. In addition, the Bi-based superconductors do not contain Ba, while it is included in other high-T_c compounds. Impurity phases and grain boundaries of cuprate superconductors containing Ba as a constituent element are quite sensitive to moisture and carbon dioxide in air even at room temperature, leading to partial decomposition of superconducting phase. In particular, degradation of superconducting grain coupling is fatal for current carrying applications. From this viewpoint, the Bi-based superconductors are desirable for achieving strong grain coupling by the conventional synthesis process for ceramic materials.

2. General Features of Bi-based Superconductors

2.1. *Crystal structures of Bi-based superconductors and related physical properties*

The ideal chemical compositions of a homologous series of Bi-based superconductors are expressed as $Bi_2Sr_2Ca_{n-1}Cu_nO_y$ with $n = 1, 2, 3, \ldots$ and these are called as "Bi22$(n-1)n$". Bi-2201 discovered in 1987[1,2] got less attention due to its low T_c, ~22 K, however, discoveries of Bi-2212 and Bi-2223[3] had triggered eager studies for development of these materials. Crystal structures of Bi-2212 ($n = 2$) and Bi-2223 ($n = 3$) are shown in Fig. 1. As clearly seen, n corresponds to the number of CuO_2 planes, which is responsible for high-T_c superconductivity, in their half unit cell. Similar to the other layered cuprate superconductors, crystal structures of the Bi-based superconductors can be regarded as alternate stacking of superconducting layers and blocking layers as illustrated in the right side of Fig. 1. In general, the blocking layer has an important role to appearance of superconductivity by providing sufficient carriers to the CuO_2 plane. In the case of the Bi-based superconductors, excess oxygen is located in between double BiO planes. Since the amount of the excess oxygen continuously varies as a function of temperature and partial pressure of oxygen[4,5], carrier concentration of the Bi-based superconductors is controlled by post-annealing and/or cooling conditions. Such non-stoichiometric oxygen composition is the common feature of the layered cuprates.

The blocking layer of the Bi-based superconductor consists of a stacking of SrO–BiO–BiO–SrO planes and its thickness is approximately 1.2 nm, which is longer than other cuprate superconductors. In addition, the blocking layer of Bi-based superconductor is almost

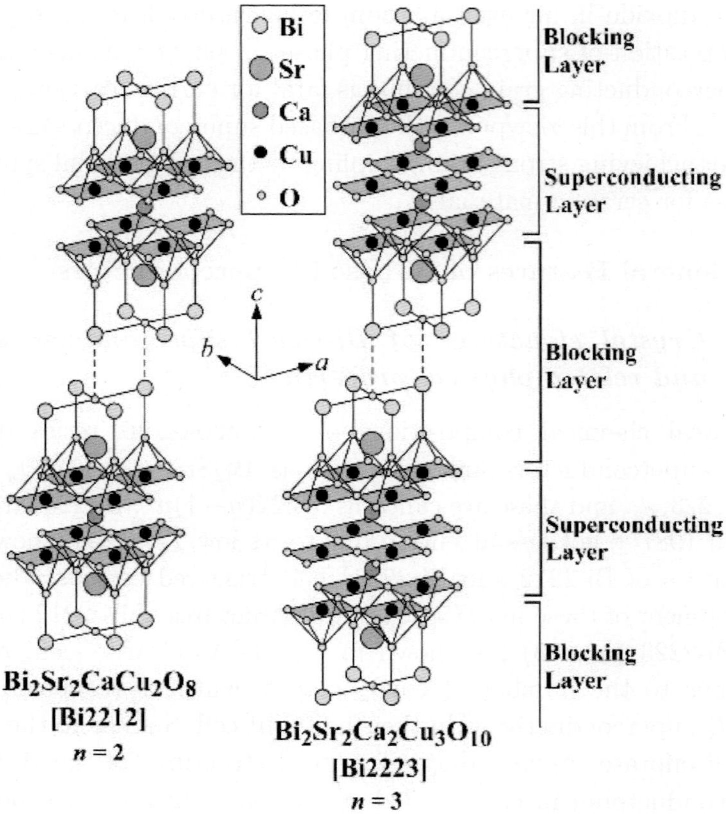

Fig. 1. Crystal structures of Bi-2212 and Bi-2223.

insulating, resulting in quite large electromagnetic anisotropy in both normal and superconducting states. The electromagnetic anisotropy parameter, γ, is defined as $(m_c*/m_{ab}*)^{1/2} \sim H_{c2(H//c)}/H_{c2(H//ab)} = \xi_{ab}^2/\xi_c^2$($H_{c2}$: upper critical field, ξ: superconducting coherence length), which is roughly estimated to be 100 for the Bi-based superconductors as shown in Table 1. Although γ can be controlled by changing carrier doping levels and element substitution, γ of the Bi-based superconductors are intrinsically much larger than that of famous cuprate superconductors $REBa_2Cu_3O_y$ (RE123: RE=rare earth elements) with γ ~7 at carrier optimally-doped state. ξ in the *ab*-plane direction, ξ_{ab}, at far below T_c of Bi-based superconductor is estimated to be ~2 nm, which is a typical value of cuprate superconductors. However, ξ along the c-axis direction, ξ_c, is less than 0.02 nm. Since this value is much shorter than the thickness of the blocking layer, coupling of superconducting layer separated by the blocking layer is quite weak. Furthermore, pinning volume proportional to $\xi_{ab}^2\xi_c$ under $H//c$ is much smaller in the Bi-based superconductors than that of RE123. This means that pinning characteristics, i.e. critical current properties under magnetic field is inherently poor in the Bi-based superconductors. Such large anisotropy of Bi-based superconductors also leads to very large difference in J_c between the ab-plane and the c-axis directions. The ratio $J_c(//ab)/J_c(//c)$ is considered to be ~100, meaning strongly c-axis orientated microstructure along the current carrying direction must be formed in the polycrystalline materials for achieving high J_c performance.

Other important structural features of the Bi-based superconductor are weak bond between BiO double planes and their modulations along the b-axis. The former feature enables us to cleave crystals of Bi-based superconductors along the ab-plane like mica and latter one result in wavy crystal structure including the CuO_2 planes. Position of Cu varies ~0.1 nm along the c-axis direction due to the modulation.

T_cs of undoped Bi-2201, Bi-2212, and Bi-2223 are ~22, ~95, and ~110 K, respectively, at each carrier optimally-doped state. Compounds with $n > 3$ can be synthesized by thin film deposition method, however their T_cs are lower than 80 K. As clearly, Bi-2223 is the most promising compound for applications at high temperatures, however, phase formation conditions of Bi-2223 quite narrow and no one has ever succeeded in the synthesis of Bi-2223 sintered bulks with almost single phase thus far. Note that the highest T_c of Tl- and Hg-based superconductors also observed in compounds with $n = 3$. Flatness of inner CuO_2 plane is believed to desirable for high T_c and decrease in T_c for $n > 3$ is explained by carrier underdoping.

2.2. *Pb-doping effects for Bi-based superconductors*

For Bi-based superconductors, Pb is the most popular dopant to control superconducting properties and/or to promote phase formation. Since Bi and Pb ions are trivalent and divalent, respectively, in the Bi-based superconductors, Pb doping to the Bi site tends to accompany hole doping. Furthermore, lead doping expands structural modulation period. For Bi-2212, the modulation period is 4.8 b for lead-free composition and that is ~10 b for ~15% lead doped composition. Further lead doping causes phase separation with modulation-free lead-rich region (>20%) and modulated region (~16%)[6]. The expansion of modulation period corresponds to flattening of local crystal structure. A dramatic effect of Pb-doping to the Bi-based superconductor is large reduction of electromagnetic anisotropy. For example, γ of Bi-2212 single crystal is ~90 at carrier overdoped state, while that of Pb-doped Bi-2212 single crystal is ~40[7,8]. A decrease in γ corresponds to an increase in pinning strength through an increase in pinning volume and enhancement of J_c along the c-axis. Enhancement of pinning strength by Pb-doping also accompanies large increases in J_c under magnetic field and irreversibility field H_{irr}[9,10]. Therefore, Pb-doping is quite effective for improving intrinsic critical current properties of the Bi-based superconductors.

In the synthesis of Bi-2223, Pb-doping is indispensable to obtain sintered bulks[11,12] or tapes with almost single phase of Bi-2223[13]. The formation mechanism of Bi-2223 from a precursor containing Bi-2212 as a main phase is considered to be varied somehow by coexistence of Pb. Flattening of the crystal structure by Pb-doping is considered to be partially contributed to form the Bi-2223 phase. However, phase formation of Bi-2223 is quite sluggish and sintering for a long time over 50 h even at optimized temperature under a moderate atmosphere with partial pressure of oxygen p_{O2} ~8 kPa, while other cuprate superconductors including Bi-2201 and Bi-2212 can be prepared within 10 h at the optimized synthesis conditions in general.

2.3. *Dependence of superconducting properties of Bi-based superconductors on their non-stoichiometric chemical composition*

As mentioned before, Bi-based superconductors have oxygen non-stoichiometry, which changes carrier concentration. Although the

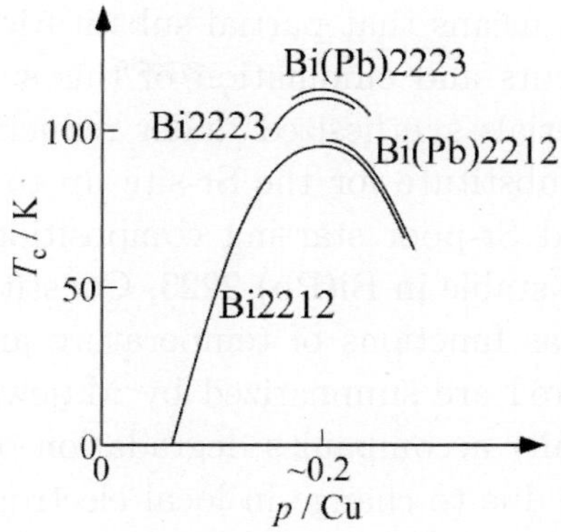

Fig. 2. Schematic illustration showing the relationship between T_c and p/Cu.

compositional range of the oxygen is much smaller than RE123, superconducting properties of Bi-based superconductors are sensitive to the oxygen content. Figure 2 shows schematic illustration of the relationship between T_c and hole concentration per Cu for Bi-based superconductors, Bi-2212, Bi-2223, Bi(Pb)-2212, and Bi(Pb)-2223. In cuprate superconductors, carrier lightly overdoped state is generally desirable to enhance critical current properties in magnetic field due to high T_c and reduced γ. For Bi-2212, materials of carrier overdoped state with $T_c \sim 80\,K$ are usually obtained when they are prepared in air. In the case of Bi(Pb)-2223, it is quite difficult to synthesize materials, such as tapes and sintered bulks, with $T_c < 108\,K$. This means that the obtained Bi(Pb)-2223 materials always have moderate carrier doping state with nearly optimal even for Ag-sheathed tapes. However, precise control of oxygen content is crucially important even if the change of T_c is small, because γ monotonically decreases with an increase in carrier concentration, i.e. oxygen content. Importance of oxygen content control of Bi(Pb)-2223 tapes are described in another chapter.

Large cation non-stoichiometry is also an important characteristics of Bi-based superconductors. Materials synthesized in atmospheres containing relatively high partial oxygen pressure (>several kPa) usually have Bi-rich and Sr-poor composition, such as $Bi_{2.1}Sr_{1.9}CaCu_2O_y$, even when the starting composition is the integral ratio. This means that partial substitution of Bi for Sr-site unintentionally occurs and elimination of this substitution is difficult even the materials synthesized under reducing atmosphere. In addition, Ca can substitute for the Sr-site up to ~40% in Bi-2212 from a Ca-rich and Sr-poor starting composition. Similar Ca-rich composition is also stable in Bi(Pb)-2223. Constituent phases of Bi-2212 and Bi-2223 as functions of temperature and nominal cation composition prepared are summarized by Majewski[14]. Such cation substitution generally accompanies degradation of superconducting properties possibly due to change in local electronic state and local lattice distortion. The degradation is more pronounced in critical current properties, suggesting a decrease in superconducting condensation energy, which affects the pinning force. Figure 3 shows J_c, H_{irr}, maximum pinning force density $F_{p(max)}$ at 20 K under $H//c$, and T_c of Bi-2212 single crystals with carrier lightly overdoped state having various cation compositions. A crystal with $c \sim 3.089$ nm has a cation composition close to Bi:Sr:Ca:Cu = 2:2:1:2, which was grown under a reducing atmosphere with an oxygen partial pressure of 1 kPa. This crystal apparently show largely improved critical current properties

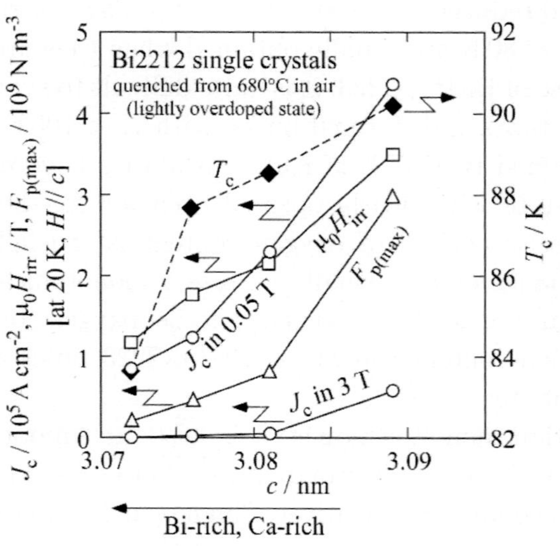

Fig. 3. J_c, H_{irr}, $F_{p(max)}$ at 20 K under $H//c$ and T_c of Bi-2212 single crystals.

compared to the crystals having shorter *c*-axis length due to Bi-rich and/or Ca-rich composition[15]. In the case of Bi(Pb)-2223, very few attentions have been paid for a control of cation composition, because synthesis of materials with almost single phase needs much attentions. Similarly adjusting cation composition close to the integral ratio, (Bi;Pb):Sr:Ca:Cu=2:2:2:3, is considered to be a promising way to improve critical current properties of Bi(Pb)-2223 materials.

3. Advantageous Points of Bi(Pb)-2223

Thus far, Bi(Pb)-2223 materials have been developed as Ag-sheathed multi-filamentary tapes fabricated by two-step sintering with an intermediate pressing by metallic rolling. The oxide core after re-sintering is composed of plate-like Bi(Pb)-2223 crystals with several tens μm in size and they are strongly oriented with their *c*-axes vertical to the tape surface. The *c*-axis oriented microstructure of Bi(Pb)-2223 crystals are formed by two origins. One is the *c*-axis oriented crystal growth from the flat interface between Ag and the oxide core, and the other is mechanical orientation at the middle part of the oxide core by applying metallic rolling after phase formation of Bi(Pb)-2223. The rolling process is also effective for an increase in density of oxide core, because large pores generate during the phase formation of wide plate-like crystals of Bi(Pb)-2223. After the rolling process, re-sintering process is indispensable for recovering micro-cracks formed in the oxide core by the rolling process. However, Bi(Pb)-2223 crystals grow again during the re-sintering process, resulting in degradation of *c*-axis orientation and generation of pores accompanying expansion of oxide cores. The overpressure sintering[16,17] suppresses the generation of pores and results in pore-free microstructure. In addition, homogeneities of phase purity and grain orientation in the oxide core are largely improved by applying the overpressure sintering, leading to high homogeneity of superconducting properties. In particular, fluctuation of critical current properties are dramatically reduced in the long tapes produced in the same batch. Improvement of homogeneity of the tapes enabled us to perform systematic studies for optimizing synthesis conditions.

Ag-sheathed Bi(Pb)-2223 long tapes with cross-sectional area of $\sim 4\,\mathrm{mm}^w \times \sim 0.25\,\mathrm{mm}^t$ are currently produced as industrial scale at Sumitomo Electric Industries, Ltd. High $I_c \sim 200$ A at 77 K in self-field[18] has already achieved in commercial tapes and they have been extensively used for power cables, current leads, and various kinds of magnets. The I_c value of ~ 200 A corresponds to $J_c \sim 60\,\mathrm{kA\ cm^{-2}}$ for the Bi(Pb)-2223 oxide core and engineering critical current density $J_e \sim 20$ kA cm^{-2}. This J_c value is almost two orders of magnitude lower than that of superconducting layer of RE123 tapes, which are called coated conductors, having bi-axially oriented thin RE123 film layer on the metal substrate, however, J_e for these superconducting tapes are comparable because of much larger volume fraction ($\sim 40\%$) of superconducting oxide in for Bi(Pb)-2223 tapes than that (1–2%) in RE123 tapes.

Applicable conditions for various superconducting wire or tapes with long length are summarized in Fig. 4. The applicable limit for each conductor is defined as $J_e = 10\,\mathrm{kA\ cm^{-2}}$ in this plot. It is apparent that the RE123 tape has larger applicable conditions in particular under high magnetic fields. However, Bi(Pb)-2223 tapes are the most popular high-T_c superconducting tapes even including

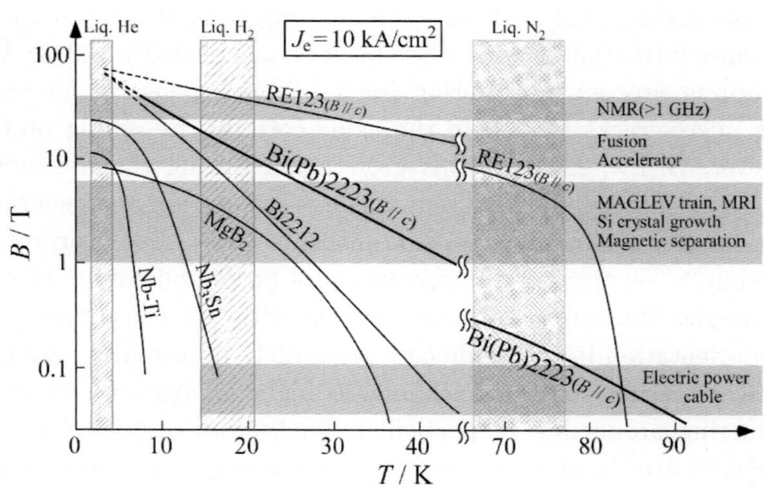

Fig. 4. Applicable conditions for superconducting wires and tapes.

high field generating magnets, because of high reliability in critical current performance, which originates in multi-filament conductors more than 100 oxide filaments in a tape, relatively simple and well established fabrication process. It was confirmed that local J_c of Bi(Pb)-2223 tape is much higher than the average J_c. This means that optimization of microstructure toward high phase purity and complete *c*-axis orientation will contribute to further improvement of I_c and J_c of the tapes.

4. Derivable Potential of Bi-2223 Materials

Intrinsic superconducting properties of Bi(Pb)-2223 will be also improved by several approaches to control the chemical composition. As described before, Bi-based superconductors have large cation non-stoichiometry and the partial substitution of cations occurs spontaneously corresponding to the synthesis conditions, temperature, and p_{O2}. In the case of Bi(Pb)-2223, narrow conditions, 830−840°C and p_{O2} ~8 kPa, for its phase formation prevented precise control of the cation composition in resulting tapes. In addition, the Bi(Pb)-2223 phase had been considered to be a meta-stable phase below the synthesis temperatures. However, recent studies revealed that the Bi(Pb)-2223 phase is stable at 600−800°C, where cation composition is controlled by applying post-annealing. For example, T_c of Bi(Pb)-2223 tape was enhanced above 116 K by addition of post-annealing process at ~700°C[19,20]. Such post-annealing process changes lattice constants of the Bi(Pb)-2223 phase, indicating changes of cation composition. When the post-annealing at ~700°C is performed in high p_{O2} atmosphere, such as in air, Pb-containing impurities generates and, hence, substitution level of Pb in Bi(Pb)-2223 decreases. This process is confirmed to have an effect on increasing hole concentration, while the amount of Pb decreases. This suggests that substitution level of Bi at the Sr-site is reduced. Since the fixed concept on T_c of Bi(Pb)-2223 was broken, synthesis conditions for commercial Bi(Pb)-2223 tapes have been reconsidered and the T_c of the current Bi(Pb)-2223 tape is ~112 K, which is approximately 3 K higher than that of former products. It should be noted that critical current properties have been also improved with an increase in T_c. Therefore, if the high T_c above 116 K would be achieved in the Bi(Pb)-2223 tapes through optimization of synthesis process, large enhancement of J_c and H_{irr} will be expected due to an increase in superconducting condensation energy as confirmed in Bi-2212 single crystals. Another promising way for improving critical current properties of Bi(Pb)-2223 tapes is increasing substitution level of Pb, which must be effective for increasing J_c along the c-axis, i.e. improvement of J_c between stacked plate-like Bi(Pb)-2223 crystals and enhancement of J_c in field and H_{irr} due to a decrease in γ. Recently, Bi(Pb)-2223 phase was found to form under relatively low p_{O2} of 2–3 kPa[21], where the solubility limit of Pb for the Bi-site increases. Sintering under such low p_{O2} reduces the optimal sintering temperature and sintering time for phase formation of Bi(Pb)-2223 within 10 h. These are preferable to suppress generation of large impurity grains.

By combining strategies on optimizations of microstructure and chemical compositions, large improvement of critical current properties of Bi(Pb)-2223 tapes can be expected, resulting in the shift of applicable limit line shown in Fig. 4 toward higher temperature and high field directions in future.

References

1. C. Michel, M. Hervieu, M. M. Borel, A. Grandin, F. Deslandes, J. Provost, and B. Raveau, Superconductivity in the Bi–Sr–Cu–O system, *Z. Phys. B* **68**, 421–423 (1987).
2. J. Akimitsu, A. Yamazaki, H. Sawa, and H. Fujiki, Superconductivity in the Bi–Sr–Cu–O system, *Jpn. J. Appl. Phys.* **26**, L2080 (1987).
3. H. Maeda, Y. Tanaka, M. Fukutomi, and T. Asano, A new high-T_c oxide superconductor without a rare earth element, *Jpn. J. Appl. Phys.* **27**, L209 (1988).
4. Y. Idemoto, S. Ichikawa, and K. Fueki, Oxygen nonstoichiometry of 2223 phase Bi–Pb–Sr–Ca–Cu–O system superconducting oxide, *Physica C* **181**, 171–178 (1991).
5. J. Shimoyama, J. Kase, T. Morimoto, J. Mizusaki, and H. Tagawa, Oxygen nonstoichiometry and phase instability of $Bi_2Sr_2CaCu_2O_{8+\delta}$, *Physica C* **185–189**, 931–932 (1991).
6. Z. Hiroi, I Chong, and M. Takano, Two-phase microstructures generating efficient pinning centers in the heavily Pb-substituted $Bi_2Sr_2CaCu_2O_{8+\delta}$ single crystals, *J. Solid State Chem.* **138**, 98 (1998).

7. T. Motohashi, Y. Nakayama, T. Fujita, K. Kitazawa, J. Shimoyama, and K. Kishio, Systematic decrease of resistivity anisotropy in $Bi_2Sr_2CaCu_2O_y$ by Pb doping, *Phys. Rev. B* **59**, 14080–14086 (1999).
8. T. Motohashi, J. Shimoyama, K. Kitazawa, K. Kishio, K. M. Kojima, S. Uchida, and S. Tajima, Observation of the Josephson plasma reflectivity edge in the infrared region in Bi-based superconducting cuprate, *Phys. Rev. B* **61**, R9269–R9272 (2000).
9. I. Chong, Z. Hiroi, M. Izumi, J. Shimoyama, Y. Nakayama, K. Kishio, T. Terashima, Y. Bando, and M. Takano, High critical current density in the heavily Pb-doped $Bi_2Sr_2CaCu_2O_{8+\delta}$ generation of novel pinning centers, *Science* **276**, 770–773 (1997).
10. J. Shimoyama, Y. Nakayama, K. Kitazawa, K. Kishio, Z. Hiroi, I. Chong, and M. Takano, Strong flux pinning up to liquid nitrogen temperature discovered in heavily Pb-doped and oxygen controlled Bi2212 single crystals, *Physica C* **281**, 69–75 (1997).
11. S. M. Green, C. Jiang, Yu Mei, H. L. Luo, and C. Politis, Zero resistance at 107 K in the (Bi,Pb)–Ca–Sr–Cu oxide system, *Phys. Rev. B* **38**, 5016 (1988).
12. M. Takano, J. Takada, K. Oda, H. Kitaguchi, Y. Miura, Y. Ikeda, Y. Tomii, and H. Mazaki, High-T_c phase promoted and stabilized in the Bi,Pb–Sr–Ca–Cu–O system, *Jpn. J. Appl. Phys.* **27**, L1041 (1988).
13. T. Hikata, K. Sato, and H. Hitotsuyanagi, Ag-sheathed Bi–Pb–Sr–Ca–Cu–O superconducting wires with high critical current density, *Jpn. J. Appl. Phys.* **28**, L82 (1989).
14. P. Majewski, Phase diagram studies in the system Bi–Pb–Sr–Ca–Cu–O–Ag, *Supercond. Sci. Technol.* **10**, 453–468 (1997).
15. Y. Yuan, J. Jiang, X. Y. Cai, S. Patnaik, A. A. Polyanskii, E. E. Hellstrom, D. C. Larbalestier, R. K. Williams, and Y. Huang, Microstructure and J_c improvements in overpressure processed Ag-sheathed Bi-2223 tapes, *IEEE Trans. Appl. Supercond.* **13**, 2921–2925 (2003).
16. S. Kobayashi, K. Yamazaki, T. Kato, K. Okura, E. Ueno, K. Fujino, J. Fujikami, N. Ayai, M. Kikuchi, K. Hayashi, K. Sato, and R. Hata, Controlled over-pressure sintering process of Bi2223 wires, *Physica C* **426–431**, 1132–1134 (2005).
17. S. Uchida, J. Shimoyama, T. Makise, S. Horii, and K. Kishio, Dramatically enhanced flux pinning properties of cation composition controlled Bi(Pb)2212 single crystals, *J. Phys. Conf. Series* **43**, 231–234 (2006).
18. N. Ayai, S. Kobayashi, K. Yamazaki, S. Yamade, M. Kikuschi, E. Ueno, J. Fujikami, T. Kato, K. Hayashi, K. Sato, R. Hata, H. Kitaguchi, H. Kumakura, and J. Shimoyama, The current transport properties of a 200 A-class Bi-2223 superconducting wire at various temperatures and magnetic fields, *IEEE Trans. Appl. Supercond.* **17**, 3113–3116 (2007).
19. J. Shimoyama, T. Kato, S. Kobayashi, K. Yamazaki, K. Hayashi, and K. Sato, Possible breakthrough for higher T_c discovered in Bi(Pb)2223 practical tapes, *Jpn. J. Appl. Phys.* **44**, L1525–1528 (2005).

20. J. Shimoyama, A. Tanimoto, T. Nakashima, T. Asanuma, S. Horii, K. Kishio, T. Kato, S. Kobayashi, K. Yamazaki, K. Hayashi, and K. Sato, Possibilities of T_c enhancement for already known HTSC compounds, *Physica C* **460–462**, 1405–1406 (2007).
21. R. Tajima, J. Shimoyama, A. Yamamoto, H. Ogino, K. Kishio, T. Nakashima, S. Kobayashi, and K. Hayashi, Synthesis of Bi(Pb)2223 tapes by low P_{O2} sintering, *IEEE Trans. Appl. Supercond.* **23**, 6400604 (2013).

Chapter 1.2

Oxygen Doping of Bi-2223 Wires

Takayoshi Nakashima

Sumitomo Electric Industries, Ltd.,
1-1-3 Shimaya Konohana-ku,
Osaka, 554-0022, Japan

The control of carrier doping level is essential for Bi-2223 wires to ensure high critical current (I_c) performance. Non-stoichiometric oxygen content is one of the most crucial factors to determine the carrier doping level. Bi-2223 wires were post-annealed in order to adjust the oxygen contents and their superconducting properties were investigated. The carrier doping level varied from under-doped to slightly over-doped states with the oxygen contents. A change of I_c limitation mechanisms in magnetic fields was confirmed at over-doped state.

1. Introduction

$(Bi,Pb)_2Sr_2Ca_2Cu_3O_{10+x}$(Bi-2223) has high superconducting transition temperature (T_c) $\sim$ 110 K, which is appealing to wire applications at 77 K such as power cable. Bi-2223 wire has also been studied in terms of coil applications. For such applications, enhancement of critical current (I_c) or critical current density (J_c) performances at lower temperatures and in magnetic fields are necessary. Carrier doping is effective way for improving flux pinning properties through a decrease of an electromagnetic anisotropy.[1] Bi-2223 has non-stoichiometries in cation compositions and oxygen content. As for the former, Pb^{2+} substitution for Bi^{3+} site can change carrier doping level; however, excess Pb addition may lead to degradation of

Bi-2223 grain boundaries with Pb-rich phases. On the other hand, the non-stoichiometric oxygen content, which is correlated with carrier doping level too, is adjustable by post-annealing without any other changes in Bi-2223 wire. In this chapter, the relationships between the non-stoichiometric oxygen and superconducting properties of our standard Bi-2223 wires[2] will be presented.

2. Carrier Doping Levels of Bi-2212 and Bi-2223

At first, we compared the carrier doping level between Bi-2212 and Bi-2223 using Bi-2223 wires. Some Bi-2212 are seen in the wires, while Bi-2212 is never confirmed in the magnetization measurement. Since the surfaces of the filaments are covered with well-aligned Bi-2223 layers, the magnetic fields are shielded by the layers. In order to measure the magnetization of Bi-2212, we need to remove them. For the purpose, the edges of the wire were cutoff and Ag sheaths of the both sides were peeled off. A photograph and scanning electron microscopy (SEM) image of the wire in peeling were shown in Fig. 1. The broken microstructure in the filaments and many Bi-2223 grains on the peeled Ag sheaths were seen.

The temperature dependences of the zero-field-cooled (ZFC) magnetizations of the original and the peeled wires were measured (Fig. 2). The field of 1 Oe was applied normal to the wire surface. The magnetization was normalized by that at 50 K. The original wire showed T_c(onset) of 110 K and superconducting transition width was ~2K. The peeled wire maintained the original T_c(onset) of 110 K and small transition width ~3K. The magnetization of Bi-2212 was not seen, though the microstructure was significantly ruined. Subsequently, we hammered the original wire to break the microstructure further without peeling. The magnetization of the hammered wire was also shown in Fig. 2. The wire exhibited T_c(onset) of 110 K, however, transition width became much broader. A cusp at around 82 K is due to the residual Bi-2212. If Bi-2212 and Bi-2223 grains exist separately in the filaments, the magnetization of Bi-2212 should be observed for the peeled wire. The comparative results with the peeled and the hammered wires indicate that Bi-2212 is encased in Bi-2223

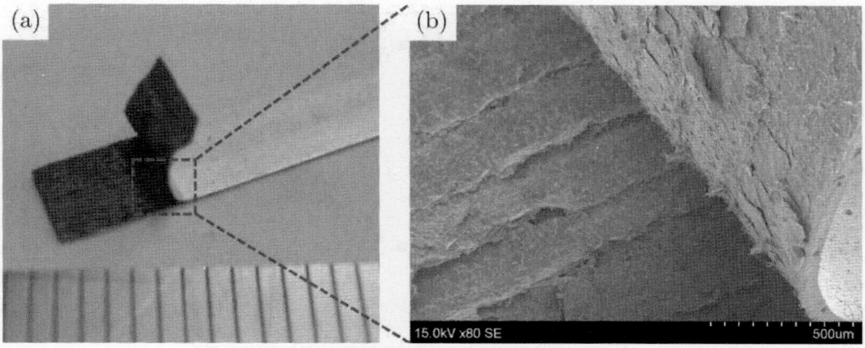

Fig. 1. Photograph of the wire in peeling (a) and the SEM image (b).

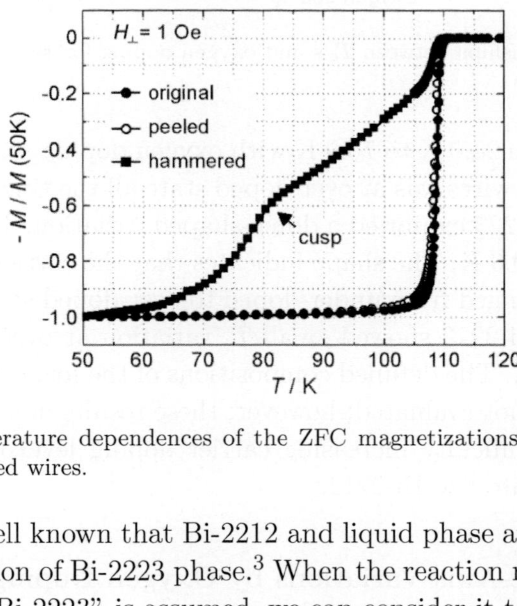

Fig. 2. Temperature dependences of the ZFC magnetizations for the original and the damaged wires.

grain. It is well known that Bi-2212 and liquid phase are required for quick formation of Bi-2223 phase.[3] When the reaction route "Bi-2212 + Liquid → Bi-2223" is assumed, we can consider it to be peritectic reaction. The magnetization results support the assumption.

The relationships between oxygen doping and T_c were investigated using the hammered wires. T_c of Bi-2212 was defined as the temperature at the cusp. The wires were post-annealed in the same manner as our previous report. The T_cs were plotted in regard to oxygen doping conditions in Fig. 3. Whether the oxygen contents were adjusted as intended will be discussed in Sec. 3. The T_cs of Bi-2212 monotonically

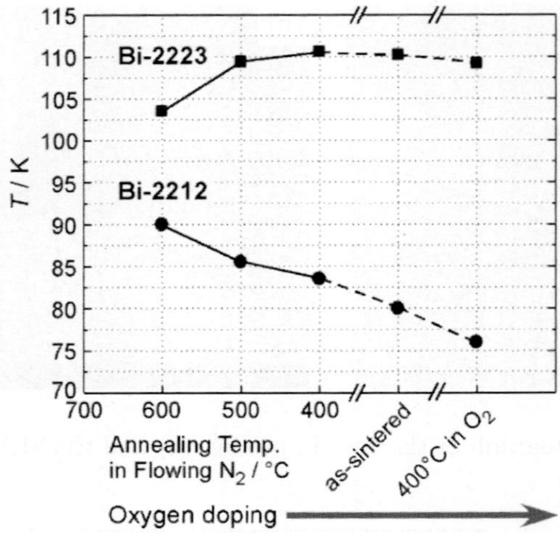

Fig. 3. Relationships between T_c's and oxygen doping for Bi-2223 and Bi-2212 in the wires.

decreased from 90.0 K to 76.0 K with oxygen doping, suggesting that Bi-2212 in the wires was at over-doped state all the time. In contrast, the T_cs of Bi-2223 exhibited a dome-shaped variation. The maximum T_c reached 110.6 K. The shape indicates that the carrier doping level of Bi-2223 changed from under-doped to over-doped states with oxygen doping. Bi-2223 showed small T_c variation at over-doped state, 109.2−110.6 K. The detailed compositions of the formed Bi-2212 and Bi-2223 were not evaluated; however, these results demonstrate that Bi-2223 has difficulty increasing carrier doping level up to heavily over-doped state like Bi-2212.

3. T_c and *c*-Axis Variations by Oxygen Doping

To confirm that the above T_c variation was caused by the changes of oxygen contents, we measured *c*-axis length of Bi-2223 after post-annealing. The *c*-axis length was calculated with 00*l* reflections in X-ray diffraction (XRD) patterns. The *c*-axis length and T_c variations with respect to oxygen doping conditions are shown in Fig. 4. In the case of Bi-based superconducting materials, excess oxygen is incorporated between Bi−O layers in their crystal structures. The oxygen

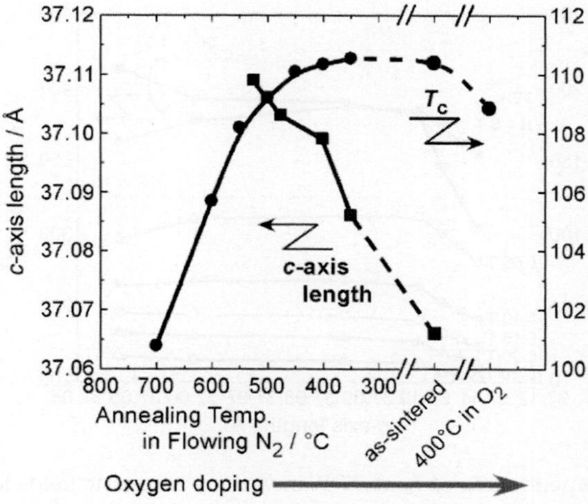

Fig. 4. Relationships between T_c's, c-axis lengths of Bi-2223 and oxygen doping. Fig. 6 in Ref. 4 is altered for this chapter.

strengthens the coupling between the layers, leading to shrinkage of c-axis. The decrease of the c-axis lengths observed in Fig. 4 demonstrates that oxygen contents were adjusted as intended. We have found that the carrier-controlled wires showed a positive relationship between I_c at 77 K in self-field and T_c regardless of whether they were at under-doped or over-doped states.[4]

4. Influence of Oxygen Doping on In-Field I_c Performance

The behaviors of carrier-controlled I_c in magnetic fields are different from that observed in self-field. Figure 5 shows the c-axis length dependences of I_c at 77 K in various magnetic fields applied normal to the wire surfaces. The dome-shaped dependences changed with magnetic fields and disappeared in over 0.15 T. In high magnetic fields, the I_cs increased with a decrease of the c-axis length. The dominant factors of I_c appears to be dependent on the applied magnetic fields.

It has been often pointed out that I_cs or J_cs of HTS wires are determined by intergrain J_c in lower magnetic fields and by intragrain J_c in higher magnetic fields.[5,6] In RE−Ba−Cu−O (RE=Rare

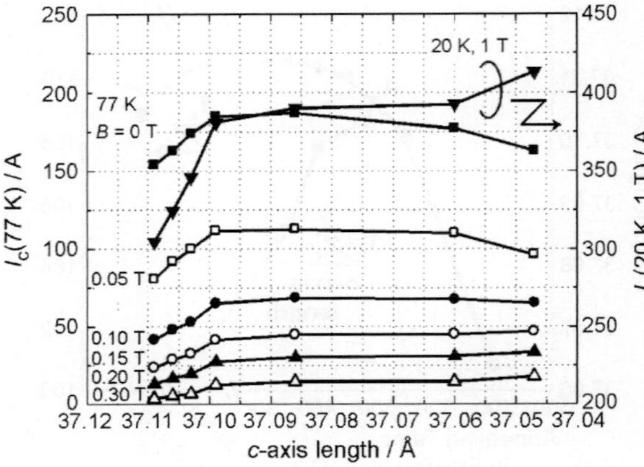

Fig. 5. c-axis dependences of I_c at 77 K in various magnetic fields for the present wire. The magnetic fields were applied normal to the wire surfaces. The data at 20 K in 1 T was also plotted.

Earth element) system, grain boundary can be arbitrarily designed with thin films, thus flux motions and J_c limitation mechanisms at the boundary have been studied.[7] According to the studies, $J-V$ curve shows the non-Ohmic linear differential behavior when intergrain J_c limits the overall J_c in low magnetic fields. And then the shape is getting close to a power-law behavior in increasing fields. The change indicates that intragrain J_c is dominant in higher fields. The discussion of switching between intergrain J_c–intragrain J_c limitations in magnetic fields is common to HTS wires, hence we checked the shapes of $I-V$ curves for Bi-2223 wires. Figure 6 shows $I-V$ curves at 77 K in various magnetic fields applied normal to the wire surfaces. The wire at over-doped state was used for this measurement. The $I-V$ curves were convex upward in low magnetic fields, and turned into straight with an increase of magnetic field. Finally, the curves were convex downward in high magnetic fields. These results were similar to those observed in RE–Ba–Cu–O system, suggesting that the dominant factors of I_c changed from intergrain J_c to intragrain J_c with increasing magnetic fields. The curvatures of the $I-V$ curves changed in 0.10–0.14 T. This field range was consistent with the dome-shaped dependences which disappeared in Fig. 5. The carrier doping generally enhances the pinning potential

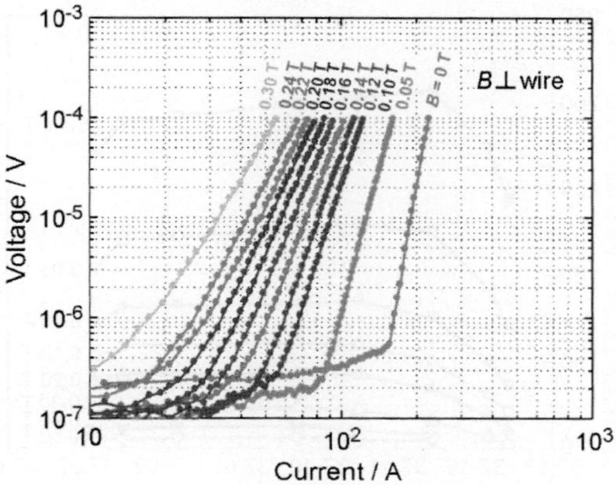

Fig. 6. $I-V$ curves measured at 77 K in various magnetic fields. The magnetic fields were applied normal to the wire surfaces.

in superconducting materials, leading to improvement of intragrain J_c. At over-doped state, the improved intragrain J_c seem to boost up I_c in over 0.10 – 0.14 T in spite of low T_c.

The *c*-axis length dependence of I_c at 20 K in 1 T was also shown in Fig. 5. The I_c's increased with a decrease of the *c*-axis length. The intragrain J_c was considered to limit I_cs in that measurement environment.

5. In-Field I_c Behaviors of the Present Wires

The results in Fig. 4 were reported for the first time in 2010. We have developed wire processing and improved I_c. In order to see if the above discussion is the case with the present wires, the control of the non-stoichiometric oxygen contents was performed. The wires were post-annealed in the same manner as before.

Figure 7 shows the *c*-axis length dependences of I_c at 77 K in various magnetic fields. The I_cs in self-field exhibited a dome-shaped dependence. The maximum I_c reached 208 Å, which was ~20 Å higher than that in Fig. 5. The dome-shaped dependences were observed in magnetic fields of up to 0.3 T. The I_cs in 0.4 and 0.5 T increased along with the decreased *c*-axis lengths. According to the discussion in Sec. 4, these results suggest that the dominant

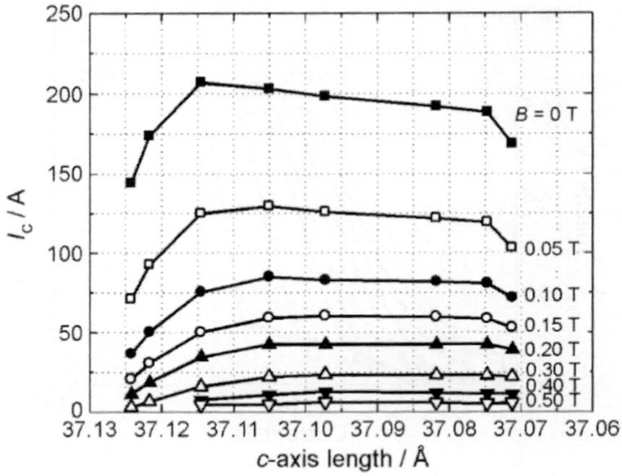

Fig. 7. c-axis dependences of I_c at 77 K in various magnetic fields for the present wire. The magnetic fields were applied normal to the wire surfaces.

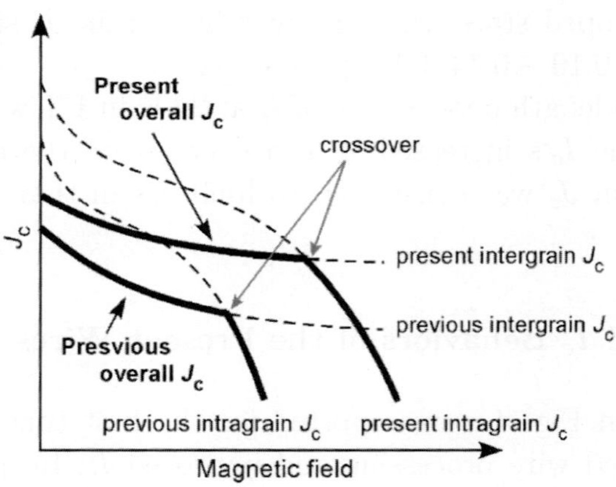

Fig. 8. Schematic image of the J_c limitations for the previous and the present wires.

factor of I_c at 77 K of the present wires changed from intergrain J_c to intragrain J_c in magnetic fields of 0.3–0.4 T. The crossover fields were higher than those of the previous wires, 0.10–0.15 T. A schematic illustration of J_c–B curves is shown in Fig. 8 to explain the enhancement in the crossover fields. Based on the report by Horide *et al.*,[8] intergrain J_c and intragrain J_c curves can be drawn separately. In the lower magnetic fields, intergrain J_c curve is dominant. On the other hand, the overall J_c is determined by intragrain J_c curve over the magnetic field where those curves are crossing. Since the present wire exhibits higher I_c at 77 K in self-field than the previous one, intergrain J_c is considered to be enhanced. If only the intergrain J_c curve is raised up, the crossover field should decrease. The intragrain J_c curve also need to move upward to increase the crossover field. Therefore, it is indicated that the enhanced I_c performances of the present wires have been brought about by both better texturing of microstructure and modification of Bi-2223 grains. The latter factor may be responsible for the difference of the *c*-axis lengths at the maximum I_c at 77 K in self-field between the previous (Fig. 5) ~37.09 Å and the present (Fig. 7) wires ~37.11 Å. We have improved the entire wire processing, thus it is difficult to specify the beneficial processes.

Figure 9 shows the *c*-axis length dependences of I_c at 20 K and 30 K in 5 T, and at 77 K in self-field. The I_cs at 20 K and 30 K in 5 T exhibited dome-shaped dependences, indicating that intergrain J_cs were still dominant. The previous wires showed the intragrain J_c-dominant behavior at 20 K in 1 T (Fig. 5), thus the crossover field was at least increased more than 4 T. At 30 K in 5 T, the dome-shaped dependence was almost collapsing. The crossover field at 30 K appears to be 6−7 T. As seen in Fig. 9, the *c*-axis lengths at the

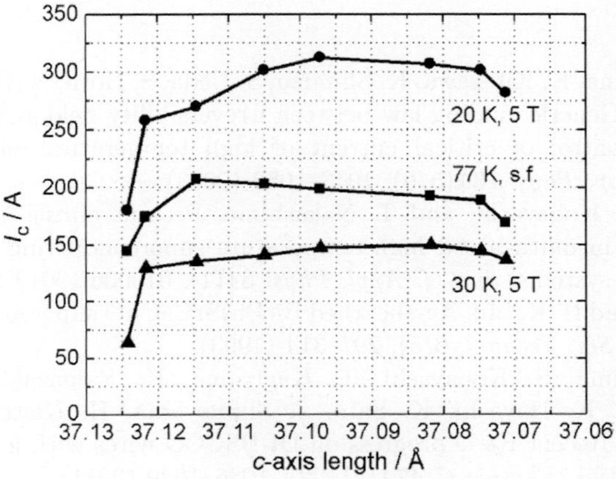

Fig. 9. *c*-axis dependences of I_c at 77 K in self-field, 30 K in 5 T, 20 K in 5 T for the present wire. The magnetic fields were applied normal to the wire surfaces.

maximum I_cs varied with the temperatures and the magnetic fields. This means that the carrier doping level should be tailored to fit the usage environment of the wires.

6. Conclusion

It was demonstrated that the carrier doping level of Bi-2223 wire easily changed from under-doped to slightly over-doped states with non-stoichiometric oxygen contents. I_c at 77 K in self-field was highly dependent on T_c and determined by intergrain J_c. For the over-doped wire, the change in the shapes of $I - V$ curves implied that I_c limitation factor tuned into intragrain J_c in higher magnetic fields. Based on these discussions, the enhanced I_c of our present wires was assumed to be due to improvement of both intragrain J_c and intergrain J_c, though we have not identified what process has benefited it.

The Bi-2223 wires have some residual Bi-2212 phases in them. Their carrier doping level is much higher than Bi-2223 matrix. At 4.2 K in high fields where intragrain J_c is dominant, Bi-2212 phases may contribute to enhancement of overall I_c performances.

References

1. J. Shimoyama, K. Kitazawa, K. Shimizu, S. Ueda, S. Horii, N. Chikumoto, and K. Kishio, Generic guiding law between irreversibility field and anisotropy — chemical control of critical current of high temperature superconductors, *J. Low Temp. Phys.* **131**(5/6), 1043–1052 (2003).
2. K. Sato, S. Kobayashi, and T. Nakashima, Present status and future perspective of bismuth-based high-temperature superconducting wires realizing application systems, *Jpn. J. Appl. Phys.* **51**(1), 010006 (2012).
3. S. X. Dou and H. K. Liu, Ag-sheathed Bi(Pb)SrCaCuO superconducting tapes, *Supercond. Sci. Technol.* **6**(5), 297–314 (1993).
4. T. Nakashima, S. Kobayashi, T. Kagiyama, K. Yamazaki, M. Kikuchi, S. Yamade, K. Hayashi, K. Sato, J. Shimoyama, H. Kiataguchi, and H. Kumakura, Recent R&D progress on DI-BSCCO wires with high critical current properties, *Physica C* **471**(21–22), 1086–1089 (2011).
5. D. C. van der Laan, J. Schwartz, B. ten Haken, M. Dhallé, and H. J. N. van Eck, Limits to the critical current in $Bi_2Sr_2Ca_2Cu_3Ox$ tape conductors: The parallel path model, *Phys. Rev. B* **77**(10), 104514 (2008).

6. L. Fernandez, B. Holzapfel, F. Schindler, B. de Boer, A. Attenberger, J. Hänisch, and L. Schultz, Influence of the grain boundary network on the critical current of $YBa_2Cu_3O_7$ films grown on biaxially textured metallic substrates, *Phys. Rev. B* **67**(5), 052503 (2003).
7. D. Dimos, P. Chaudhari, J. Mannhart, and F. K. LeGoues, Orientation dependence of grain boundary critical currents in $YBa_2Cu_3O_{7-\delta}$, *Phys. Rev. Lett.* **61**(2), 219–222 (1988).
8. T. Horide, K. Matsumoto, Y. Yoshida, M. Mukaida, A. Ichinose, and S. Horii, The limitation mechanism of J_c-θ characteristics in $YBa_2Cu_3O_{7-x}$ thin film with a single low angle grain boundary, *Physica C* **463**–**465**, 678–681 (2007).

Chapter 1.3

Magnetic Field and Temperature Dependence of Critical Currents of Bi-2223 Wires

Hitoshi Kitaguchi
Superconducting Wire Unit,
National Institute for Materials Science (NIMS)
1-2-1, Sengen, Tsukuba 305-0047, Japan

The critical current (I_c) of Bi-2223 wires depends on magnetic field (B) and temperature (T). I_c depends on the history of external magnetic field (so called "hysteresis") reflecting inter-grain weak coupling in Bi-2223 filament. I_c depends also on magnetic field direction (so called "anisotropy") reflecting microstructure of Bi-2223 filament where plate-like Bi-2223 grains are aligned. In this chapter, typical I_c−B−T characteristics of Bi-2223 wires are described.

1. Introduction

Magnetic field (B) and temperature (T) dependence of critical currents (I_c) is a key information to design superconducting devices. It is well known that I_c of superconductor is a function of external magnetic field strength and temperature. In addition to this feature which is common in all superconductors, I_c of Bi-2223 shows two other characteristic dependence on magnetic field. The first one is hysteresis. Because of weak coupling at grain boundaries in Bi-2223 filament, I_c depends on the history of external magnetic field. I_c observed at certain magnetic field in field increasing sequence is lower than that in field decreasing sequence. The other is anisotropy. As Bi-2223 crystal has highly anisotropic layer structure, superconducting parameters along c-axis and in ab-plane are much different. Therefore I_c of Bi-2223 depends strongly not only on magnetic field strength but also on its direction.

In this chapter, typical $I_c - B - T$ characteristics of Bi-2223 wires are summarized. (I_c means transport I_c and all I_c values mentioned and shown were measured using the four probe transport method).

2. Hysteretic Behavior of I_c

Figure 1 shows I_c values of a Bi-2223/Ag multi-filamentary tape measured in liquid helium (4.2 K). External magnetic fields were applied in parallel to tape surface (i.e. perpendicular to c-axis of Bi-2223 grains). $I_c$values in field increasing sequence from zero external field (closed circles in Fig. 1) stay lower than I_c values in field decreasing sequence (open circles in Fig. 1) from the field higher than 15 T.

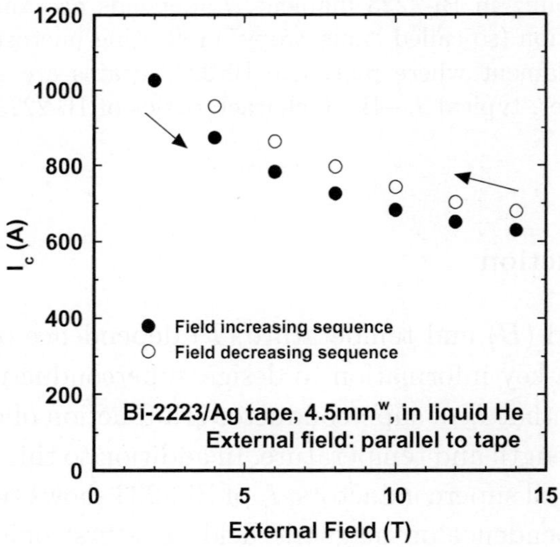

Fig. 1. An example of hysteretic behavior in $I_c - B$ relation of Bi-2223/Ag multi-filamentary tape. External magnetic fields were applied in parallel to tape surface (i.e. perpendicular to c-axis of Bi-2223 grains).

Inter-grain weak-coupling gives an explanation for this hysteresis. In polycrystalline Bi-2223, superconductivity (superconductive coupling) at and near grain-boundaries is weaker than that inside Bi-2223 grains.[1–3] Therefore, I_c of the wire is controlled by inter-grain I_c (I_c at grain boundary) rather than intra-grain I_c (I_c inside grain) in most cases. In field increasing sequence, magnetic flux cannot easily go inside of Bi-2223 grains, thus they are concentrated near grain-boundaries due to strong pinning at low temperature. Therefore, the local magnetic fields near grain-boundaries become higher than the external field and suppress superconducting current. On the contrary, the local magnetic field near grain-boundaries in field decreasing sequence is lower than the external field because magnetic flux introduced in higher fields are pinned inside Bi-2223 grain and cannot easily move out of the grains. This difference in the local magnetic field near grain-boundaries results in hysteretic $I_c - B$ relation.

3. Magnetic Field Angular Dependence of I_c

Figures 2(a) and (b) show magnetic field dependence of I_c of Bi-2223/Ag multi-filamentary tape at (a) 4.2 K (in liquid helium) and (b) 77.3 K (in liquid nitrogen), respectively. At both temperatures, I_c values in magnetic fields parallel to tape surface (i.e. perpendicular to *c*-axis of Bi-2223 grains) stay higher than those in perpendicular fields (i.e. parallel to *c*-axis of Bi-2223 grains).[3–5] This anisotropic I_c–B relation is observed at any temperature. Superconducting parameters of Bi-2223 are much different along *c*-axis and in *ab*-plane directions due to highly anisotropic layer structure; Superconducting Cu-O layer spreads in *ab*-plane and is not strongly coupled to next superconducting Cu-O layer because of the existence of non-superconducting Bi-O layer in between. This layer structure can pin magnetic flux along *ab*-plane but is not effective to pin the flux in *c*-axis direction. On the other hand, Bi-2223 filament in wires has highly textured microstructure where *c*-axis aligns in one direction (for example, *c*-axis aligns in perpendicular to tape surface). Therefore, the intrinsic anisotropy of Bi-2223 crystal leads to anisotropic angular dependence of I_c.

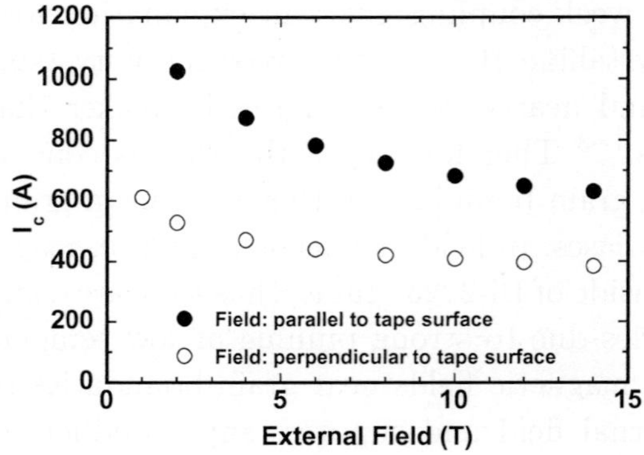

(a) 4.2K

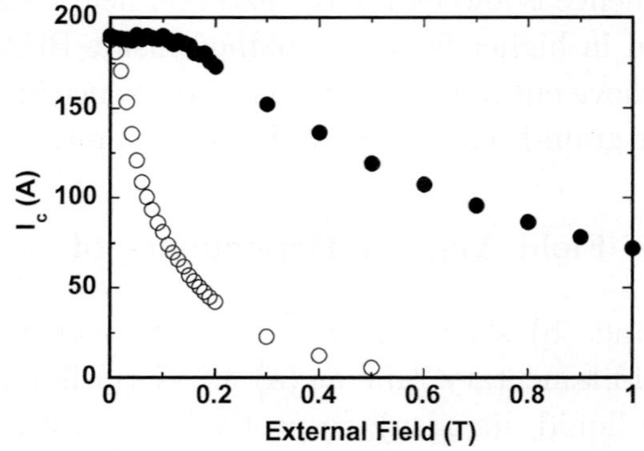

(b) 77.3K

Fig. 2. Effect of magnetic field direction on I_c of Bi-2223/Ag multi-filamentary tape (width: 4.5 mm) in (a) liquid helium (4.2 K) and (b) liquid nitrogen (77.3 K). External magnetic fields were applied in parallel to tape surface (i.e. perpendicular to c-axis of Bi- 2223 grains) and perpendicular to tape surface (i.e. parallel to c-axis of Bi-2223 grains). All I_c values were measured in field increasing sequence.

In low fields parallel to tape surface [see Fig. 2(b), below 0.1 T], I_c is almost constant. Superconducting current creates magnetic field (so called "self-field") that has field component perpendicular to tape surface. In this regime where I_c–B shows a plateau, I_c is not controlled by external fields but by self-field. This indicates that the higher I_c can be expected when this self-field is cancelled or suppressed.

Figure 3(a) shows angular dependence of I_c in detail at 30 K, 6 T and 77.3 K, 0.3 T. Magnetic field angle θ, is defined to be 0 and 90°

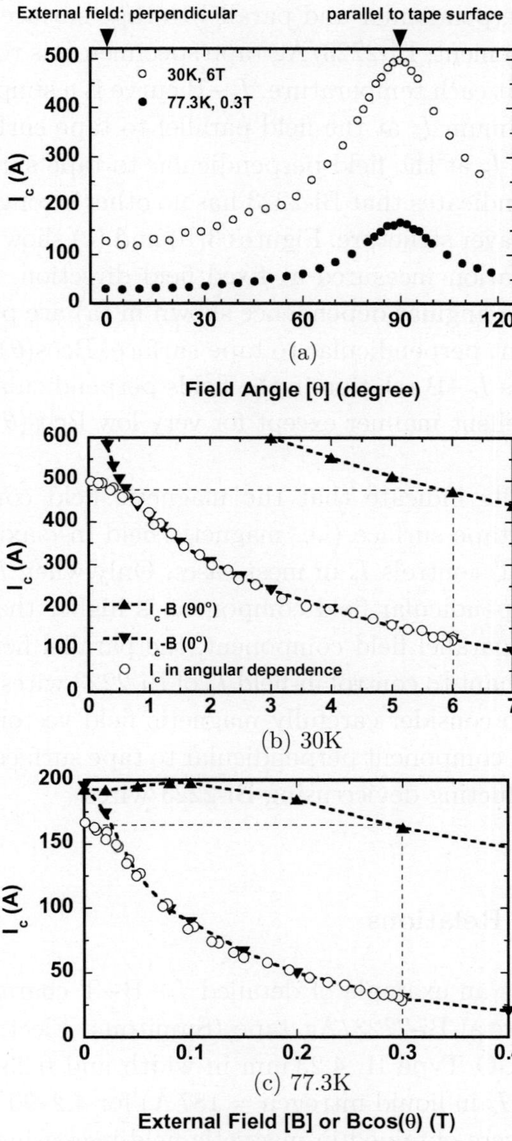

Fig. 3. (a) Magnetic field angular dependence of I_c at 30 K, 6 T and 77.3 K, 0.3 T. Tape specimen was rotated in fixed external field. Comparison with I_c–B relation measured at fixed field direction are shown in (b) 30 K and (c) 77.3 K. In these panels, I_c values shown in (a) are plotted against field component perpendicular to tape surface, Bcos(θ).

for the field perpendicular and parallel to tape surface, respectively. In this measurement, Bi-2223/Ag tape specimen was rotated in fixed external field at each temperature. I_c–B curve is a simple single peak one with maximum I_c at the field parallel to tape surface ($\theta = 90°$) and minimum I_c at the field perpendicular to tape surface ($\theta = 0°$). This relation indicates that Bi-2223 has no other strong pinning than its intrinsic –layer structure. Figures 3(b) and (c) show a comparison with I_c–B relation measured at fixed field direction. In these panels, I_c values in angular dependence shown in (a) are plotted against field component perpendicular to tape surface, Bcos(θ). I_c – Bcos(θ) relation follows I_c–B relation in the fields perpendicular to tape surface in an excellent manner except for very low Bcos(θ) part at each temperature.

These results indicate that the magnetic field component perpendicular to tape surface (i.e. magnetic field in c-axis direction of Bi-2223 crystal) controls I_c in most cases. Only when I_c corresponding to the perpendicular field component is higher than that determined by the parallel field component, the parallel field component becomes dominant to control in-field I_c of Bi-2223 wires. Therefore, it is inevitable to consider carefully magnetic field vector distribution, especially field component perpendicular to tape surface, in designing any superconducting device using Bi-2223 wires.

4. I_c–B–T Relations

Figure 4 shows an example of detailed I_c–B–T characteristics of a recent commercial Bi-2223/Ag tape (Sumitomo Electric Industries, Ltd., DI-BSCCO Type H, 4.23 mm in width and 0.23 mm in thickness, self-field I_c in liquid nitrogen = 187 A) for 4.2–90 K and 0-10 T. All I_c values were obtained in magnetic field increasing sequence. In the fields parallel to tape surface, plateau like behavior mentioned before can be seen in low field regime. This plateau becomes wider with lowering of temperature. I_c in the parallel fields at 12 K and 4.2 K exceeded measurement capability because I_c in the fields parallel to tape surface are much higher than in the perpendicular fields

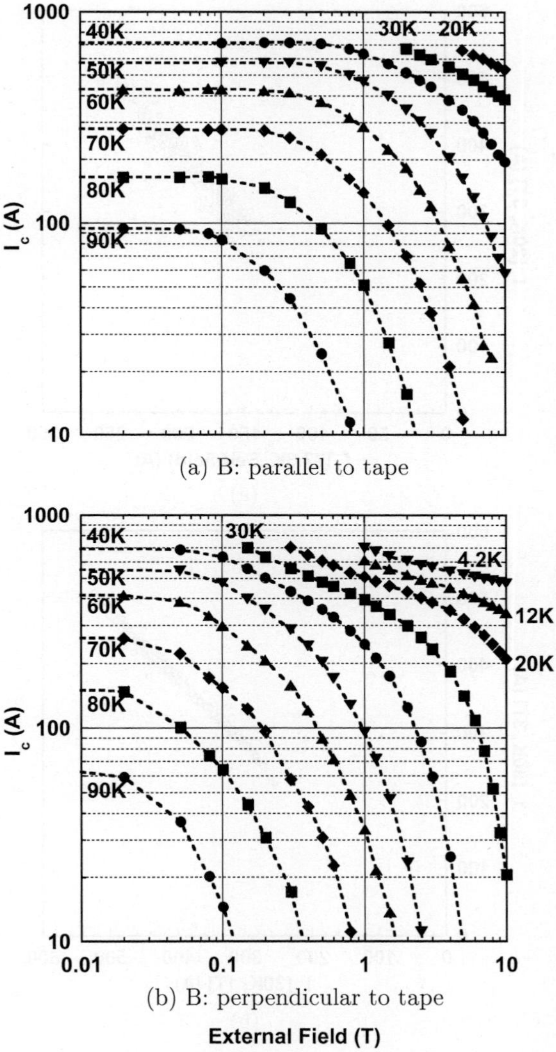

Fig. 4. An example of $I_c - \mathrm{B} - \mathrm{T}$ characteristics of Bi-2223/Ag tape in magnetic fields (a) parallel and (b) perpendicular to tape surface.

as described before. In lower temperatures, no more than 20 K, Bi-2212 shows an excellent in-field I_c performance even in high fields perpendicular to tape surface. For example, I_c at 20 K, 10 T perpendicular to tape surface exceeds 200 A.

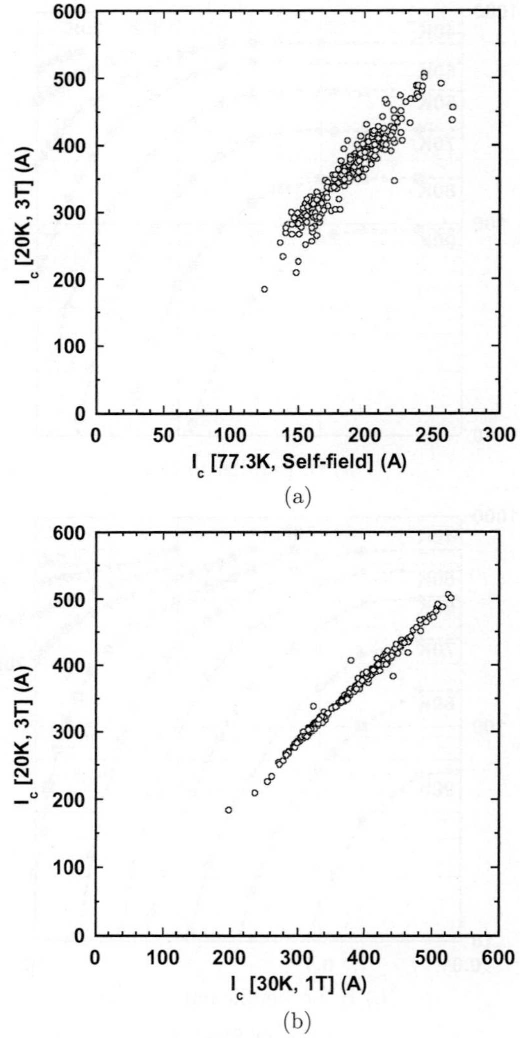

Fig. 5. Relation between I_c values at different temperatures, (a) I_c [20 K, 3 T] vs. I_c [77.3 K, self-field] and (b) I_c [20 K, 3 T] vs. I_c [30 K, 1 T]. Magnetic field direction is perpendicular to tape surface at 20 and 30 K.

Superconducting characteristics of Bi-2223 are much influenced by its oxygen content.[6] Temperature dependence of I_c is affected by oxygen content which varies depending on fabrication process conditions. Relations between I_c values at different temperatures, I_c [20 K,

3 T] *vs.* I_c [77.3 K, self-field] and I_c [20 K, 3 T] *vs.* I_c [30 K, 1 T], are shown in Fig. 5(a) and (b), respectively. Magnetic field direction is perpendicular to tape — surface at 20 and 30 K. I_c values for more than 200 kinds of Bi-2223 tape samples including commercial and R&D products fabricated in 2011–2014 are plotted in this figure. Self-field I_c at 77.3 K varies from 125 to 265 A. I_c [20 K, 3 T] is positively correlated with I_c [77.3 K, self-field] [see Fig. 5(a)], however the correlation is not strong. For example, I_c [20 K, 3 T] values for I_c [77.3 K, self-field] = 200 (+/−2) A are scattered from 353 to 431 A. On the contrary, the correlation between I_c [20 K, 3 T] and I_c [30 K, 1 T] is very strong. I_c determining mechanism at high and low temperatures might be different. The ratios of I_c [20 K, 3 T] to I_c [77.3 K, self-field] and I_c [30 K, 1 T] are 1.92 ± 0.112 and 0.95 ± 0.017 (coefficient of variation: 5.8% and 1.8%), respectively. These results suggest that self-field I_c value at liquid nitrogen temperature is not an ideal indicator of I_c property at low temperatures.

5. Summary

In this chapter, typical I_c–B–T characteristics of Bi-2223 wires were described. In summary,

(1) I_c–B relation shows a hysteretic behavior due to inter-grain weak coupling in polycrystalline Bi-2223 filament. At each external magnetic field, I_c in field increasing sequence is lower than that in field decreasing sequence.

(2) Relation between I_c and magnetic field angle is an anisotropic one, reflecting *c*-axis oriented microstructure of Bi-2223 filament and intrinsic anisotropy of Bi-2223 crystal. I_c becomes maximum and minimum in the field parallel and perpendicular to tape surface, respectively.

(3) A typical I_c–B–T property for 4.2–90 K and 0–10 T was shown in detail. Bi-2223/Ag tape with self-field I_c in liquid nitrogen of 187 A carries more than 200 A at 20 K, in 10 T perpendicular to tape surface.

(4) Strong correlation exists between I_c values at low temperature, however less strong for low temperature I_c to I_c at liquid nitrogen temperature. The ratios of I_c [20 K, 3 T] to I_c [77.3 K, self-field] and I_c [30 K, 1 T] are 1.92 ± 0.112 and 0.95 ± 0.017 (coefficient of variation: 5.8% and 1.8%), respectively.

Acknowledgments

Part of I_c measurements were performed at Grenoble High Magnetic Field Laboratory (GHMFL)/CNRS, Grenoble, France and Robinson Research Institute (RRI)/Victoria University of Wellington, Gracefield, New Zealand. The author would like to thank measurement collaborators; G. Nishijima (NIMS), Y. Miyoshi (NIMS), X. Chaud (GHMFL), N. Strickland (RRI), and S. Wimbush (RRI).

References

1. K. Sato, T. Hikata, and Y. Iwasa, *Appl. Phys. Lett.* **57**, 19284d (1990).
2. T. Matsushita, B. Ni, K. Yamafuji, K. Watanabe, K. Noto, H. Morita, H. Fujimori, and Y. Muto, *Advances in Superconducting.* Tokyo: Springer-Verlag, 393 (1989).
3. Q. Y. Hu, R. M. Schalk, H. W. Weber, C. Czurda, and S. X. Dou, *J. Appl. Phys.* **78**, 1123 (1995).
4. H. Kitaguchi, K. Takahashi, H. Kumakura, T. Hayashi, K. Fujino, N. Ayai, and K. Sato, *Supercond. Sci. Technol.* **22**, 045005 (2009).
5. S. Choi, T. Kiyoshi, S. Matsumoto, K. Itoh, T. Hase, K. Zaitsu, M. Hamada, and M. Sugano, *IEEE Trans. Appl. Supercond.* **18**, 1159 (2008).
6. T. Nakashima, S. Kobayashi, T. Kagiyama, K. Yamazaki, M. Kikuchi, S. Yamade, K. Hayashi, K. Sato, J. Shimoyama, H. Kitaguchi, and H. Kumakura, *Physica C* **471**, 1086 (2011).

Chapter 1.4

Electro-Magnetic Properties of Bi-2223 Wires

Teruo Matsushita

Department of Electronics and Computer Science,
Kyushu Institute of Technology
680-4 Kawazu, Iizuka, Fukuoka 820-8502, Japan

The electromagnetic properties such as critical current density and irreversibility field are described for superconducting Bi-2223 wires fabricated by the controlled over-pressure (CT-OP) process. The improvement of the critical current density was more significant in the magnetic field parallel to the tape surface than in the normal field. This improvement is limited in the low magnetic field region and the irreversibility field was not improved by this process. The observed properties are analyzed using the theoretical model of flux creep and flow. It is found that the improvement of the properties by introducing the CT-OP process is mainly ascribed to the improvement of the crystal alignment by the high pressure, while the pinning mechanism is unchanged.

1. Introduction

Bi-2223 superconductor has a thick insulating blocking block, and for this reason it has a fairly large anisotropy with respect to the direction of the magnetic field. This restricts the condensation energy density to a relatively low value especially at high temperatures in spite of the high critical temperature. That is, the thick insulating blocking block lowers the superconducting property. Figure 1 shows the temperature dependence of the condensation energy density of

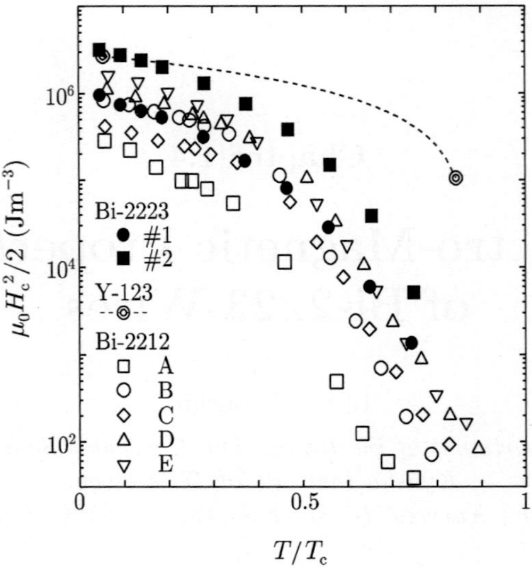

Fig. 1. Reduced temperature dependence of condensation energy density of Bi-2223 superconductors.[1] The characteristics are compared with other high-temperature superconductors.

Bi-2223 single crystals.[1] As a result, the flux pinning property is strongly affected by thermal agitation and the critical current density decreases quickly with increasing magnetic field especially along the *c*-axis at temperatures such as the liquid nitrogen temperature. In addition, the in-field critical current density is determined only by the magnetic field component along the *c*-axis.[2] In particular, the irreversibility field in the direction of the *c*-axis is much lower than that of more three-dimensional (3D) REBCO superconductors. This limits the application of Bi-2223 wires to the low magnetic field region. In the low temperature region, however, these wires can be applied to equipment used at high magnetic fields because of the enhanced condensation energy density as shown in Fig. 1.

The critical current properties of superconducting Bi-2223 wires have been significantly improved by introducing the high pressure annealing process under pressurized oxygen called the controlled over-pressure (CT-OP) process. In this chapter the electromagnetic properties of Bi-2223 wires fabricated by this process are shown. The theoretical analysis using the flux creep and flow model is also shown to figure out the reason for the improvement.

2. Fundamental Properties

Sample #1 was recently fabricated by the CT-OP process and sample #2 was fabricated just after the introduction of this process. These are 200 A and 130 A class wires, respectively. The specifications of these samples are listed in Table 1. Figure 2(a) and 2(b) shows the critical current density in the magnetic field parallel to the c-axis at various temperatures using the conventional four terminal

Table 1. Specifications of CT-OP processed Bi-2223 wires.

Sample	Width (mm)	Thickness (mm)	No. of filaments	T_c [K]	I_c [A]*
#1	0.21	4.0	121	112.5	197
#2	0.23	4.2	61	109.5	126

Note: *I_c is the critical current at 77.3 K in the self-field.

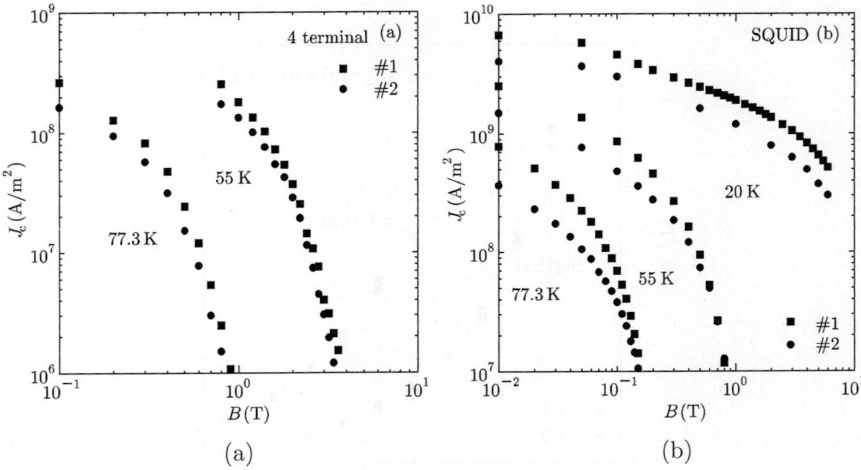

Fig. 2. Critical current density of the two samples at various temperatures in the magnetic field parallel to the c-axis using (a) conventional four terminal method and (b) SQUID magnetometer.[3]

method and SQUID magnetometer.[3] It is found that the critical current density measured using the SQUID magnetometer is appreciably lower than that obtained with the four terminal method. This difference comes from the large difference in the electric field level between the two measurement methods. That is, the electric field is of the level of 10^{-9} or 10^{-10} V/m for the magnetization measurement using the SQUID magnetometer, while the critical current density is determined at 10^{-4} V/m for the four terminal method. If the n-value is about 10; it brings about a difference of a factor about 3 between the two J_c values. From the observed results in Fig. 2, the improvement of J_c from sample #2 to sample #1 is found to be appreciable in the low magnetic fields. This improvement is more significant in the low electric field region for the magnetic measurement using the SQUID, indicating the improvement of the n-value. Figure 3 shows the irreversibility field defined using the critical current density criterion of 10^7 A/m^2 as a function of the temperature for the two measurement methods. The irreversibility field in the low electric field region is much lower than that in the usual electric field region due to the significant influence of the flux creep. The irreversibility field is almost the same between the two samples.

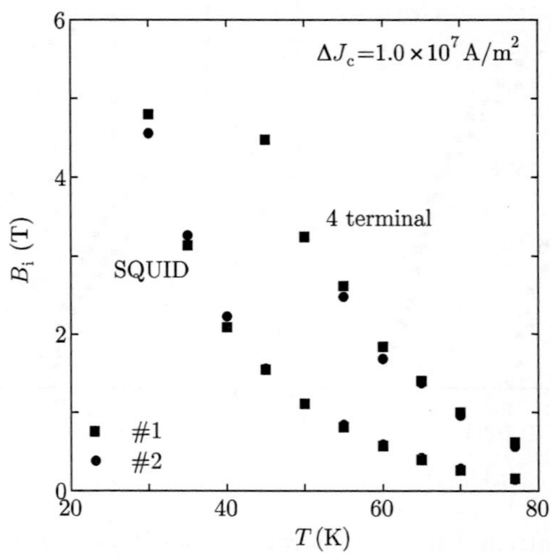

Fig. 3. Irreversibility field along the c-axis for the two measurement methods.

Figure 4(a) and 4(b) shows the transport critical current density at 77.3 K in the in-plane and normal magnetic fields, respectively. The critical current density has a large anisotropy due to the two-dimensional (2D) superconducting property of Bi-2223. This is more clearly seen from Fig. 5 showing the anisotropy of the

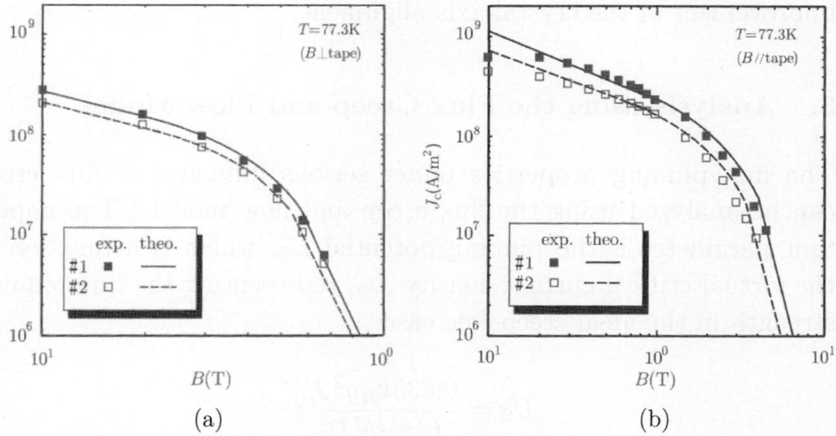

Fig. 4. Transport critical current density of the two samples at 77.3 K in the magnetic field (a) parallel and (b) normal to the c-axis.[4] The lines show the theoretical predictions of the flux creep and flow model.

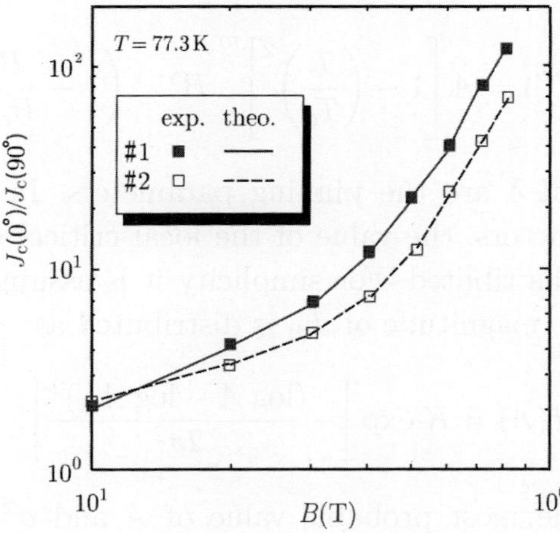

Fig. 5. Anisotropy of transport critical current density at 77.3 K.[4] The lines show the theoretical predictions of the flux creep and flow model.

transport critical current density. The anisotropy increases simply with increasing magnetic field and is larger for sample #1. This field dependence is simply attributed to the difference in the irreversibility field between the two field directions.The improvement from sample #2 to sample #1 shown in Fig. 4 is more significant for the magnetic field parallel to the *a*-*b* plane. This is speculated to be caused by the improvement of the crystal-axis alignment.

3. Analysis using the Flux Creep and Flow Model

The flux pinning properties under serious influence of flux creep can be analyzed using the flux creep and flow model.[5] The important parameter is the pinning potential U_0, which is a function of the virtual critical current density J_{c0}, representing the flux pinning strength in the ideal creep-free case:

$$U_0 = \frac{0.835 k_B g^2 J_{c0}^{1/2}}{(2\pi)^{3/2} B^{1/4}} \tag{1}$$

for the 3D pinning. In Eq. (1), k_B is the Boltzmann constant and g^2 is the number of flux lines in the flux bundle. The magnetic field and temperature dependence of J_{c0} is assumed to be given by

$$J_{c0}(B,T) = A\left[1 - \left(\frac{T}{T_c}\right)^2\right]^m B^{\gamma-1}\left(1 - \frac{B}{B_{c2}}\right)^\delta, \tag{2}$$

where m, γ and δ are the pinning parameters. In high temperature superconductors, the value of the local critical current density is statistically distributed. For simplicity it is assumed that only A representing the magnitude of J_{c0} is distributed as

$$f(A) = K \exp\left[-\frac{(\log A - \log A_m)^2}{2\sigma^2}\right], \tag{3}$$

where A_m is the most probable value of A and σ^2 is the parameter representing the distribution width. If we assume a sinusoidal washboard potential, the activation energy is given as function of the current density J by

$$U(j) = U_0[(1 - j^2)^{1/2} - j\cos^{-1} j] \tag{4}$$

with $j = J/J_{c0}$ denoting the normalized current density. The electric field due to the flux flow is given by

$$\begin{aligned} E_{\mathrm{ff}} &= 0; \quad j \le 1, \\ &= \rho_{\mathrm{f}} J_{c0}(j-1); \quad j > 1, \end{aligned} \tag{5}$$

where ρ_{f} is the flux flow resistivity. On the other hand, the component of the electric field due to the flux flow is

$$\begin{aligned} E_{\mathrm{fc}} &= Ba_{\mathrm{f}}\nu_0 \exp\left[-\frac{U(j)}{k_{\mathrm{B}}T}\right]\left[1 - \exp\left(-\frac{\pi U_0 j}{k_{\mathrm{B}}T}\right)\right]; \quad j \le 1, \\ &= Ba_{\mathrm{f}}\nu_0 \left[1 - \exp\left(-\frac{\pi U_0 j}{k_{\mathrm{B}}T}\right)\right]; \quad j > 1, \end{aligned} \tag{6}$$

where a_{f} is the flux line spacing and ν_0 is the attempt frequency. The total electric field from the two contributions is approximately given by

$$E' = (E_{\mathrm{ff}}^2 + E_{\mathrm{fc}}^2)^{1/2} \tag{7}$$

Then, the electric field E is estimated as the function of J from

$$E(J) = \int_0^\infty E'(J)f(A)\mathrm{d}A. \tag{8}$$

The parameters m, γ, A_{m} and σ^2 are determined so as to get good agreement with experimental results of the E-J characteristics. We used $\delta = 2$ because of relatively weak pinning and high κ value (Ginzburg–Landau parameter). In principle, g^2 can be obtained analytically in the case where the statistical distribution width can be neglected.[6] However, the distribution width is relatively wide for Bi-2223 superconductors. Hence, g^2 is also treated as a fitting parameter. The critical current density was determined using the corresponding electric field criterion, and the irreversibility field was obtained with the same criterion of the critical current density as in experiments.

The determined pinning parameters are listed in Table 2. The theoretical prediction is compared with the experimental results in Figs. 4 and 5 for the transport critical current density at 77.3 K. It is

Table 2. Parameters of the theoretical model of flux creep and flow.

	Sample #1		Sample #2	
	$(\|a-b)$	$(\|c)$	$(\|a-b)$	$(\|c)$
m	3.53	3.53	3.49	3.49
γ	0.46	0.46	0.51	0.51
A_{m} [$\mathrm{AT}^{1-\gamma}$]	8.79×10^9	3.85×10^9	7.02×10^9	3.59×10^9
σ^2	0.011	0.019	0.013	0.021
g^2	1.71	1.06	1.69	1.09

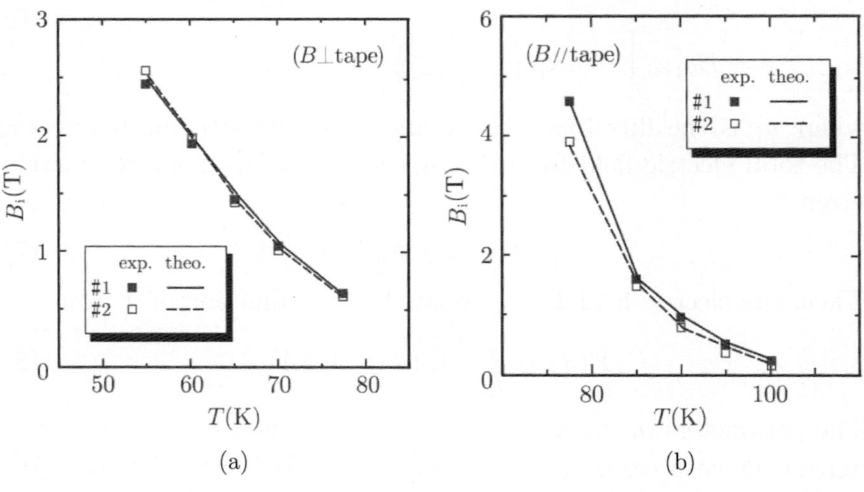

Fig. 6. Irreversibility field along the c-axis for (a) transport measurement and (b) magnetic measurement.

found that the agreement is satisfactory. The theoretical prediction of the irreversibility fields along the c-axis for the two measurement methods is compared with the experiment in Fig. 6(a) and 6(b). The good agreement is obtained again. This shows that the theoretical model of flux creep and flow describes correctly the electromagnetic phenomena over a wide range of electric field.

Here, we discuss the nature of the improvement of the pinning property by introduction of the CT-OP process through the change in the pinning parameters. The value of A_{m} is appreciably increased for the both field directions, while the values of m and γ are unchanged during the development of the CT-OP process. This indicates that the flux pinning mechanism does not change by introduction of this process. This is consistent with the unchanged irreversibility field. The improvement of the critical current density is attributed to the increase in the condensation energy that can be seen from the improvement of the critical temperature and to the improvement of the percolation property influenced by the crystalline alignment and defects such as different phase particles. The latter point can be seen from the decrease in σ^2. That is, the poor J_c region is eliminated, and this contributes to the increase in the n-value. This explains the slight increase in the irreversibility field at low electric field region as shown in Fig. 6(b).

The improved crystal alignment can be directly seen from the increase in the anisotropy: The anisotropy of A_m increased from 1.96 (sample #2) to 2.28 (sample #1) and that of g^2 increased from 1.55 (sample #2) to 1.61 (sample #1). The anisotropy of g^2 is directly associated with that of B_{c2}.[7] The high pressure during the final heat treatment is considered to be effective to align the crystal axes in Bi-2223 phase.

For a further improvement of Bi-2223 wires, the optimization of the CT-OP process is important. It is expected, however, that the improvement of the J_c property in the normal field may be limited. Hence, introduction of pinning centers will be a key point in the future. This seems to be also useful for applications of Bi-2223 wires in the lower temperature region, since the condensation energy approaches to that of the 3D YBCO as shown in Fig. 1.

4. Summary

The influence of the CT-OP process on the electromagnetic properties of superconducting Bi-2223 wires was theoretically analyzed using the model of flux creep and flow. This process is useful to improve the critical current density through reduction of defects and achievement of better crystalline alignment along the direction normal to the tape surface. As a result, the improvement of the critical current density is more significant in the magnetic field parallel to the tape surface than in the magnetic field normal to the tape surface. Thus, the anisotropy increased. Since the pinning mechanism was essentially unchanged, the irreversibility field was also unchanged. For further improvement of the electromagnetic properties, introduction of pinning centers is desirable.

References

1. E. S. Otabe, I. Kohno, M. Kiuchi, T. Matsushita, T. Nomura, T. Motohashi, M. Karpinen, H. Yamauchi, and S. Okayasu, *Adv. Cryog. Eng.* (American Inst. Phys.) **52**, 805 (2006).
2. T. Matsushita, *Flux Pinning in Superconductors, 2nd Ed.* Springer, Berlin (2014), 371.
3. M. Kiuchi, S. Takayama, E. S. Otabe, T. Matsushita, J. Fujikami, K. Hayashi, and K. Sato, *Physica C*, **463–465**, 825 (2007).
4. S. Takayama, M. Kiuchi, E. S. Otabe, T. Matsushita, N. Ayai, J. Fujikami, K. Hayashi, and K. Sato, *Physica C*, **468**, 1779 (2008).
5. M. Kiuchi, K. Noguchi, T. Matsushita, T. Kato, T. Hikata, and K. Sato, *Physica C*, **278**, 62 (1997).
6. T. Matsushita, *Physica C*, **217**, 461 (1993).
7. T. Matsushita, *Flux Pinning in Superconductors, 2nd Ed.* Springer, Berlin (2014), 428.

Chapter 1.5

AC Loss of Bi-2223 Wires

Ryoji Inada

Department of Electrical and
Electronic Information Engineering,
Toyohashi University of Technology
1-1 Hibarigaoka, Tempaku-cho,
Toyohashi, Aichi 4418580, Japan

In most alternating current (AC) power devices, Bi-2223 wire should carry an AC transport current under an AC external magnetic field, and transport losses due to transport current and magnetization losses due to external field are simultaneously generated in the wire. The magnitude of losses and difficulty of the loss reduction are strongly influenced by the direction of external field against the broader face of a tape-form Bi-2223 wire. In this article, the AC loss characteristics of Bi-2223 wire carrying a current in a transverse external field is shown and contribution of both transport and magnetization losses in total losses is briefly discussed. In addition, demonstration of the magnetization loss reduction under a transverse field in perpendicular to the broader face of the wire is described.

1. Introduction

Powder-in-tube processed Ag-sheathed $(Bi,Pb)_2Sr_2Ca_2Cu_3O_x$ (Bi-2223) wires with high and uniform critical current I_c above 200 A (77 K, self-field) and the length above 1 km are commercially available at present.[1] In addition to the enhancement of the current capability by improving the fabricating process, the reduction of alternating current (AC) losses in Bi-2223 wire is one of the key issues for the application to power devices such as cables, transformers and motors, because the magnitude of losses directly influences on the efficiency and the cooling cost of the devices. In most power devices, the wires must carry an AC transport current in an AC external magnetic field with various directions, so that transport losses Q_t due to transport current and magnetization losses Q_m due to external field are simultaneously generated in the wire.[2–6] Moreover, Bi-2223 wire has anisotropic cross sectional geometry with its aspect ratio (= width/thickness) of 10 − 20 and the loss properties are strongly influenced by the direction of an external field against the broader face of the wire.[3,7]

In typical Bi-2223 wire, superconducting filaments are embedded in Ag-matrix with low electric resistivity ρ (= 0.3×10^{-8} Scm^{-1} at 77 K). It is widely known that the Bi-2223 filaments are electromagnetically coupled *via* Ag-matrix and the bundle of the filaments behave as a bulk superconductor at operating frequency around 50 Hz.[2–7] Suppression of the electromagnetic filament coupling should be necessary, particularly for reducing Q_m. As well as the case for metal-based superconducting wires with low critical temperature T_c, twisting the filaments with a suitable pitch length and/or increasing the matrix resistivity is necessary to suppress the inter-filamentary coupling and achieve noticeable Q_m reduction, but the difficulty of loss reduction are strongly influenced by the direction of external field against the broader face of a tape-form wire.[3,7–9]

This article presents AC loss properties of Bi-2223 wires and the method for reducing the losses at specific electromagnetic operating condition. First, the loss properties of Bi-2223 wire carrying a transport current in a transverse external field (in perpendicular to the current transport direction) are presented and contribution of both Q_t and Q_m in total losses Q_{total} in the wire is discussed briefly. Next, demonstration of Q_m reduction under a transverse field in perpendicular to the broader face of the wires are described, by applying the filament twisting and introduction of thin oxide layers around the Bi-2223 filaments as resistive barriers to enhance the matrix resistivity.

2. AC Losses of Bi-2223 Wire Carrying AC Transport Current under AC Transverse External Field

In case of cables with multi-layer structures, the wire carries current in transverse field mainly in parallel to the broader face of the wire, with the field amplitude depending on the layer positions. On the other hand, for winding in transformers or motors, the wire is subjected to a transverse field with both the field amplitude and field angle against the broader wire surface depending on the positions in winding. Therefore, it is important to understand the AC loss characteristics in the wire carrying a transport current in an external transverse field. This section presents the loss properties of Bi-2223 wire carrying a transport current in a transverse external field in parallel and perpendicular to the wire surface [Fig. 1(a)].

PIT-processed Bi-2223 wire (width = 3.60 mm, thickness = 0.30 mm, volume fraction of Bi-2223 filaments = 24%) used for loss

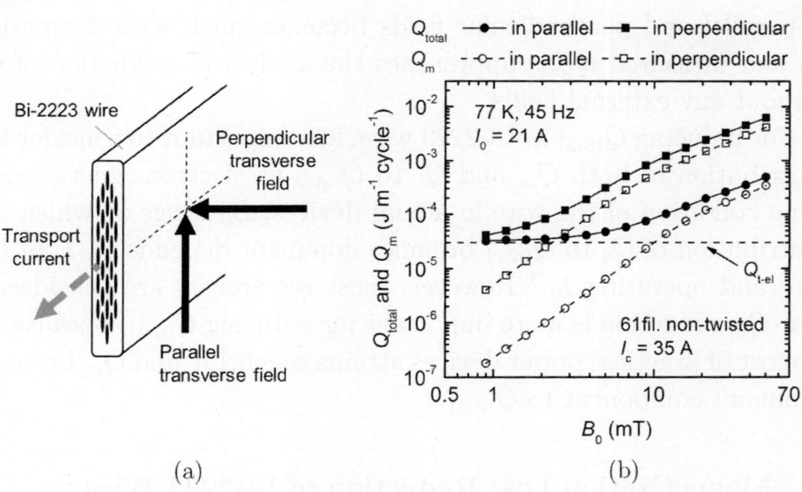

Fig. 1. (a) Classification of parallel or perpendicular transverse fields and transport current for loss measurement and (b) Total losses Q_{total} and magnetization losses Q_{m} at 77 K and 45 Hz for 61-filamentary Bi-2223 wire (width = 3.6 mm, thickness 0.30 mm, I_{c} = 35 A at 77 K, self-field) carrying a transport current (current amplitude I_0 = 21 A) in a parallel or perpendicular transverse field plotted against field amplitude B_0. Analytical predictions of transport loss without external field ($Q_{\mathrm{t-el}}$) using elliptical model approximation.[10]

measurements. The wire has non-twisted 61 Bi-2223 filaments in pure Ag-matrix as well as commercially available ones. The total loss Q_{total} in a wire carrying transport current in transverse field was measured by the electromagnetic method.[2–6] In the measurements, the phase difference between transport current and external field was exactly adjusted to 0°.

Figure 1(b) shows Q_{total} and Q_{m} per cycle per unit length of Bi-2223 wire at 77 K and 45 Hz plotted against the amplitude B_0 of parallel or perpendicular transverse field. Note that the amplitude of transport current $I_0 = 21$ A, which is 60% of self-field I_c (= 35 A) at 77 K of the wire. For comparison, analytical prediction of Q_{t} without any external fields is shown for fully coupled filaments using an elliptical model approximation[10]. As can be seen, measured Q_{m} of Bi-2223 wire in perpendicular transverse field is one order or more than those in parallel one at same B_0. The difference in Q_{m} due to external field direction is caused by the anisotropic wire geometry. On the other hand, due to the contribution of Q_{t}, the difference in Q_{total} in parallel and perpendicular fields becomes small with decreasing B_0 and measured Q_{total} approaches the analytical prediction of Q_{t} without any external fields.

For reducing Q_{total} in Bi-2223 wire, it is important to consider the contribution of both Q_{m} and Q_{t} to Q_{total} at electromagnetic operating condition of the wire in actual devices. B_0 range at which the contribution of Q_{t} to Q_{total} becomes dominant depends on I_{c} of the wire and operating I_0.[6] However, most researchers are considering that Q_{m} reduction is more important for reducing Q_{total}, because B_0 in most of practical power devices attains to ~0.1 T and Q_{m} becomes dominant component to Q_{total}.

3. Magnetization Loss Reduction of Bi-2223 Wire in Perpendicular Transverse Field

As mentioned above, twisting the superconducting filaments with a suitable pitch length and/or increasing the matrix resistivity is necessary to suppress the inter-filamentary coupling and achieve noticeable Q_{m} reduction in Bi-2223 wire. However, due to the anisotropic cross sectional geometry of Bi-2223 wire, the difficulty of Q_m reductions are strongly influenced by the direction of external field against the broader wire surface.[3,7–9] It has been demonstrated that Q_m in parallel transverse field can be reduced to acceptable level by only filament twisting, even in the wire with pure Ag matrix with low resistivity.[3–5,7] On the other hand, due to the large aspect ratio of Bi-2223 wire, both the hysteresis loss Q_h in superconductor and the coupling loss Q_c in matrix in a perpendicular transverse field becomes much larger than in a parallel one, and the conditions for filament decoupling become more restrictive.[3,7–9] Not only twisting the filaments but also increasing the matrix resistivity by introducing oxide layers among the filaments as resistive barriers should be necessary to suppress the filament coupling and achieve noticeable Q_m reduction in a perpendicular transverse field.[11–21]

Coupling frequency f_c, which is related to coupling time constant τ_c as the expression of $f_c = (2\pi\tau_c)^{-1}$, is one of the important parameters to determine the AC operating condition for filament decoupling. To achieve a significant loss reduction by decoupling the filaments, f_c should at least be higher than operating frequency f_{op}. In the previous works for the development of Bi-2223 wires with oxide barriers, f_c under a perpendicular transverse field was increased above 100 Hz by introducing $BaZrO_3$[11–13] or $SrZrO_3$[13–16,18–21] barriers combined with filament twisting. The higher f_c of 400 – 500 Hz was also achieved in twisted wires with $SrZrO_3$ with small amount of $SrCO_3$[14–16] or $Bi_2Sr_2CaCu_2O_x$ (Bi-2212)[18,19] as barriers. However, J_c of those barrier wires with $f_c > 100$ Hz were limited to far below 10 kAcm^{-2} at 77 K and self-field.

Recently, we have succeeded in achieving both $J_c \sim 20$ kA/cm^2 in self-field and significant loss reduction in an AC perpendicular field at 77 K around 50 Hz simultaneously, by controlling the thickness of oxide (85wt% $SrZrO_3$ + 15wt% Bi-2212) barriers and geometrical parameters such as wire widths and twist pitch lengths.[20,21] Figure 2 shows optical images of cross section and broader face (after removing sheath part) of Bi-2223 wire with barriers and twist pitch length L_t = 4 mm. The wire width is reduced to 2.7 mm, which is approximately 30% narrower than commercially available Bi-2223

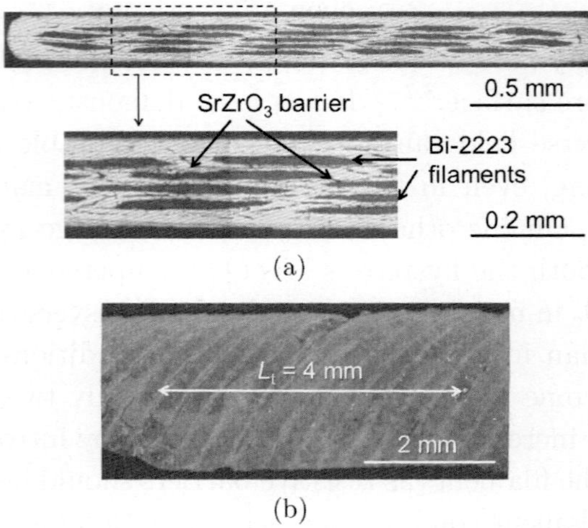

Fig. 2. Optical images of (a) cross section and (b) broader face (after removing sheath part) of Bi-2223 wire with $SrZrO_3$barrier and twist pitch length L_t = 4 mm.

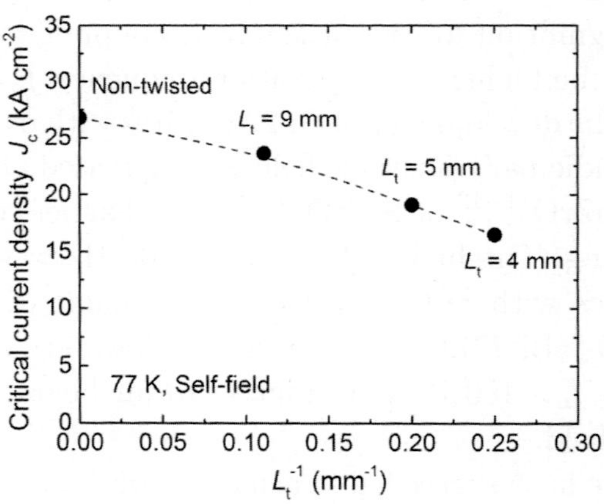

Fig. 3. Transport critical current densities J_c of Bi-2223 wire with $SrZrO_3$ at 77 K and self-field plotted against inverse of twist pitch length L_t.

wire (= 4 mm). Even in tightly-twisted wire, each filament is confirmed to be partitioned by barrier layers.

Transport J_c at 77 K and in self-field for Bi-2223 wires with barriers are shown in Fig. 3, as a function of inverse of L_t. By optimizing $SrZrO_3$ barrier thickness, wire geometry and deformation method, self-field J_c attained to 27 kA/cm^2 for a non-twisted wire and $17-24$ kA/cm^2 for twisted ones with $L_t = 4-9$ mm, respectively. Although these J_c are still lower than those for commercially available Bi-2223 wires, it is worth to note that the value for our non-twisted barrier wire of 27 kA/cm^2 and degradation behavior of J_c by reducing L_t is nearly the same as wires without barriers fabricated in our laboratory.[22]

For a composite multi-filamentary wire with a normal metal matrix, frequency dependence of Q_c per cycle at fixed B_0 is expressed as[23]

$$Q_c \propto B_0^2 \frac{\omega\tau_c}{1+(\omega\tau_c)^2} = B_0^2 \frac{f_{op}/f_c}{1+(f_{op}/f_c)^2} \tag{1}$$

where $\omega = 2\pi f_{op}$. As expected from Eq. (1), Q_c per cycle included in Q_m shows the maximum at $f_{op} = f_c$, so that f_c can be determined experimentally by frequency dependence of Q_m particularly in lower B_0 range. Figure 4(a) shows the frequency dependence of losses Q_m per-cycle for twisted wires with $SrZrO_3$ barriers and different L_t at 77 K and $B_0 = 0.5$ mT. As can be seen, Q_m data shows the maximum at different frequencies among the wires with different L_t. These specific frequencies correspond to f_c at which Q_c per-cycle included in Q_m shows the maximum for each wire. It is confirmed that f_c increases monotonically with decreasing L_t and attains to 270 Hz for the wire with the shortest $L_t = 4$ mm, while self-field J_c is well above 10 kA cm^{-2} at 77 K.

Coupling time constant τ_c in perpendicular transverse field are related to L_t and transverse resistivity $\rho_{t\perp}$ as[8, 11–16, 18–21]

$$\tau_c = \frac{1}{2\pi f_c} \propto \frac{L_t^2}{\rho_{t\perp}}. \tag{2}$$

By using Eq. (2), τ_c of twisted wires with $SrZrO_3$ barriers were calculated and plotted as a function of L_t in Fig. 4(b). It is confirmed that τ_c is nearly proportional to L_t^2, suggesting that $\tau_{t\perp}$ of twisted wires with $SrZrO_3$ barriers is not changed remarkably by difference in L_t. The wire with the shortest $L_t = 4$ mm has the highest $\tau_c = 0.59$ ms among the samples. To estimate the transverse resistivity $\rho_{t\perp}$ of the wire with $SrZrO_3$ barrier, we also measured f_c of twisted wire

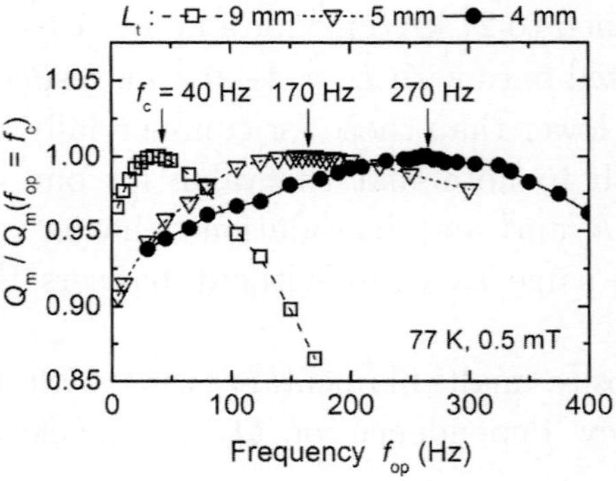

(a)

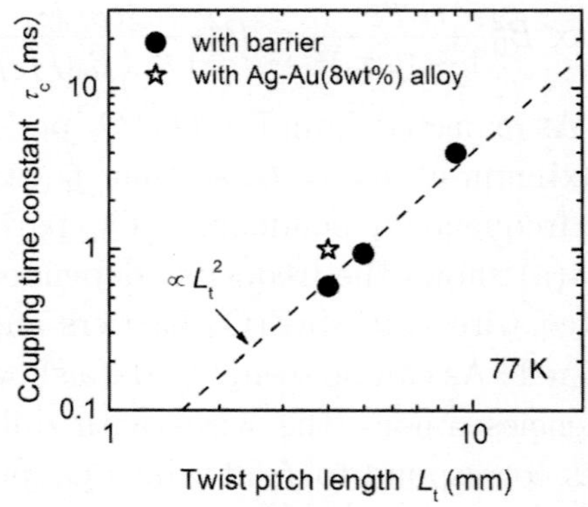

(b)

Fig. 4. (a) Frequency dependence of magnetization loss $Q_{\rm m}$ per cycle per unit length of Bi-2223 wire with $SrZrO_3$ barriers and different twist pitch lengths $L_{\rm t}$. $Q_{\rm m}$ data for each sample is normalized by maximum Q_m in the measurement range. (b) Coupling time constant $\tau_{\rm c}$ of Bi-2223 wire with $SrZrO_3$ barriers plotted against $L_{\rm t}$. $\tau_{\rm c}$ of Bi-2223 wire with Ag–Au (8wt%) as matrix and $L_{\rm t} = 4$ mm is shown for comparison.

with resistive Ag-Au (8 wt%) alloy matrix and the similar geometrical parameters (width = 2.5 mm, $L_{\rm t}$ = 4 mm)[20,21]. It was found that Ag-Au (8 wt%) sheath with resistivity 8 times higher than pure Ag at 77 K give $f_{\rm c}$ = 160 Hz and τ_c = 0.99 ms. From the extrapolation using Eq. (2), $\rho_{t\perp}$ of the wire with barrier is expected to be approximately 14 times higher than a pure Ag-sheathed Bi-2223 wire without barrier.

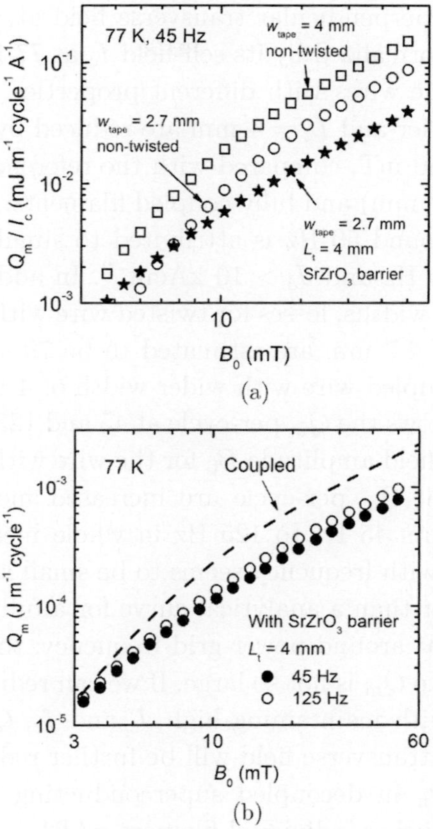

Fig. 5. (a) Comparison of magnetization loss Q_m at 77 K and 45 Hz normalized by self-field I_c of 2.7 mm-width Bi-2223 wire with $SrZrO_3$ barrier and $L_t = 4$ mm, non-twisted 2.7 mm-width wire without barrier and non-twisted 4 mm-width wire without barrier. (b) Perpendicular transverse field amplitude B_0 dependence of magnetization losses Q_m per- cycle per unit length for twisted barrier tape TW3 ($L_t = 4$ mm) at 45 and 125 Hz. Analytical prediction of Q_m for fully coupled filaments is also shown.

In order to confirm Q_m reduction around 50 Hz in perpendicular field for twisted wire with $SrZrO_3$ barrier $L_t = 4$ mm and $f_c = 270$ Hz, B_0 dependence of Q_m at 77 K and 45 Hz are shown in Fig. 5(a). As references, the data for non-twisted wires with the wire widths of 4 mm and 2.7 mm are also plotted.[20,21] These two reference wires have no barrier layers so that all filaments are electromagnetically coupled under a perpendicular transverse field at 45 Hz. Here, Q_m for each wire is normalized by its self-field I_c at 77 K for direct comparison among the wires with different properties. Q_m for the wire with $SrZrO_3$ barrier and $L_t = 4$ mm are reduced by 35 – 40% at B_0 between 10 and 50 mT, compared with the reference one with same wire width (= 2.7 mm) and fully coupled filaments. Such remarkable loss reduction around 50 Hz is attributed to simultaneous achievement of $f_c > 250$ Hz and $J_c > 10$ kAcm^{-2}. In addition, due to the difference in wire widths, losses for twisted wire with $SrZrO_3$ barriers and the width of 2.7 mm are estimated to be 70 – 80% lower than those for fully coupled wire with wider width of 4 mm.

Figure 5(b) shows the Q_m per-cycle at 45 and 125 Hz as a function of perpendicular field amplitude B_0 for the wire with $SrZrO_3$ barriers and $L_t = 4$ mm. Q_m per-cycle are increased monotonically with increasing f_{op} from 45 Hz to 125 Hz in whole measured B_0 range, but the increase with frequency seems to be small and Q_m at 125 Hz is still much lower than a analytical curve for a fully-coupled wire[24]. This suggests that around power-grid frequency, the contribution of coupling loss Q_c to Q_m is not so large. If we can reduce the individual filament width with maintaining high J_c and f_c, Q_m around 50 Hz in perpendicular transverse field will be further reduced because the hysteresis loss Q_h in decoupled superconducting filaments can be reduced with reducing individual filament width.

4. Conclusions and Outlook

With increasing the amplitude of AC external transverse magnetic field, magnetization loss Q_m becomes dominant factor to total loss Q_{total} of Bi-2223 wire carrying transport current in an external transverse field. Particularly for reducing Q_m in transverse field in perpendicular to broader wire surface, not only twisting the filaments but also increasing matrix resistivity should be necessary. Noticeable Q_m reduction in a perpendicular transverse field at 77 K around 50 Hz is successfully demonstrated in tightly twisted Bi-2223 wire with inter-filamentary $SrZrO_3$ layers as resistive barriers, which is attributed to simultaneous achievement of self-field $J_c > 10$ kAcm^{-2} at 77 K and coupling frequency $f_c > 250$ Hz.

Acknowledgments

This work was partially supported by the Grant-in-Aids for Scientific Research from MEXT (Grant No. 20686020) and JSPS (Grant No. 23760257) of Japan.

References

1. T. Nakashima, S. Kobayashi, T. Kagiyama, K. Yamazaki, M. Kikuchi, S. Yamade, K. Hayashi, K. Sato, J. Shimoyama, H. Kitaguchi, and H. Kumakura, *Physica C* **471**, 1086 (2011).
2. J. J. Rabbers, D. C. van der Laan, B. ten Haken, and H. H. J. ten Kate, *IEEE Trans. Appl. Supercond.* **9**, 1185 (1999).
3. N. Amemiya, Z. Jiang, N. Ayai, and K. Hayashi, *Physica C* **392–396**, 1083 (2003).
4. Z. Jiang, N. Amemiya, N. Ayaiand, and K. Hayashi, *Supercond. Sci. Technol.* **17**, 1311 (2004).
5. Z. Jiang, N. Amemiya, T. Nishioka, and S. S. Oh, *Cryogenics* **45**, 29 (2005).
6. R. Inada, K. Tateyama, Y. Nakamura, A. Oota, C. S. Li, and P. X. Zhang, *Supercond. Sci. Technol.* **20**, 138 (2007).
7. H. G. Knoopers, J. J. Rabbers, B. ten Haken, and H. H. J. ten Kate, *Physica C* **372–376**, 1784 (2002).
8. E. Martínez, Y. Yang, C. Beduz, and Y. B. Huang, *Physica C* **331**, 216 (2000).
9. K. Funaki, Y. Sasashige, H. Yanagida, S. Yamasaki, M. Iwakuma, N. Ayai, T. Ishida, Y. Fukumoto, and Y. Kamijo, *IEEE Trans. Appl. Supercond.* **19**, 3053 (2009).
10. W. T. Norris, *J. Phys. D: Appl. Phys.* **3**, 489 (1970).
11. K. Kwasnitza, St. Clerk, R. Flükiger, and Y. B. Huang, *Physica C* **299**, 113 (1998).
12. M. Dhallé, A. Polcari, F. Marti, G. Witz, Y. B. Huang, R. Flükiger, St. Clerck, and K. Kwasnitza, *Physica C* **310**, 127 (1998).
13. K. Kwasnitza, S. Clerk, R. Flükiger, and Y. B. Huang, *Cryogenics* **39**, 829 (1999).
14. H. Eckelmann, J. Krelaus, R. Nast, and W. Goldacker, *Physica C* **355**, 278 (2001).
15. H. Eckelmann, R. Nast, C. Schmidt, and W. Goldacker, *IEEE Trans. Appl. Supercond.* **11**, 2955 (2001).
16. R. Nast, H. Eckelmann, O. Zabara, S. I. Schlachter, and W. Goldacker, *Physica C* **372–376**, 1777 (2002).

17. N. Ayai, K. Hayashi, and K. Yasuda, *IEEE Trans. Appl. Supercond.* **15**, 2510 (2005).
18. R. Inada, Y. Mitsuno, K. Soguchi, Y. Nakamura, A. Oota, C. S. Li, and P. X. Zhang, *IEEE Trans. Appl. Supercond.* **19**, 3018 (2009).
19. R. Inada, Y. Nakamura, A. Oota, C. S. Li, and P. X. Zhang, *Supercond. Sci. Technol.* **20**, 085014 (2009).
20. R. Inada, Y. Okumura, R. Ohtsu, A. Oota, C. S. Li, and P. X. Zhang, *Proceedings of ICEC 24-ICMC 2012*, 857 (2013).
21. R. Inada, Y. Okumura, R. Ohtsu, and A. Oota, *J. Cryo. Super. Soc. Jpn.* **49**, 62 (2014).
22. Y. Nakamura, Y. Fujiwara, A. Nagaoka, T. Machida, R. Inada, and A. Oota, *Physica C* **470**, 1369 (2010).
23. A. M. Campbell, *Cryogenics* **22**, 3 (1982).
24. B. ten Haken, J. J. Rabbers, and H. H. J. ten Kate, *Physica C* **377**, 156 (2002).

Chapter 1.6

Mechanical Properties of Bi-2223 Wires

Kozo Osamura

Research Institute for Applied Sciences
Sakyo-ku, Tanaka Ohi-cho 49
Kyoto 606-8202, Japan

The characteristics of tensile properties of commercial Bi-2223 wires were explained by using the rule of mixture below the reversible strain limits. The reversible tensile strain limit coincides well with the start of degradation in critical current, because both phenomena have the same origin, at which the SC filaments start to fracture macroscopically. In order to improve both properties, it is crucial to maximize the reversible strain region. Its strategy was discussed in terms of thermal strain and pre-strain dependence through the strengthening by lamination.

1. Introduction

Bi-2223 wires as a typical high temperature superconductor are now manufactured on industrial scale, which has high current density and small cross-sectional area. Its major application is to build electrical power devices such as superconducting cable and magnet.

When the power device is being manufactured, complicated stresses/strains are applied to the wires. In order to manufacture sound products, therefore, the assessment of mechanical properties of the wires is indispensable. Further when the power device is energized in practice, a large electromagnetic force is applied to the superconducting wires because of its high current density. In order to design the power device, it is necessary to determine beforehand the mechanical properties of the practical SC wires at operation temperature.

Here the characteristics of tensile properties of commercial Bi-2223 wires were explained together with the strain dependence of critical current. The definition of reversible strain region was given, of which optimization was discussed in terms of thermal strain and pre-strain dependence through the strengthening by lamination.

2. Tensile Properties of Commercial Bi-2223 Wires

A typical stress *versus* strain relationship for a commercial Bi-2223 wire (NX3ply) laminated by nickel alloy sheet is shown in Fig. 1, and also the strain dependence of the normalized critical current is depicted. The normalized critical current decreased gradually with increasing applied strain (A_a). Beyond a critical strain (A_{95}), the critical current decreased abruptly due to the start of SC filament

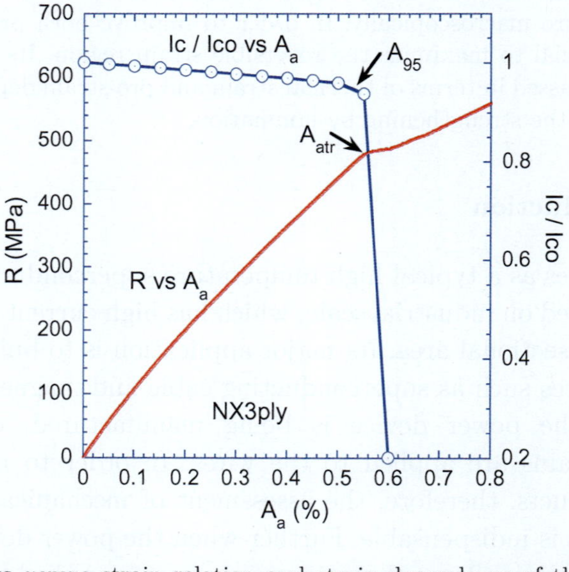

Fig. 1. Stress *versus* strain relation and strain dependence of the normalized critical current at 77 K for the NX3ply Bi-2223 wire.

fracture. The strain A_{95} is defined in the practical viewpoint, where the normalized critical current still retains 95%.

The stress–strain curve shows a quasi-elastic behavior.[1] The slope decreased gradually with increasing strain up to the shoulder indicated by A_{atr}, where a macroscopic yielding takes place due to the start of BSCCO filament fracture.[2] Thus both parameters of A_{95} and A_{atr} have the same origin, in which the SC filaments start to fracture.

The commercial Bi-2223 wire is a typical composite material which includes five components of BSCCO filaments (1), Ag matrix (2), Ag alloy (3), solder (4) and laminated metallic sheet (5). The stress (R_c) of the wire is given by the sum of those of constituent components through the rule of mixture as given by the equation,

$$R_c = V_{f1}R_1 + V_{f2}R_2 + V_{f3}R_3 + V_{f4}R_4 + V_{f5}R_5, \qquad (1)$$

where V_{fi} and R_i are the volume fraction and stress of component i, respectively. Within the external strain region up to 1%, it is presumed that BSCCO filaments behave elastic, while other components of Ag matrix and Ag alloy, solder and laminated metallic sheet change elasto-plastically. Figure 2 shows the stress–strain curves at

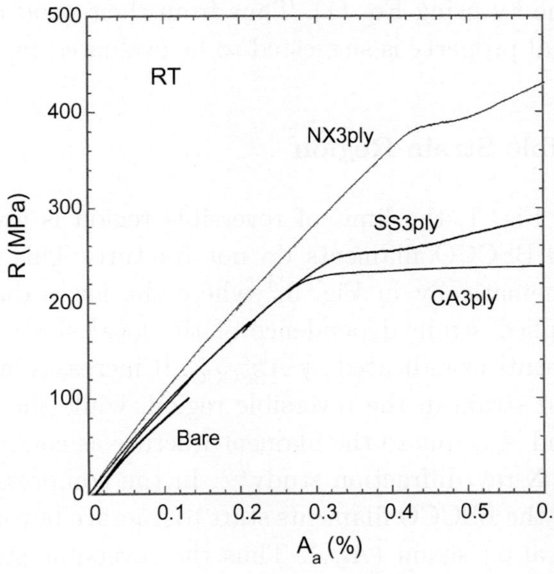

Fig. 2. Stress *versus* strain curves at room temperature for four kinds of Bi-2223 wires.

Table 1. Modulus of elasticity and force free strain at room temperature for three kinds of commercial Bi-2223 wires.[1]

	E_c(GPa)		A_{aff}(%)	
	Obs	Cal	Obs	Cal
Bare	89	89.2	−0.0069*	−0.059
Brass3ply	88	83.8	−0.110	−0.086
SUS3ply	95	95.7	−0.059	−0.057

room temperature for some Bi-2223 wires, where the bare means the original wire before laminating the metallic sheet. CA3ply and SS3ply indicate the laminated metallic sheet of copper alloy (CA) and stainless steel (SS), respectively. The behavior of stress–strain curve depends on the laminated materials, while the bare wire fractures at low strain region.

In order to determine the mechanical properties of Bi-2223 wires, the international standard IEC 61788-18[3] can be applied. The observed modulus of elasticity is listed in Table 1 together with the calculated one by using Eq. (1). Thus from their good coincidence, the mechanical property is suggested to be evaluated by calculation.

3. Reversible Strain Region

As shown in Fig. 1, the limit of reversible region is given by A_{atr} within which BSCCO filaments do not fracture. This situation is explained schematically in Fig. 3,[4] where the lower diagram indicates the applied strain dependence of the local strain exerted on BSCCO filaments as indicated by $A_{\mathrm{BSCCO}}^{\mathrm{loc}}$. It increases in proportion to the applied strain in the reversible region, while the local strain relaxes beyond A_{atr} due to the filament fracture as confirmed experimentally by X-ray diffraction study[2,5]. In the compressive applied strain region, the BSCCO filaments start to fracture beyond the compressive relaxation strain (A_{acr}). Thus the reversible strain region, in which the BSCCO filaments do not fracture, exists in between A_{atr} and A_{acr}.

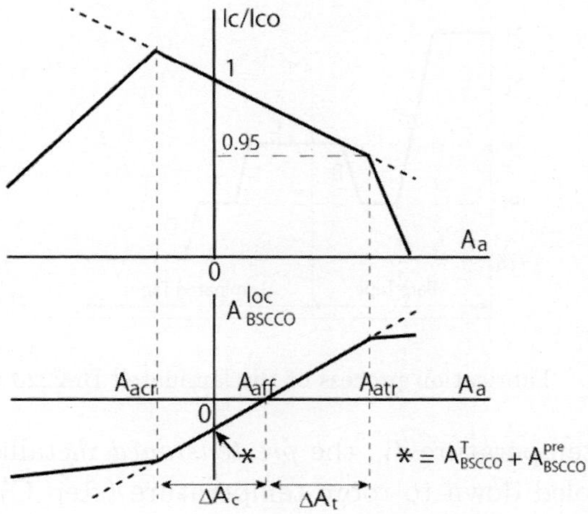

Fig. 3. Schematic illustration of the applied strain dependence of the normalized critical current and the local strain exerted on BSCCO filaments.[4]

As indicated in Fig. 3, the reversible strain region for the critical current expands to the compressive strain until A_{acr}, beyond which the critical current decreased. As made clear recently,[4] the critical current increases linearly with decreasing applied strain in the reversible strain region as shown in Fig. 3. Therefore the apparent critical current maximum appears at around the compressive reversible strain limit (A_{acr}).

4. Thermal Strain and Pre-Strain Exerted on BSCCO Filaments

Due to different coefficient of thermal expansion (CTE) of constituent components, the thermal strain generates in each component, when the sample is cooled down from high temperature. The principal procedure to evaluate the thermal strain has been discussed previously.[1] Figure 4 explains the fabrication process of the laminated wire. After the final heat treatment at T_f for optimizing the SC property, the wire is cooled down to room temperature (step A). In order to laminate the metallic sheet, the wire is re-heated to the soldering temperature

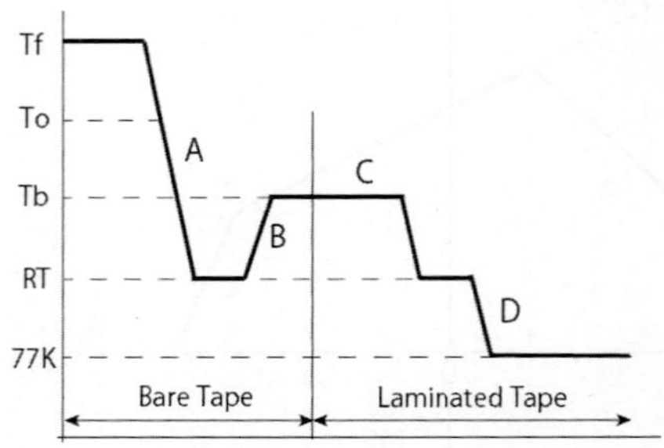

Fig. 4. Fabrication process of the laminated Bi-2223 wires.

(step B). At temperature T_{b}, the *pre-tensioned* metallic sheet is soldered and cooled down to room temperature (step C). Further the wire is cooled down to 77 K (step D). In each step, the local strain exerted on constituent components changes due to the thermal strain and the applied pre-strain.

During cooling, the wire shrinks as a whole. In the constraint shrinkage, the whole components shrink with the same amount giving by the average CTE (α_{c}). The thermal strain of component i during the constraint shrinkage is given by the term $(\alpha_{\mathrm{i}} - \alpha_{\mathrm{c}})$. In general, the CTE is non-linear function with temperature as described previously in the literature.[6] The thermal strain exerted on BSCCO filaments cooled at temperature T ($A^{\mathrm{T}}_{\mathrm{BSCCO}}$) from T_o is expressed by the expression,

$$A^{\mathrm{T}}_{BSCCO} = \int_{T}^{T_o} (\alpha_{BSCCO} - \alpha_c)dT, \tag{2}$$

where T_o is a specific temperature, at which the thermal strain exerted on BSCCO filament becomes zero.[1] When cooling down to T, the thermal stress ($R^{\mathrm{T}}_{\mathrm{i}}$) generates in each component. Their corresponding thermal stresses balance among them through the following constraint condition under the rule of mixture,

$$V_{f1}R_1^T + V_{f2}R_2^T + V_{f3}R_3^T + V_{f4}R_4^T + V_{f5}R_5^T = 0. \tag{3}$$

Only an unknown parameter is α_{c} in this equation. After getting α_{c} at any temperature, it is possible to integrate Eq. (2) and to get the thermal strain of each component. The estimation of thermal strain for the BSCCO component is very important in order to discuss both mechanical and superconducting properties. The procedure to get exactly the residual strain is very time consuming. So an approximate expression has been proposed to evaluate the thermal strain.[1] The thermal strain at temperature T was driven for the five-component system as follows,

$$A^T_{BSCCO} \cong \frac{(\alpha_3 - \alpha_1)E_3V_{f3}(T - T_o) + (\alpha_5 - \alpha_1)E_5V_{f5}(T - T_b)}{E_c}, \tag{4}$$

where E_i is the modulus of elasticity of component i. In order to look for different lamination material, the above expression is applicable as far as the composite structure does not change.

When the bare tape was laminated with metallic sheet at temperature T_{b}, the metallic sheet was stretched by the pretension (P_{p}). By means of this process, a compressive residual strain was added to the BSCCO filaments. In practice, the additional strain ($A^{\mathrm{pre}}_{\mathrm{BSCCO}}$) in proportion to P_{p} was observed in the increase of A_{atr}.[5]

Before applying an external strain, that is, $A_{\mathrm{a}} = 0$, the local strain exerted on BSCCO filaments is compressive as given by $A^{\mathrm{T}}_{\mathrm{BSCCO}} + A^{\mathrm{pre}}_{\mathrm{BSCCO}}$ as shown in Fig. 3.[2] When applying external strain, the local strain increases and changes sign over crossing a specific strain, A_{aff}, which is called the force free strain.[1] The local stress exerted on BSCCO filaments becomes zero at A_{aff}. Still the local strain changes linearly with increasing the applied strain. This linearity ensures the elastic behavior of BSCCO filaments against the applied strain. Then the following relation was held;

$$A_{\mathit{aff}} = -k(A^T_{BSCCO} + A^{pre}_{BSCCO}), \tag{5}$$

where k is the proportional constant.[4]

As discussed previously,[4] the BSCCO filaments start to fracture at the reversible strain limit (A_{atr}), which consists of the following two terms as shown in Fig. 3,

$$A_{atr} = A_{\mathit{aff}} + \Delta A_t \tag{6}$$

ΔA_{t} indicates an intrinsic tensile strain to fracture, that is, the tensile strength of BSCCO filaments themselves. On the other hand, ΔA_{c} is the intrinsic compressive strain.

Thus, we get important suggestions for strengthening the Bi-2223 wires. One of them is to increase the force free strain. As given by Eq. (5), the thermal strain can be increased by selecting the metallic sheet, of which CTE is larger than that of BSCCO. The pre-strain shall be applied as much as possible. Further, the intrinsic tensile strain can be increased by the microstructure control during fabrication as well as the selection of lamination material.

5. Effect of Pretension

In order to examine the effect of pretension, several laminated Bi-2223 wires were fabricated, which included one original bare tape and four tapes laminated by stainless steel (SS) sheet on both sides with thickness of 50 μm. When the bare tape was laminated with SS sheet by the Pb-free solder at temperature between 480 and 520 K, the SS sheet was stretched by the pretension (P_p) between 44 and 161 N. Stress–strain behavior at room temperature is shown in Fig. 5 for

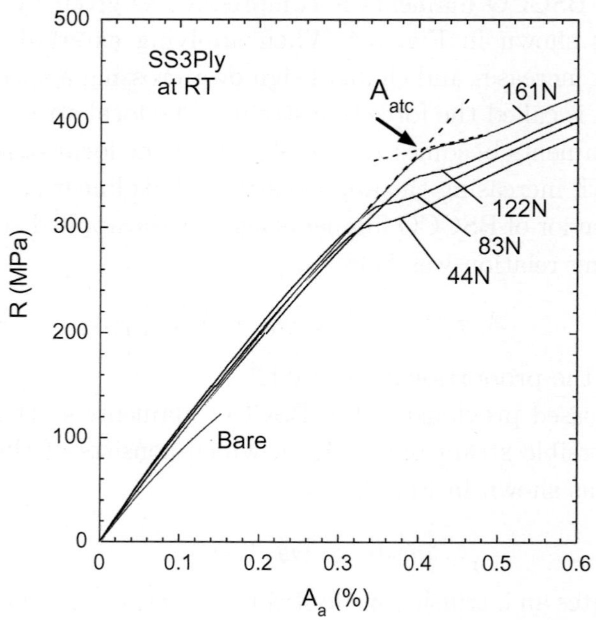

Fig. 5. Stress–strain curve at room temperature for four Bi-2223 wires, tapes laminated with SS sheet and one bare tape.

Table 2. Change of characteristic strains as a function of pretension measured at room temperature.[2]

Symbol	Pretension P_p (N)	A_{95} (%)	A_{atc} (%)	A_{aff} (%)	ΔA_t (%)
Bare	0	0.218	0.148	0.041	0.081
44 N	44	0.410	0.318	0.138	0.162
83 N	83	0.430	0.342	0.169	0.157
122 N	122	0.467	0.383	0.192	0.158
161 N	161	0.506	0.405	0.239	0.161

all the wires. The strain indicated by A_{atc}, increased with increasing pretension. Table 2 gives the change of characteristic strains as a function of pretension. Over the shoulder, the tapes did not fracture as a whole, but the plastic deformation was observed successively accompanying with the work hardening of laminated metallic sheet, while the bare tape fractured as a whole. The stress corresponding to the shoulder R_{tc} was 364 MPa for the 161 N tape. It was also reported that R_{tc} reached 440 MPa for the similar tape laminated by SS sheet with thickness of 100 μm[7].

6. Strengthening by Lamination

As we understand from Eq. (1), the modulus of elasticity of the wire becomes higher by choosing the laminated metallic sheet with the higher modulus and with the larger thickness. However when the thicker sheet is selected, the engineering critical current, defined as the critical current divided by the total cross section of tape becomes lower.

In general, the material with high modulus of elasticity has low CTE. As given in Eq. (4), the selection of the metallic sheet with high modulus of elasticity results in the low thermal strain. Thus the optimization of material selection is important.

Figure 6 shows the reversible limits for various Bi-2223 wires at room temperature and 77 K. Here SS and CA indicate that the laminated metallic sheet is SS and CA, respectively. They were fabricated by means of the pretensioned lamination technique under various pretension levels. The reversible stress limit for the SS laminated wires are higher than those for the CA laminated ones. This fact is mainly

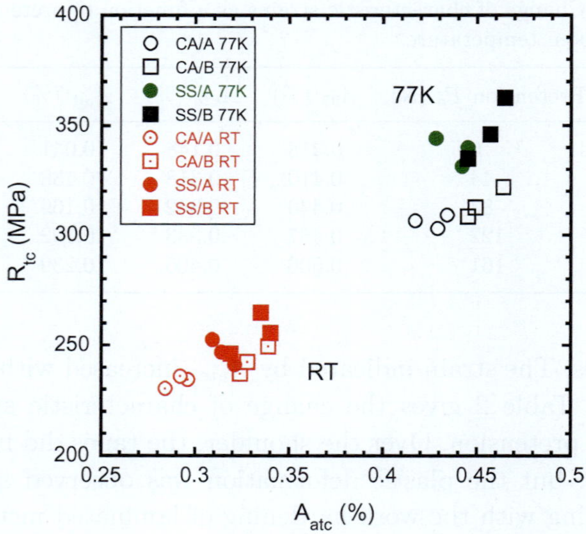

Fig. 6. Reversible stress and strain limits for various Bi-2223 wires laminated by CA and SS sheets and pretensioned.

attributed to the higher modulus of elasticity for the SS laminated wires. On the other hand, the reversible limit for the bare wire was typically $R_{\mathrm{tc}} = 120$ and 152 MPa, and $A_{\mathrm{atc}} = 0.15$ and 0.23% at room temperature and 77 K, respectively.

7. Summary

The general behavior of tensile properties of commercial Bi-2223 wires could be explained by using the rule of mixture below the reversible strain limits under both tensile and compressive applied strains. The reversible tensile strain limit coincides well with the start of degradation in critical current, because both phenomena have the same origin, at which the SC filaments start to fracture. In order to improve the mechanical and superconducting properties, it is crucial to maximize the reversible strain region. The strategy was discussed in terms of thermal strain and pre-strain dependence through the strengthening by lamination.

Acknowledgments

This work was supported in part by a grant-in-aid of the Ministry of Education, Culture, Sports, Science and Technology, Japan (Grant No. 26420669).

References

1. K. Osamura, S. Machiya, H. Suzuki, S. Ochiai, H. Adachi, N. Ayai, K. Hayashi, and K. Sato, Mechanical behavior and strain dependence of the critical current of DI-BSCCO tapes, *Supercond. Sci. Technol.* **21**, 054010–054018 (2008).
2. K. Osamura, S. Machiya, S. Ochiai, G. Osabe, K. Yamazaki, and J. Fujikami, Direct evidence of the high strain tolerance of the critical current of DI-BSCCO tapes fabricated by means of the pretensioned lamination technique, *Supercond. Sci. Technol.* **26**, 045012–045016 (2013).
3. K. Osamura, A. Nyilas, K.-P. Weiss, H.-S. Shin, K. Katagiri, S. Ochiai, M. Hojo, M. Sugano, and K. Ohsawa, International round robin test for mechanical properties of BSCCO-2223 superconductive tapes at room temperature, *Cryogenics* **51**, 21–26 (2011).
4. K. Osamura, S. Machiya, D. P. Hampshire, Y. Tsuchiya, T. Shobu, K. Kajiwara, G. Osabe, K. Yamazaki, Y. Yamada, and J. Fujikami, Uniaxial strain dependence of the critical current of DI-BSCCO tapes, *Supercond. Sci. Technol.* **27**, 085005–085015 (2014).
5. K. Osamura, S. Machiya, S. Ochiai, G. Osabe, K. Yamazaki, and J. Fujikami, High strength/high strain tolerance DI-BSCCO tapes by means of pretensioned lamination technique, *IEEE Trans. Appl. Supercond.* **23**, 6400504–6400507 (2013).
6. K. Osamura, S. Machiya , Y. Tsuchiya, H. Suzuki, T Shobu, M Sato, S. Harjo, K. Miyashita, Y. Wadayama, S. Ochiai, and A. Nishimura, Thermal strain exerted on superconductive filaments in practical Nb_3Sn and Nb_3Al strands, *Supercond. Sci. Technol.* **26**, 094001–094008 (2013).
7. N. Ayai, K. Kato, J. Fujikami, S. Kobayashi, M. Kikuchi, K. Yamazaki, S. Yamade, T. Ishida, K. Tatamidani, K. Hayashi, K. Sato, K. Hata, H. Kitaguchi, H. Kumakura, K. Osamura, and J. Simoyama, DI-BSCCO wire with I_c over 200 A at 77 K, *J. Phys.* **97**, 012112–012117 (2008).

Chapter 1.7

Thermal Properties of Bi-2223 Wires

Hiroyuki Fujishiro and Tomoyuki Naito

Department of Materials Science and Engineering, Iwate University, 4-3-5, Ueda Morioka 020-8551, Japan

We introduced the thermal conductivity $\kappa(T)$ of Bi-2223 wires sheathed with Ag-based alloys. The thermal conductivity of several sheathed Ag-based alloys was also summarized. For several kinds of DI-BSCCO tapes, $\kappa(T)$ was reported and the anisotropic thermal conductivities of the stacked or sandwiched DI-BSCCO bundles were measured and analyzed using the heat flow model. The $\kappa(T)$ data are useful and valuable for calculating the heat intrusion and thermal stability for the application to the superconducting coils and power current leads.

1. Introduction

In the various wiring processes for oxide superconductors, the choice of metal stabilizer seriously influences superconducting characteristics such as critical current density J_c and critical temperature T_c. Silver (Ag) was widely used as the metal stabilizer at the early stage in order to realize a high crystal orientation of the oxide compounds and excellent superconducting characteristics.[1,2] In cases where Ag-sheathed oxide superconducting tapes are applied to power current leads for cryogenic apparatus such as a high field superconducting magnet (SCM), the Ag sheath acts as a heat conduction medium, and the intrusion of heat becomes a serious problem. For this reason, it is necessary to reduce the thermal conductivity of the metal stabilizer instead of pure Ag. On the other hand, in cases where the tapes are used to wind the SCM themselves, the sheathed pure Ag is an ideal metal because of higher thermal conductivity even at low temperatures. However, it is necessary to enhance the mechanical strength of the sheath metal to endure the hoop stress because Ag is not mechanically strong. A popular solution to reduce the thermal conductivity and/or to enhance the mechanical strength of the metal is alloying; that is, in the solid solution, the thermal conductivity decreases with the increase in the mixing. For the development of wiring processing of Bi-based superconductors such as $Bi_2Sr_2Ca_1Cu_2O_{8+x}$ (Bi-2212) and $(Bi,Pb)_2Sr_2Ca_2Cu_3O_{10+x}$ (Bi-2223), searches for Ag-based alloys have been widely performed.

The Bi-2223 tapes sheathed with Ag-based alloy (Bi-2223/Ag-alloy) are popular at present, and have been developed and investigated by many researchers and companies. Sumitomo Electric Industries (SEI), Ltd. developed a new sintering technique with the controlled overpressure (CT-OP),[3] which enables to fabricate the Bi-2223/Ag-alloy tape including very small amount of voids and non-superconducting regions and having a high Young's modulus in comparison with the conventional Bi-2223/Ag-alloy tapes. Using this technique, SEI succeeded in fabricating a Bi-2223/Ag-alloy tape with high critical current I_c, over 200 A at 77 K and under the self-field.[4] This I_c value gives us the possibility that the superconducting applications, such as high-field SCMs and power cables, are practicably operated at 77 K. Since the Bi-2223/Ag-alloy tapes made by CT-OP technique surpass other tapes in all the properties, SEI named them Dynamically Innovative BSCCO (DI-BSCCO).

The thermal conductivity data of the tapes are necessary to calculate the heat flow through the tapes, and thermal conductivity and electrical resistivity are important in determining the dynamic stability of the current leads and SCMs. When the superconducting materials are used in the practical devices, we need to estimate the amount of heat intrusion through them.[5] The heat intrusion $Q_{\rm int}$ in the materials from high temperature ($T_{\rm H}$) side to low temperature side (T_L) is in general given by

$$Q_{\text{int}} = \frac{S}{l} \int_{T_L}^{T_H} \kappa(T) dT, \tag{1}$$

where $\kappa(T)$ is the thermal conductivity, l the length and S the cross section of the material. In the case of the Bi-2223/Ag-alloy tapes, the absolute value and temperature dependence of $\kappa(T)$ are mainly dominated by the sheath material, the volume ratio of sheath material to the Bi-2223 and the fabrication process.[5–7] However, $\kappa(T)$ of DI-BSCCO had not been measured yet just after the development. The thermal dilatation $\delta L/L$ is also the important physical property, because the change in the length of the tape would give itself a mechanical stress and suppress the superconducting properties.

We introduce the measuring technique of thermal properties for materials in Sec. 2. The thermal conductivity of Ag-based alloys such as Ag–Au, Ag–Cu is summarized in Sec. 3, including our data. Thermal conductivity of BSCCO tapes fabricated in 1990s is presented in Sec. 4. The thermal properties of several types of DI-BSCCO tapes, which are the most popular commercial tapes in the world, are summarized in Sec. 5. The anisotropic thermal properties of stacked and sandwiched DI-BSCCO tapes for the double-pancake coil application are presented in Sec. 6 using model analysis of heat equation. Finally in Sec. 7, we summarize the thermal properties of Bi-2223 wires.

2. Measurement of Thermal Properties

Thermal conductivity $\kappa(T)$ is usually measured by a steady-state heat flow method. In this section, we introduce the basic measuring technique of $\kappa(T)$ for Bi-2223 tapes in our laboratory.[8,9] The $\kappa(T)$ measurement technique for stacked or sandwiched tapes is also shown in later sections. The experimental setup of the $\kappa(T)$ measurement along the tape surface is schematically shown in Fig. 1. One end of the tape was soldered to the copper-block (sample stage) on the cold stage of a Gifford–McMahon (GM) cycle helium refrigerator. A small metal chip resistor of 1 kΩ was used as a heater, which was adhered to the other end of the tape by GE7031 varnish. κ is obtained by the

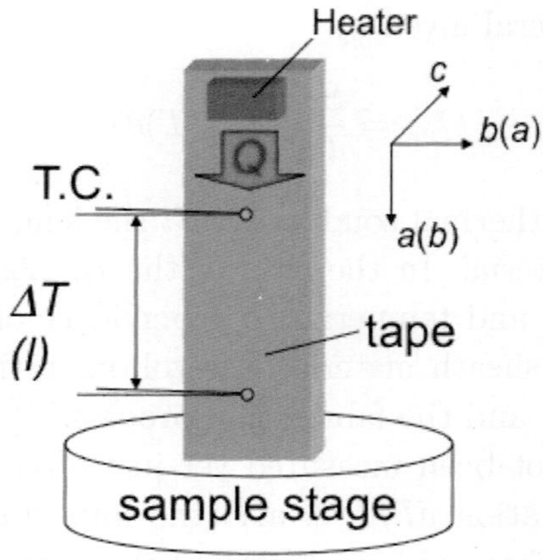

Fig. 1. Experimental setup of the thermal conductivity measurement for the Bi-2223 superconducting tapes.

following relation,

$$\kappa = \frac{Q}{\Delta T} \cdot \frac{l}{S}, \tag{2}$$

where Q is applied heat flow from the heater, ΔT is temperature difference and l is the distance between two chromel–constantan (or Au(Fe)–chromel) thermocouples (76 μm in diameter), and S is the cross section of sample. ΔT was maintained as 0.5–0.8 K by controlling the heater power automatically using a personal computer. To reduce radiation loss, $\kappa(T)$ was measured below 250 K.

Electrical resistivity $\rho(T)$ of the tape was measured by a standard four-probe method with a typical current density of about 3 A/cm^2.[10] The temperature of sample stage was controlled between 5 K and 300 K using a GM cycle helium refrigerator.

Thermal dilatation normalized by the sample length L at 300 K, $\delta L(T)/L(300\text{K}) = (L(T) - L(300\text{K}))/L(300\text{K})$, was measured from 10 K to 300 K by a strain-gauge method using a commercial strain-gauge (CFLA-1-350-11; Tokyo Sokki Kenkyujo Co., Ltd.), in which the gauge length is 1 mm, the gauge resistance is 350 Ω, and the gauge factor is 2.09.

3. Thermal Conductivity of Sheath Alloys

3.1. *Ag–Au and Ag–Cu alloys*

A popular method of reducing the thermal conductivity of metals is alloying. As Ag is generally congenial to oxide superconductors, the Ag–Au and Ag–Cu alloys were fabricated as sheath materials and the dependence of the thermal conductivity and the electrical resistivity on the Au and Cu contents in the alloys were investigated.[11]

Ag–Au and Ag–Cu alloys, containing Au and Cu from 0 to 11 at% and from 0 to 2.6 at%, respectively, were prepared by RF induction furnace and cold-rolled into tapes 50 μm in thickness. The alloy tapes were heat-treated at 840°C for 150 h in air, which was the typical condition of the Bi-2223/Ag fabricating process.

Figures 2(a) and 2(b) show the temperature dependence of the thermal conductivity $\kappa(T)$ and electrical resistivity $\rho(T)$ of Ag–Au alloy tapes with various Au contents, respectively. $\kappa(T)$ drastically decreased with increasing Au content at low temperatures. $\kappa(T)$ of the Ag-11 at% Au alloy was about three orders of magnitude smaller than that of the pure Ag tape at 20 K. The $\kappa(T)$ data were consistent with those previously reported by Crisp *et al.*,[12] in which the Ag–Au alloys were heat-treated at 900°C for 72 h in low oxygen pressure. In Fig. 2(b), $\rho(T)$ drastically increased with increasing Au content at low temperatures. As the electronic thermal conductivity ($\kappa_{\rm el}$) and the electrical resistivity ρ are related by the Wiedemann–Franz law, $\kappa_{\rm el} = L_0T/\rho$, where L_0 ($= 2.45 \times 10^{-8}$ WΩK^{-2}) is the Lorenz number, the Au content dependences in Figs. 2(a) and 2(b) are reasonable.

Figures 3(a) and 3(b) show the temperature dependence of the thermal conductivity $\kappa(T)$ and electrical resistivity $\rho(T)$ of Ag–Cu alloy tapes with various nominal Cu contents, respectively. Contrary to the Ag–Au system, $\kappa(T)$ of the Ag–Cu alloy slightly decreased with increasing Cu content at low temperatures. In accordance with the $\kappa(T)$ data, the change of $\rho(T)$ was very small with increasing Cu content.

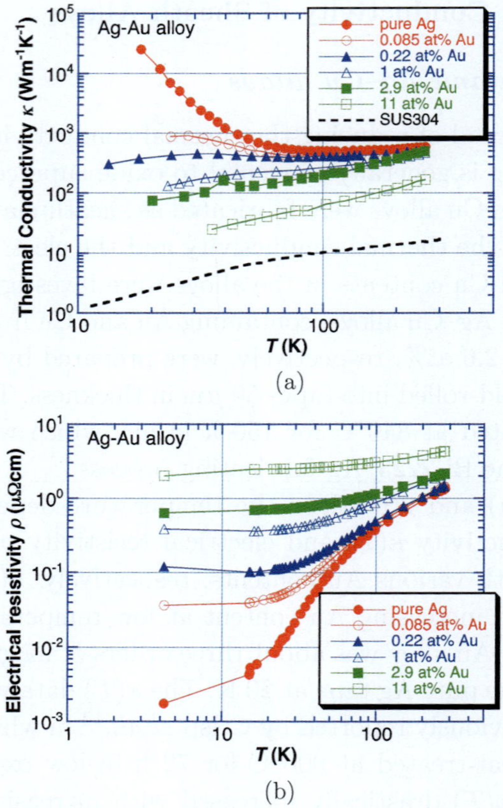

Fig. 2. Temperature dependence of (a) thermal conductivity κ and (b) electrical resistivity ρ of Ag–Au alloy tapes with various Au contents.

In order to investigate the different behavior between Ag–Au and Ag–Cu alloys, the surface observations and the compositional analysis were performed by SEM and EPMA. Figure 4 shows the typical SEM images of Ag–Au and Ag–Cu alloy tapes subjected to heat treatment in air. For comparison, the SEM image of the pure Ag tape subjected to the same heat treatments is also shown. It is found that the recrystallized grain size in the Ag–Au alloy tapes decreased drastically as the amount of Au is increased. On the other hand, the grain size in the Ag–Cu alloy tapes, which is somewhat smaller than that for pure Ag, remains almost unchanged even when the amount of Cu added is increased.

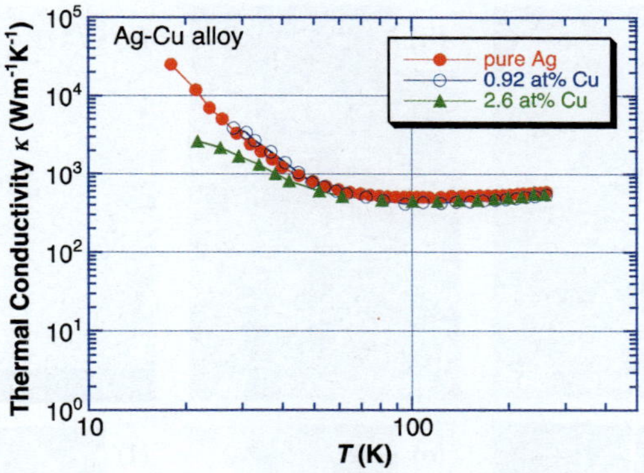

(a)

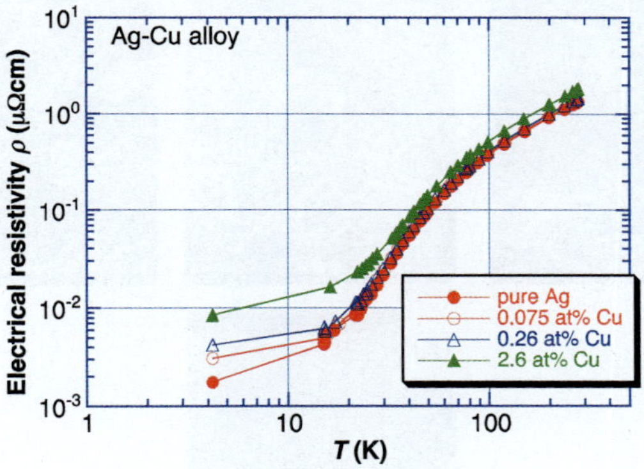

(b)

Fig. 3. Temperature dependence of the (a) thermal conductivity κ and (b) electrical resistivity ρ of Ag–Cu alloy tapes with various Cu contents.

Figure 5 presents the relation between the images from the reflection electron microscope and an area analysis of Ag, Au and Cu identified from the characteristic X-ray intensity of each element. Figure 5(a) shows the reflection electron micrograph of polished Ag-1.0 at% Au alloy tape. Figures 5(b) and 5(c) show the area analysis patterns of Au and Ag elements in the same position, respectively. Figure 5(d) shows the reflection electron micrograph of polished Ag-0.92 at% Cu alloy tape. Figures 5(e) and 5(f) show the area analysis patterns of Cu and Ag elements in the same position, respectively. In the Ag–Au alloy tape, it was found that Au atoms are homogeneously dissolved in Ag matrix. On the other hand, in the Ag–Cu

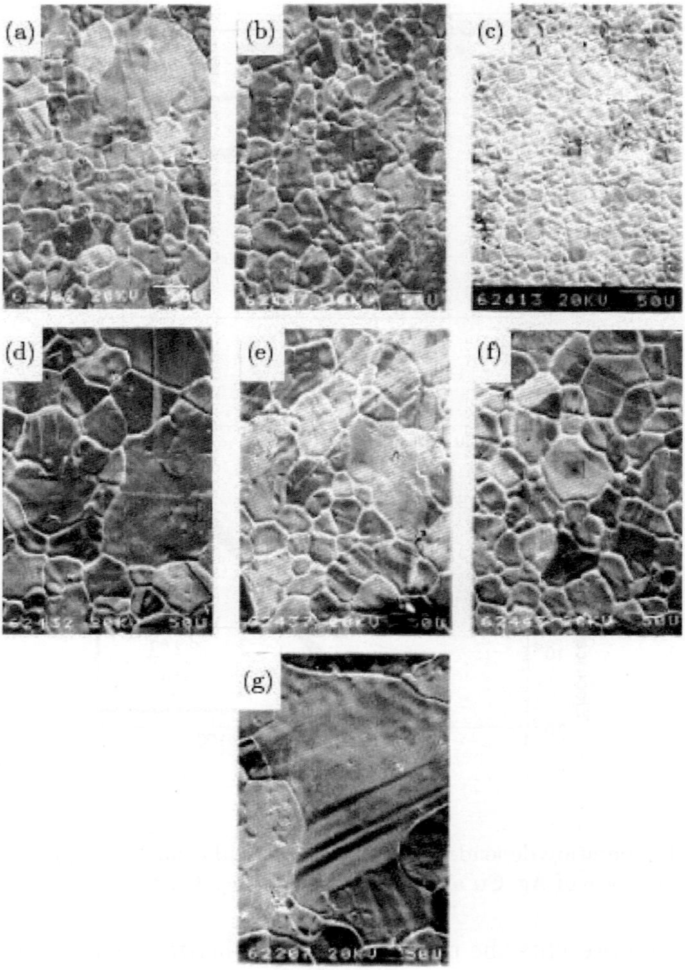

Fig. 4. SEM images of Ag–Au alloy tapes [(a) 1.0 at% Au; (b) 2.9 at% Au; (c) 11 at% Au], Ag–Cu alloy tapes [(d) 0.23 at% Cu; (e) 0.92 at% Cu; (f) 2.6 at% Cu] and (g) pure Ag alloy tape.

alloy, only a small fraction of the Cu atoms are dissolved homogeneously in the Ag matrix, accompanied by local precipitation. The Cu content in the Ag matrix was estimated to be ~0.4 at% from quantitative analysis by EPMA. The equilibrium phase diagrams of these two alloy systems show that the Ag–Au alloy is a solid solution

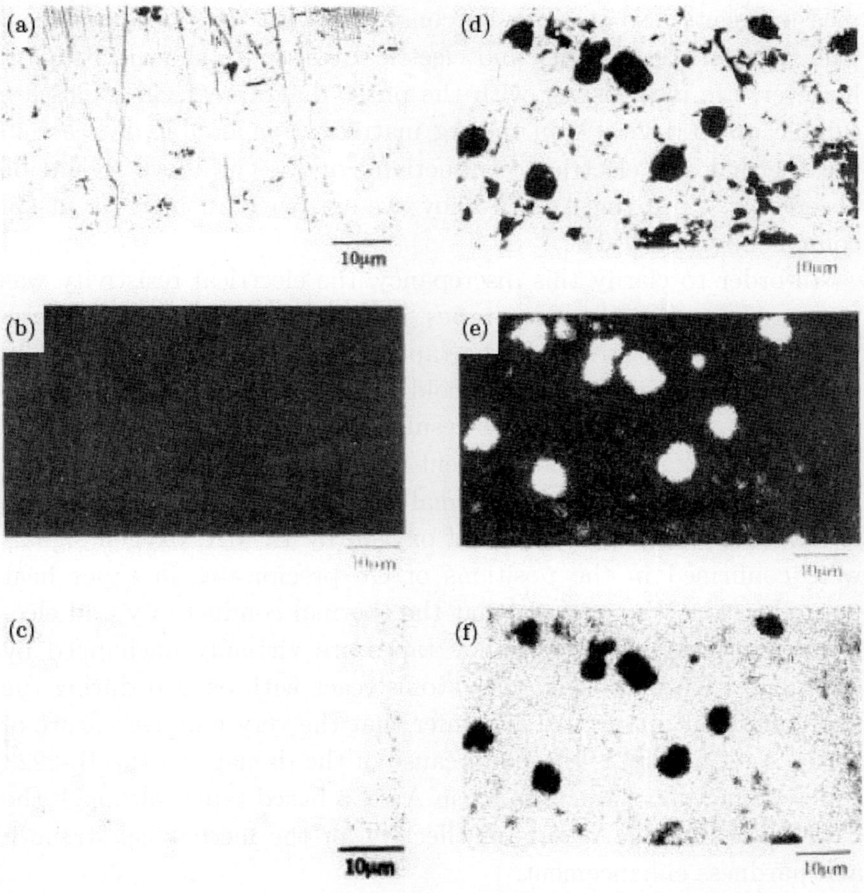

Fig. 5. (a) Reflection electron micrograph of polished Ag-1.0 at% Au alloy tape. (b), (c) Area analysis patterns of Au and Ag elements in the same position, respectively. (d) Reflection electron micrograph of polished Ag-0.92 at% Cu alloy tape. (e), (f) Area analysis patterns of Cu and Ag elements in the same position, respectively.

type,[13] and the Ag–Cu alloy is a eutectic reaction type.[14] Accordingly, Au atoms can be homogeneously dissolved in the Ag matrix at any concentration, whereas Cu atoms can be dissolved up to 0.35 at% and then the rest of the Cu atoms are precipitated. The results for the Ag–Au alloys shown in Figs. 2 and 5 are consistent with the phase diagram. For the Ag–Cu alloy system, the results of the EPMA measurements also seem to be consistent with the phase diagram. The thermal conductivity and electrical resistivity shown in Fig. 4, however, are inconsistent with the phase diagram. If Cu atoms are soluble up to 0.35 at% in the Ag matrix, a remarkable decrease in the thermal and electrical conductivity of Ag–Cu alloys should be observable, as in the Ag–Au alloy system, with an increase in Cu content up to 0.35 at%.

In order to clarify this discrepancy, the electrical resistivity was measured for Ag–Cu alloy tapes in air or nitrogen atmosphere. The electrical resistivity of the tape heat-treated in air drastically decreased, compared with the cold-rolled tape and the tape heat-treated in nitrogen gas. These results indicate that Cu atoms react with oxygen during heat treatment in air. The oxidized copper does not seriously affect both the thermal conductivity and electrical resistivity. From the area analysis of oxygen by EPMA, oxygen signals were confirmed in the positions of Cu precipitates in tapes heat treated in air. We conclude that the thermal conductivity and electrical resistivity of Ag–Cu alloy tapes are virtually unchanged by addition of Cu because copper atoms react with oxygen during the heat treatment in air. We also infer that the very reactive nature of Cu in Ag–Cu alloy is the main cause of the damage to the Bi-2223 superconducting characteristics in Au–Cu based tapes, although the Cu addition to the Ag tape is effective for the mechanical strength and hardness enhancement.

3.2. *Other Ag–based alloys*

$\kappa(T)$ decreases with the increase in the content of alloy-elements of Au, Pd, Cu and Mg *etc.*, suggesting that the alloy-elements play an important role in determining the $\kappa(T)$ value.[11,12,15–17] At the same time, these additions to the Ag sheath improved mechanical strength (Young's modulus and hardness). This improvement is related with the presence of dispersed alloying element, which leads to the smaller grain size. Table 1 shows the typical thermal conductivity $\kappa(T)$ for several Ag–Au alloys.[11,15,16] $\kappa(T)$ of all the Ag alloys decreases with increasing contents of the impurity atoms. However, at present, the main Ag-based alloy as a sheath material is Ag–Au alloy with some

Table 1. Thermal conductivity of Ag alloys.

	κ ($Wm^{-1}K^{-1}$)			
Sample	77 K	40 K	30 K	Ref.
Pure Ag	363.8	—	997.2	15
Pure Ag	—	1004.6	—	16
Pure Ag	530	1220	2867	11
Ag–Au(0.22 at%)	401	440	455	11
Ag–Au(7 at%)	—	109.7	—	16
Ag–Au(11 at%)	51.1	28.5	—	11
Ag–Mg(0.0005 at%)	266.5	—	342.6	15
Ag–Mg(0.3 at%)	—	411.4	—	16
Ag–Au(0.25 at%)-Mg(0.001 at%)	118.1	—	62.1	15
Ag–Au(0.025)-Mg(0.002 at%)	112.6	—	59.2	15
Ag–Pd(0.06 at%)-Mg(0.02 at%)	82.5	—	28.9	15
Ag–Sb(1.1 at%)	—	142.3	—	16

additional impurity such as Mg, Al, Pd *etc.* for the enhancement of the mechanical strength.

4. Thermal Conductivity of BSCCO Tapes

Using the Ag–Au alloy sheath, many BSCCO tapes were fabricated and the thermal and superconducting properties had been investigated in 1990s. In this section, the thermal properties, heat intrusion and the influence of magnetic field on the thermal conductivity of the BSCCO tapes were introduced.[11]

$Bi_{1.8}Pb_{0.4}Sr_{2.0}Ca_{2.2}Cu_{3.0}O_x$ (Bi-2223) oxide superconducting powder was wrapped with Ag-11 at% Au alloy tape and then cold-rolled into a tape, which was a single filament tape. The tape was heat-treated at the maximum temperature of 840°C in air, which was necessary for the fabrication of Bi-2223 superconducting tape. The final size of the tape was about 300 μm in thickness and about 8 mm in width.

Figure 6 shows the temperature dependence of the thermal conductivity κ_{AgAuT} of the Bi-2223 superconducting tapes sheathed with Ag-11 at% Au alloy. The superconductor cross-section ratio f_{sc} to the total cross section is 0.65. In Fig. 6, the thermal conductivities of the

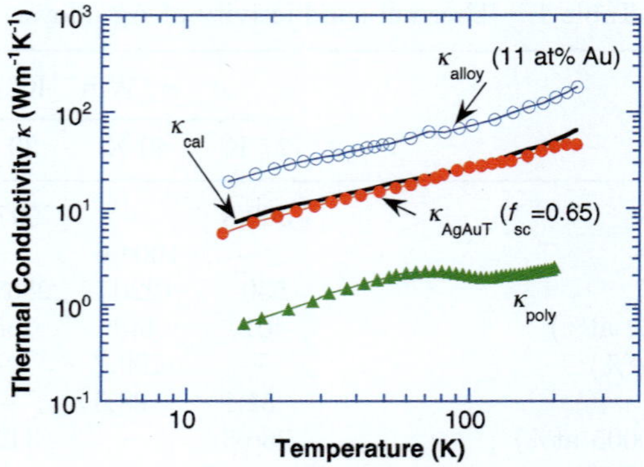

Fig. 6. Temperature dependence of the thermal conductivity κ_{AgAuT} of Bi-2223 tape sheathed with Ag-11 at% Au alloy with the ratio of $f_{\mathrm{sc}} = 0.65$. The thermal conductivity of the Ag-11 at% Au alloy κ_{alloy}, that of Bi-2223 polycrystal κ_{poly} and the calculated thermal conductivity κ_{cal} were also shown.

Ag–Au alloy κ_{alloy} and that of Bi-2223 sintered polycrystal κ_{poly} are also shown. κ_{poly} was 2.4 $\mathrm{Wm^{-1}K^{-1}}$ at 200 K, decreased monotonically with decreasing temperature, showed a significant upturn just below T_{c} (=105 K) and took a local maximum around 70 K, which is about two orders of magnitude smaller than that of Ag-11 at% Au tape and the magnitude of κ_{AgAuT} lies between κ_{alloy} and κ_{poly}. The thermal conductivity of the superconducting tape κ_{cal} can be estimated using the κ_{poly}, κ_{alloy} and f_{sc} by $\kappa_{\mathrm{cal}} = f_{\mathrm{sc}}\kappa_{\mathrm{poly}} + (1 - f_{\mathrm{sc}})\kappa_{\mathrm{alloy}}$ with thick solid line as shown in Fig. 6. As a result, κ_{cal} is very close to the experimentally obtained κ_{AgAuT}. This confirms that the thermal conductivity of a superconducting tape with arbitrary ratio f_{sc} and with various Au contents in the alloy sheath can be practically estimated by use of the relation. It is useful in a power current lead design to be able to estimate the heat flow through the lead rather than to measure it.

The heat intrusion Q_{int} through the current lead using the superconducting tapes can be calculated using Eq. (1). We used the Bi-2223 superconducting tapes sheathed with Ag–Au alloys with $f_{\mathrm{sc}} = 0.65$ and calculated the Q_{int} value for $l = 100$ cm current lead of $I = 1,000$ A. Figure 7(a) presents the results of the calculation of the relation between the heat intrusion through the lead and the

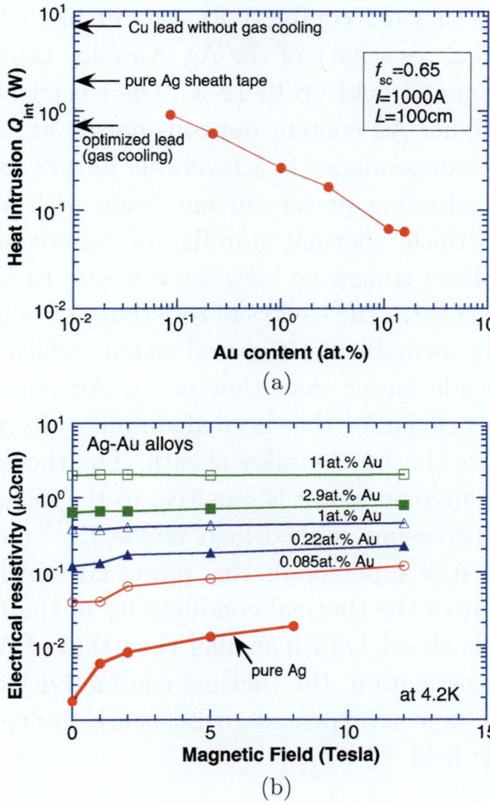

Fig. 7. (a) Relation between the heat intrusion Q_{int} through a 100 cm superconducting lead ($f_{\text{sc}} = 0.65$) of $I = 1,000$ A and the Au content in the Ag–Au alloy sheath. (b) Magnetic field dependence of the electrical resistivity of Ag–Au alloys with various concentrations of Au at 4.2 K.

Au content in the Ag–Au alloy sheath. The heat intrusion Q_{int} of the current lead decreases with the increase in the Au content. For example, the Q_{int} value of a superconducting lead sheathed with Ag–11 at% Au alloy is about two orders of magnitude smaller than that of a conventional copper lead without gas cooling or a Bi-2223 superconducting lead sheathed with pure Ag.

In many cases, power current leads for cryogenic equipment such as SCMs are used under magnetic fields. The influence of magnetic fields on the electrical resistivity of Ag–Au alloy tapes was measured and then on the thermal conductivity was estimated. Figure 7(b) shows the electrical resistivity of the Ag–Au alloy tapes at 4.2 K, as a function of magnetic field up to 12 T. The electrical resistivity of the tapes with higher Au content depends hardly at all on the magnetic field. This independence is a favorable nature for stable operation of superconducting power current leads with alloy sheathes. Although the intrinsic thermal stability of superconducting tapes sheathed with alloys somewhat deteriorates due to large electrical resistivity in the sheath, the electrical resistivity of bulk oxide superconductors in the normal state is several mΩcm, which is about three orders of magnitude larger than that of the Ag–Au alloy tapes. It should still be beneficial for the thermal stability of superconducting power leads to use the Ag–Au alloy sheath. The thermal conductivity of the oxide superconductor is sensitive to the magnetic field and decreases with increasing applied field below T_c,[18] and could cause unstable operation of superconducting power current leads. Because the absolute value of the thermal conductivity of the Bi-2223 superconducting core is about 1/15th smaller than that of Ag–Au alloy as shown in this investigation, the thermal conductivity of the Bi-2223 superconducting tape is expected to be nearly independent of the applied magnetic field.

5. Thermal Properties of DI-BSCCO Tape

In this section, we introduce the thermal properties of single DI-BSCCO tapes.[10] Table 2 summarizes the specifications of four types of DI-BSCCO tapes (types H, HT, G, and AC). Type H, whose sheath is a silver-alloy, is a standard tape. The silver-alloy consists of the silver and a small amount of other metal element. In order to achieve high yield strength, type HT was prepared, in which the type H tape was laminated from both surfaces by the SS (SUS) tapes 0.02 mm in thickness. The sheath material of type G is an Ag-5.4 wt% Au alloy. Type AC is designed for AC applications, in which the twisted Bi-2223 filaments are sheathed by the silver-alloy.

Figure 8 shows the temperature dependence of the electrical resistivity $\rho(T)$ for four DI-BSCCO tapes. The critical temperature T_c is

Table 2. Specifications of the DI-BSCCO tapes.

Type	Sheath Material	T_c(K)	Cross Section (mm^2)	Width (w) (mm)	Thickness (t) (mm)
H	Ag alloy	110.8	0.935	4.25	0.22
HT	Ag alloy + SUS	111.0	1.100	4.40	0.31*
G	Ag-5.4wt%Au	111.0	0.994	4.14	0.24
AC	Ag alloy	110.8	0.440	2.20	0.20

Note: * including the thickness of SUS parts (0.02 mm × 2) and solder layers (N.A.).

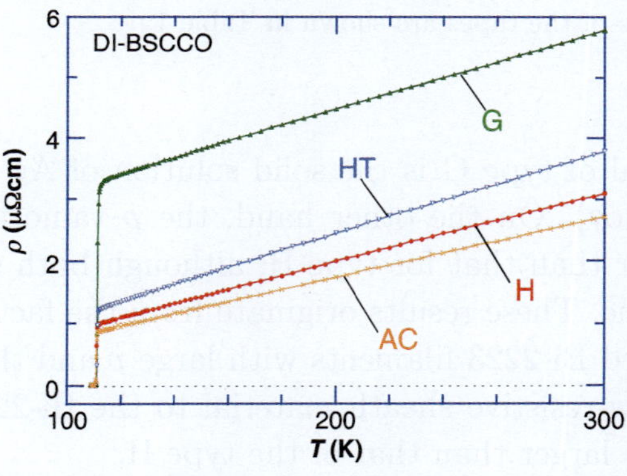

Fig. 8. Temperature dependence of the electrical resistivity $\rho(T)$ for four types of DI-BSCCO tapes (Type H, HT, G, and AC) with different sheath materials. The specifications of the tapes are shown in Table 1.

about 111 K for all the tapes, indicating good quality of the superconducting part of Bi-2223 filaments. All the $\rho(T)$ curves show a good linear dependence in the normal state. In this regime, the absolute value of $\rho(T)$ of the metal sheaths are rather smaller than that of Bi-2223 filaments. Therefore, the applied current mainly flows through the metal sheath; for instance, the ρ value for the Bi-2223 polycrystal was reported to be about 900 $\mu\Omega$cm at 295 K[19] and that for pure silver 1.6 $\mu\Omega$cm at room temperature.[20] The absolute value of the type G is rather larger than those for other three types, because the

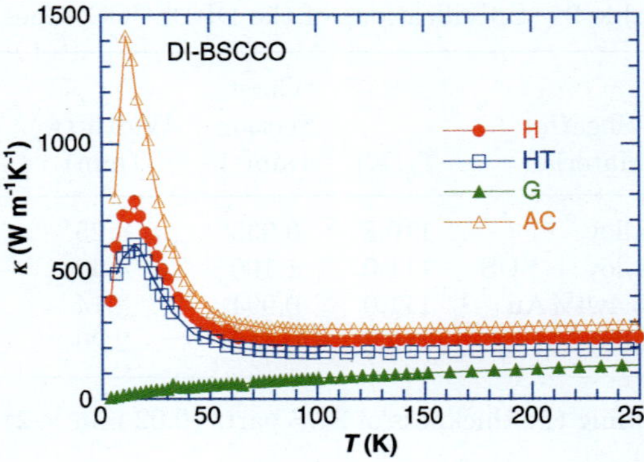

Fig. 9. Temperature dependence of the thermal conductivity $\kappa(T)$ for four types of DI-BSCCO tapes (Type H, HT, G, and AC) with different sheath materials. The specifications of the tapes are shown in Table 1.

sheath material of type G is the solid solution of Ag–Au alloy (Ag-5.4wt% Au alloy). On the other hand, the ρ value for type AC is slightly smaller than that for type H, although both sheath materials are the same. These results originate from the fact that type AC contains twisted Bi-2223 filaments with large ρ and that the volume ratio of the low-resistive sheath material to the Bi-2223 filament of the type AC is larger than that of the type H.

Figure 9 presents the temperature dependence of the thermal conductivity $\kappa(T)$ for four DI-BSCCO tapes. $\kappa(T)$ of the type H decreases quite slightly with decreasing temperature from 250 K to 100 K. It increases moderately between 100 K and 60 K, starts to increase rapidly below 60 K, and takes a peak of 780 $\mathrm{Wm^{-1}K^{-1}}$ around 17 K. The steep increase and peak structure is often observed in $\kappa(T)$ of a high-purity silver,[12] which originates from a large mean free path of the electrons at low temperatures. The behavior of $\kappa(T)$ for the type HT is basically similar to that of type H, but the absolute value for type HT is about 20–25% smaller than that for type H. Assuming that the thickness of the type H tape in the type HT tape is the same as that of the type H tape, the ratio of the cross section between a couple of SUS with solder layers and the whole tape is estimated to be 0.29:1.0. Therefore, $\kappa(T)$ of type HT should be smaller about 29% than that of type H, if the applied heat thoroughly flows through the part of Ag-alloy. SUS is a poor heat conductor compared with Ag. For instance, the absolute values of a standard sample of the austenitic stainless steel (SRM1460) provided by the National Institute of Standards and Technology (NIST) are about 12 $Wm^{-1}K^{-1}$ at 200 K and 1.5 $Wm^{-1}K^{-1}$ at 20 K.[21] On the other hand, those of pure Ag are approximately 600 $Wm^{-1}K^{-1}$ at 200 K and 20,000 $Wm^{-1}K^{-1}$ at 20 K.[12] Therefore, the most of applied heat flows through the part of Ag-alloy sheath also for type HT, giving rise to the reduction of $\kappa(T)$. $\kappa(T)$ of type G decreases monotonically with decreasing temperature. The absolute value of κ of type G near 77 K is about one-third of that of type H. No peak and a small value in the $\kappa(T)$ for type G is often observed in the Ag–Au alloys with higher Au contents,[12] which originates from the large phonon scattering by disorders. The similar $\kappa(T)$ have been observed in the Bi-2223 tapes sheathed with other metal alloy.[5–7] $\kappa(T)$ of type AC behaves similarly to that of types H and HT, but the peak value of type AC near 10 K is extremely larger than that for other two types. This large peak suggests that the volume ratio of sheath material to the Bi-2223 is large in type AC. We note that no anomaly in T_c was observed for the samples.

In a conducting solid, the measured thermal conductivity comprises both contributions due to phonons and electrons, κ_{ph} and κ_{el}. Since the $\rho(T)$ curves show the good T-linear dependence above T_c, as found in Fig. 8, the κ_{el} is expected to be T-independent. The obtained values of κ_{el} at 150 K for each sample are about 244 $Wm^{-1}K^{-1}$ (type H), 198 $Wm^{-1}K^{-1}$ (type HT), 110 $Wm^{-1}K^{-1}$ (type G), and 282 $Wm^{-1}K^{-1}$ (type AC), respectively. These values are comparable with the measured κ values in the normal state. On the other hand, the κ value of the sintered Bi-2223 polycrystal was reported to be about 3.3 $Wm^{-1}K^{-1}$ at 150 K.[19] Therefore, the contribution of the Bi-2223 filament to the heat transport is negligibly small and the most of the applied heat is carried by electrons in the metal sheath.

Figure 10 shows the temperature dependence of the thermal dilatation $\delta L(T)/L(300\ \mathrm{K})$ for various DI-BSCCO tapes. $\delta L(T)/L(300\ \mathrm{K})$ of all the samples shows a similar temperature

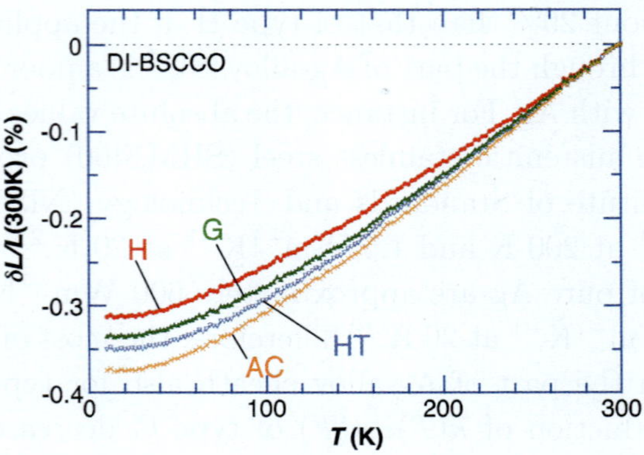

Fig. 10. Temperature dependence of the thermal dilatation $\delta L(T)/L(300\text{ K})$ for four types of DI-BSCCO tapes (Type H, HT, G, and AC) with different sheath materials.

dependence; from 300 K to 25 K, the $\delta L(T)/L(300\text{ K})$ decreases monotonically with decreasing temperature, and shows almost constant value of $-0.31 \sim -0.38\%$ below 25 K. Weiss *et al.* reported that the thermal dilatation of types H and HT at 5 K are approximately 0.27% and 0.3%, respectively,[7] which are about 10% smaller than the present results. However, we cannot discuss the discrepancy, because they did not describe the measuring method and accuracy.

6. Thermal Conductivity of DI-BSCCO Tapes with Stacked or Sandwiched Structure

6.1. *Thermal conductivity of multi-stacked DI-BSCCO tapes*

In this subsection, the thermal conductivity of stacked samples consisting of six DI-BSCCO tapes (Type H) that were joined by a commercial solder (LFM-48) was presented. The measured $\kappa(T)$ was compared with the estimated $\kappa(T)$ which was calculated using the measured $\kappa(T)$ of the single tapes and solder.[22]

Figure 11 shows the temperature dependence of the thermal conductivity, $\kappa^{(1)}_{\text{mes}}(T)$, $\kappa^{(6)}_{\text{mes}}(T)$ and $\kappa^{\text{sol}}_{\text{mes}}(T)$, for the single tape (type H), six-fold sample (type H6) and solder (LFM-48), respectively.

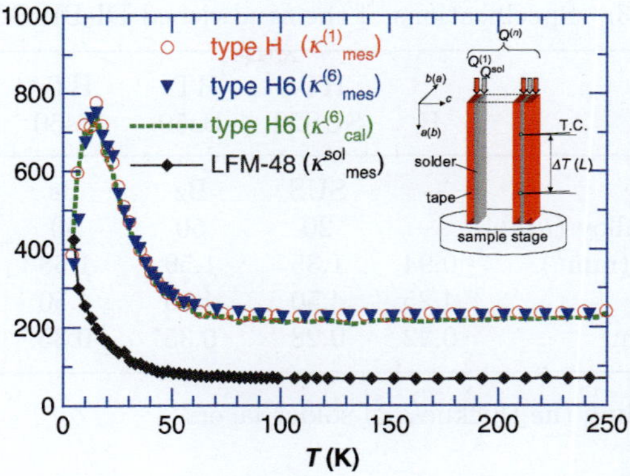

Fig. 11. Temperature dependence of the thermal conductivities, $\kappa(T)$, for the DI-BSCCO tape sheathed by Ag (H), the six-fold sample (H6), and the solder (LFM-48). The calculated $\kappa(T)$ values for the six-fold sample (H6 cal) are also shown.

$\kappa^{(6)}_{\rm mes}(T)$ almost coincides with $\kappa^{(1)}_{\rm mes}(T)$. The absolute value of $\kappa^{\rm sol}_{\rm mes}(T)$ is somewhat smaller than both $\kappa^{(1)}_{\rm mes}(T)$ and $\kappa^{(6)}_{\rm mes}(T)$, but increases very slowly with decreasing temperature from 250 K to around 40 K. Below 40 K, it increases somewhat rapidly with decreasing temperature.

$\kappa^{(6)}_{\rm cal}(T)$ was estimated using those of the single tape and the solder. In the estimation, parallel heat flow model was used. The schematic image of heat flow through the stacked sample is demonstrated in the inset of Fig. 11. The stacked six-fold sample consists of six tapes and five solder layers. The detail explanation was referred in ref. 22. $\kappa^{(6)}_{\rm cal}$ was also shown in Fig. 11, which nearly corresponds to the measured $\kappa^{(6)}_{\rm mes}(T)$, meaning that the thermal conductivity of the six-fold sample can be estimated using those of the single tape and the solder.

6.2. *Thermal conductivity of sandwiched DI-BSCCO tapes*

In order to improve the mechanical strength of the DI-BSCCO tape, the tape is usually reinforced by metal tapes such as stainless steel and brass. In this subsection, the thermal conductivity of

Table 3. Specifications of the sandwiched DI-BISCCO tapes.

Type	H	HT-SUS20	HT-Bz50	HT-Bs50	HT-Bs100	H6
Alloy tape	—	SUS	Bz	Bs	Bs	—
Thickness of alloy (μm)	—	20	50	50	100	—
Cross section (mm^2)	0.94	1.35	1.59	1.58	2.03	5.88
Width (mm)	4.25	4.50	4.55	4.50	4.50	4.42
Thickness (mm)	0.22	0.28	0.35*	0.35*	0.47*	1.33*

Note: * including the thickness of solder layers.

the DI-BSCCO tape sandwiched by various kinds of metal tapes was presented. The measured $\kappa(T)$ was compared with the estimated $\kappa(T)$ using the $\kappa(T)$ values of the bare DI-BSCCO tape, the reinforcing metal tape and the solder.[23]

The type HT tape was prepared to achieve a high yield stress, in which the type H tape was sandwiched from both surfaces by the thin metal tapes with various thicknesses using the commercial solder (LFM-48). Stainless steel (SUS), bronze (Bz) or brass (Bs) were used as the reinforcing material. The type HT tape with SUS tapes 20 μm in thickness was named HT-SUS20. Other HT samples were named by this rule as shown in Table 3, in which the specifications of the DI-BSCCO tapes studied here were summarized. A stacked bundle (H6) was also presented.

Figure 12(a) shows the temperature dependence of the thermal conductivity $\kappa(T)$ for the type H, four sandwiched type HT and type H6 tapes. On the cooling run, $\kappa(T)$ of the type H tape decreases moderately from 200 K down to 100 K and subsequently increases below 100 K. It begins to increase rapidly below 60 K, and takes a maximum of about 800 Wm^{-1}K^{-1} around 15 K. The shape of $\kappa(T)$ for all the type HT tapes is qualitatively similar to those for the type H and the type H6. However, the absolute value of the thermal conductivity, $\kappa_{\text{abs}}(T)$, depends on the species and thickness of the metal tape, because $\kappa_{\text{abs}}(T)$ of the metal strongly depends on the species of element and the composition. Figure 12(b) shows the measured $\kappa(T)$ for the SUS tape 20 μm in thickness and the solder (LFM-48). The reported data for bronze (Cu:Sn = 9:1)[24] and

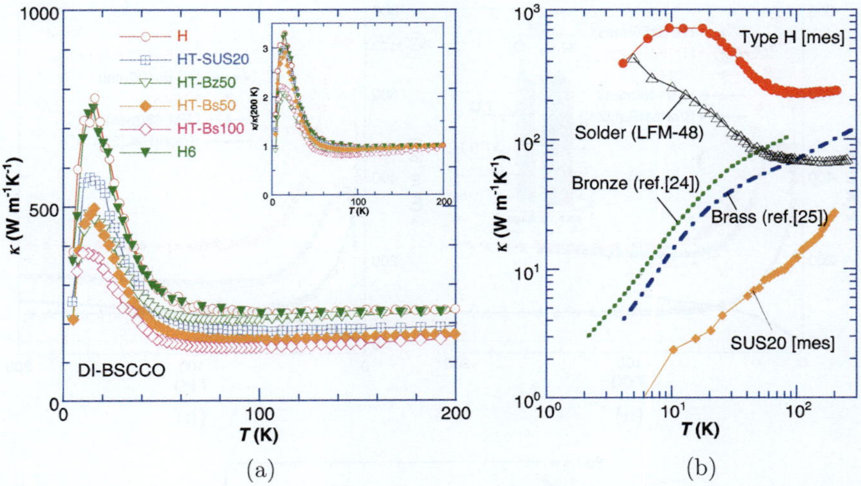

Fig. 12. (a) Temperature dependence of the thermal conductivity $\kappa(T)$ for the type H, four kinds of type HT laminated with different metal and six stacked type H (H6) tapes. The inset shows the $\kappa(T)/\kappa(200\ \mathrm{K})$ for the same samples. (b) The measured $\kappa(T)$ for the type H tape, the SUS tape 20 μm in thickness and the solder (LFM-48), the reported $\kappa(T)$ for bronze (Cu:Sn=9:1)[24] and the recommended $\kappa(T)$ for brass (Cu:Zn=7:3).[25]

the recommended data for brass (Cu:Zn = 7:3)[25] are also shown. $\kappa(T)$ for the solder increases very gradually with decreasing temperature from 200 K down to around 40 K. It begins to increase rapidly below 40 K with decreasing temperature. $\kappa(T)$ for other metals (SUS, bronze, brass) decreases monotonically with decreasing temperature. The $\kappa_{\mathrm{abs}}(T)$ for the SUS tape is about one to three orders of magnitude smaller than that for the type H tape. For bronze and brass, $\kappa_{\mathrm{abs}}(T)$ is somewhat smaller than that for the type H tape above 30 K and is one to two orders of magnitude smaller below 30 K. The inset of Fig. 12(a) shows the reduced $\kappa(T)$ value at 200 K, $\kappa(T)/\kappa(200$ K). The $\kappa(T)/\kappa(200\ \mathrm{K})$ curves for the types H, HT-SUS20, HT-Bs50 and H6 tapes almost coincide with each other, which means that the contribution of the alloy tape to the heat transport can be neglected and the applied heat almost flows through the silver sheath in the type H tape. On the other hand, the $\kappa(T)/\kappa(200\ \mathrm{K})$ for the types HT-Bz50 and HT-Bs100 tapes deviates from that for the type H

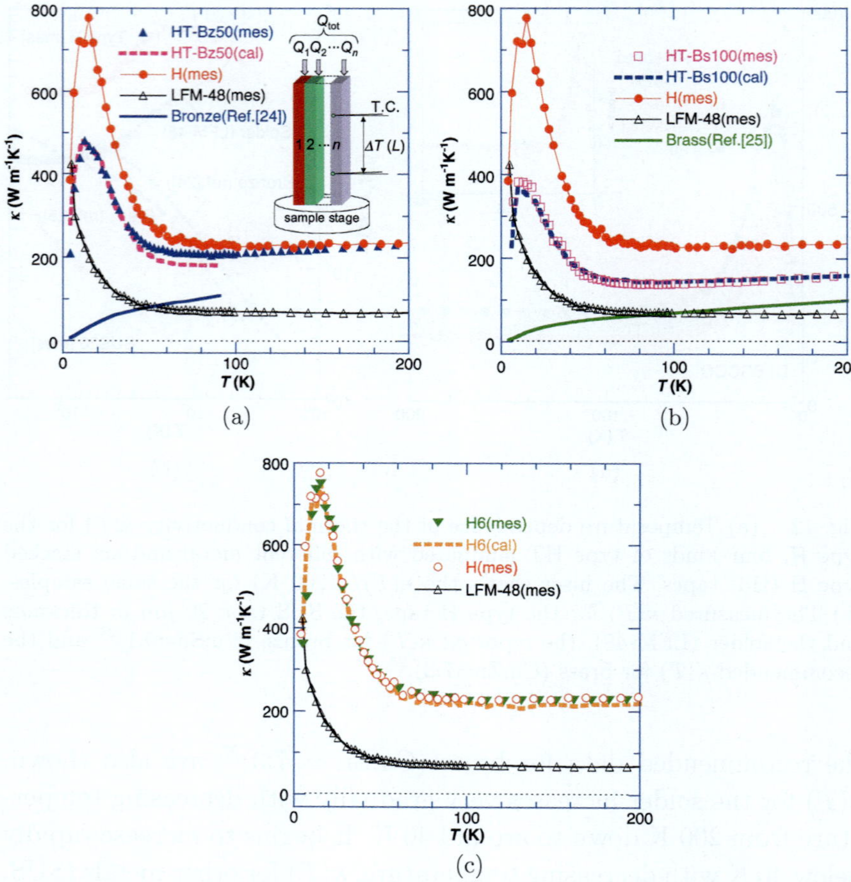

Fig. 13. The measured and calculated $\kappa(T)$ for the (a) HT-Bz50, (b) HT-Bs100 and (c) H6 tapes. The inset of (a) demonstrates a schematic image of the heat flow through the composite sample consisting of n kinds of materials.

tape, indicating that the contribution of the metal tapes cannot be ignored to the heat transport.

We try to estimate the $\kappa(T)$ for both HT-Bz50 and HT-Bs100 tapes using $\kappa(T)$ for each material. The schematic image of the heat flow through a composite sample, which consists of n kinds of materials, is illustrated in the inset of Fig. 13(a). Assuming that the applied heat flows in parallel, the total heat through the composite sample

Q_{tot} is given by

$$Q_{\text{tot}} = \sum_{i=1}^{n} Q_i, \tag{3}$$

where Q_i is the heat flow through the i-th part. Since the temperature difference ΔT and the length l are the same for all the parts, the total thermal conductivity κ_{tot} can be estimated by

$$\kappa_{\text{tot}} = \left(\sum_{i=1}^{n} (\kappa_i \times S_i) \right) \Big/ S_{\text{tot}}, \tag{4}$$

where, κ_i is the thermal conductivity and S_i is the cross section for i-th part. Figures 13(a) and 13(b) present the calculated $\kappa(T)$ for the HT-Bz50 and HT-Bs100 tapes, respectively, using this relation. The calculated $\kappa(T)$, which is depicted as the broken line, nearly corresponds to the measured one for both tapes. We note that the disagreement between the measured and calculated $\kappa(T)$ is found for the type HT-Bz50 tape above 40 K, which indicates the possibility that the amount of Cu in bronze for reported $\kappa(T)$ is somewhat smaller than that in the brass tape of the type HT-Bz50 tape. The similar estimation was performed for the type H6 tape in Fig. 13(c).

The measured $\kappa(T)$ was well reproduced by the calculated one. These results indicate that the $\kappa(T)$ for the composite sample can be estimated from $\kappa(T)$ for each part using the equivalent circuit for the heat flow with the simple sum rule.

6.3. *Anisotropic thermal conductivity of sandwiched DI-BSCCO tapes*

The thermal conductivity $\kappa(T)$ of the Bi-2223/Ag tapes have been usually measured only along the length direction, because the thickness of the tape is too thin to measure $\kappa(T)$ by a steady-state heat flow method. In this subsection, we estimated the thermal conductivity of the Bi-2223/Ag tape along three directions; the length, width and thickness ones.[26] $\kappa(T)$ along the length direction was measured using a single tape, and those along the width and thickness directions were estimated using stacked bundles which consisted of several Bi-2223/Ag tapes. We discuss the anisotropic heat flow in the Bi-2223/Ag tape using an equivalent heat current circuit.

Type H of DI-BSCCO was used, in which the sheath metal is Ag with a small amount of other metal elements. The "Ag-ratio" defined by V_{Ag}/V_{Bi2223} was estimated to be about 1.8 from the photograph of the cross section of the tape, where V_{Ag} and V_{Bi2223} were the volume of the Ag sheath and Bi-2223 filaments, respectively. $\kappa(T)$ of the Bi-2223/Ag tape was determined along three typical directions, the length (l), width (w) and thickness (t) ones in the following. As described in Fig. 14(a), $\kappa(T)$ along the l-direction was measured directly using a single Bi-2223/Ag tape (type H), which was abbreviated as the l-sample. To estimate $\kappa(T)$ of the tape along the w- and t-directions, two stacked bundles were prepared by joining several

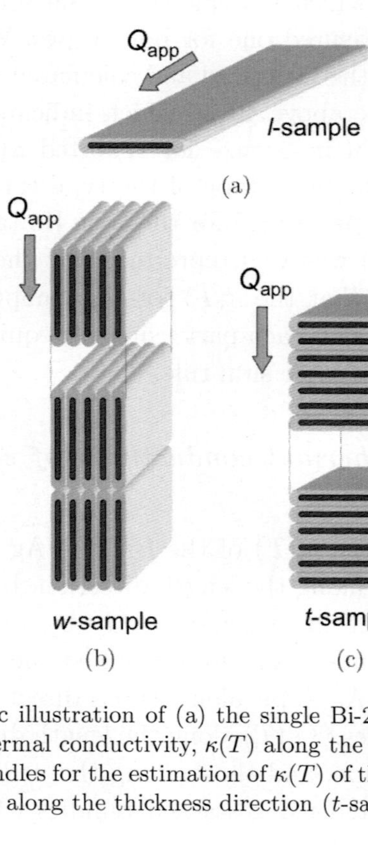

Fig. 14. Schematic illustration of (a) the single Bi-2223/Ag tape for the measurement of the thermal conductivity, $\kappa(T)$ along the length direction (l-sample) and the stacked bundles for the estimation of $\kappa(T)$ of the tape (b) along the width (w-sample) and (c) along the thickness direction (t-sample).

Table 4. Specifications of the anisotropic samples.

Sample name	l-sample	w-sample	t-sample
Distance between the thermocouples L (mm)	10.30	8.40	6.50
Cross section S (mm^2)	0.935	8.20	8.60
Width (mm)	4.25	4.10	4.30
Thickness (mm)	0.220	2.00	2.00

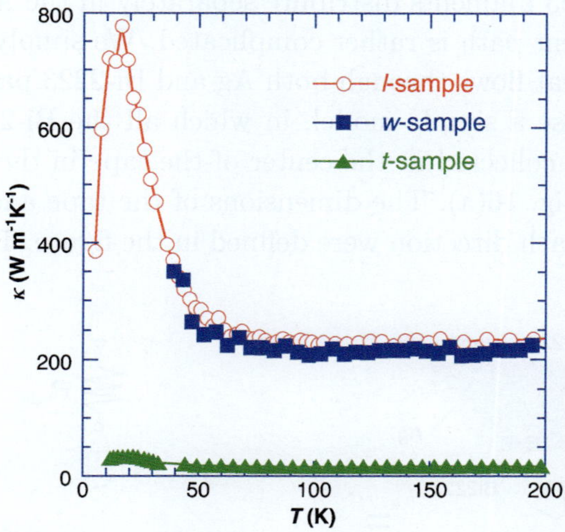

Fig. 15. Temperature dependence of the thermal conductivity $\kappa(T)$ along the three directions, the length (l), width (w) and thickness (t) ones, for Bi-2223/Ag tapes.

Bi-2223/Ag tapes using a commercial solder (LFM-48), as described in Figs. 14(b) and 14(c), which were named w- and t-sample, respectively. The dimensions of three samples are summarized in Table 4.

Figure 15 presents the temperature dependence of the thermal conductivity $\kappa(T)$ of three samples. $\kappa(T)$ of the l-sample slightly decreases with decreasing temperature, begins to increase below 60 K, and becomes maximum of about 800 $\mathrm{Wm^{-1}K^{-1}}$ around 17 K, which is similar to $\kappa(T)$ of Bi-2223/Ag tape reported previously.[22,23,26] $\kappa(T)$ of the w-sample almost coincides with that of the l-sample. For the w-sample, $\kappa(T)$ could not be measured below 40 K because of the limit of the heater power. The temperature dependence of $\kappa(T)$ of the t-sample is similar to that of other two samples. However the absolute value of $\kappa(T)$ of the t-sample is about one order of magnitude smaller than that of other samples. The thickness of the solder layer between the tapes is found to be extremely thin from the photograph and the $\kappa(T)$ value of the solder (LFM-48)[22] is rather larger than that of the t-sample. Therefore, the small $\kappa(T)$ value of the t-sample does not originate from the solder layers.

The Bi-2223 filaments distribute separately in the Ag sheath and the heat current path is rather complicated. We simply assume that the applied heat flows through both Ag and Bi-2223 parts in parallel and we propose a simple model, in which all the Bi-2223 filaments were virtually collected at the center of the tape in the cross section as shown in Fig. 16(a). The dimensions of the tape and the Bi-2223 region along each direction were defined in the figure. The equivalent

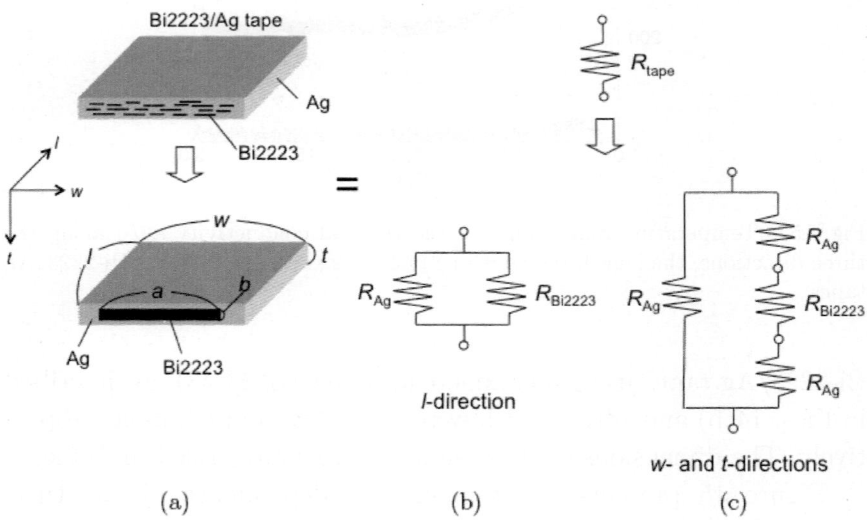

Fig. 16. (a) Schematic image of a model for the equivalent heat current circuit. The dimensions of the tape and virtually collected Bi-2223 region are illustrated. The equivalent heat current for (b) the l-direction and the (c) both w- and t-directions. R_{tape}, R_{Ag} and $R_{\mathrm{Bi\text{-}2223}}$ are the thermal resistances of the whole tape, the Ag region and the Bi-2223 filaments, respectively.

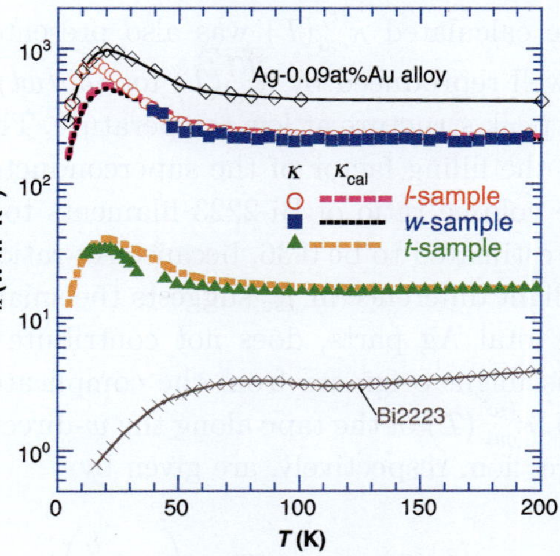

Fig. 17. Temperature dependence of the measured and calculated thermal conductivity $\kappa(T)$ and $\kappa_{\mathrm{cal}}(T)$ for three samples. The reported $\kappa(T)$ curves of Ag-0.09 at% Au alloy and Bi-2223 polycrystal are also plotted.

heat current circuits for the l-sample and both w- and t-samples are represented in Figs. 16(b) and 16(c), respectively. Here, R_{tape}, R_{Ag}, and $R_{\text{Bi-2223}}$ are the thermal resistances of the whole tape, the Ag region and the Bi-2223 filaments, respectively. The thermal resistance R is given by $\kappa^{-1} \times (l/S)$. In general, the absolute value of $\kappa(T)$ of Bi-2223 is about two orders of magnitude smaller than that of high-purity Ag, as shown in Fig. 17. The purity of the Ag sheath in the Type H tape was estimated to be about 99.9% from the $\kappa(T)$ measurement.[22] In this study, we use the reported $\kappa(T)$ of Ag-0.09 at% Au alloy[11] as $\kappa_{\mathrm{Ag}}(T)$. The $\kappa(T)$ value of the polycrystalline Bi-2223[19] is used as $\kappa_{\text{Bi-2223}}(T)$, because $\kappa(T)$ of Bi-2223 single crystal has not been reported. Consequently, the heat current path including Bi-2223 in series can be ignored because of $R_{\text{Bi-2223}} \gg R_{\mathrm{Ag}}$. Thermal conductivity along the l-direction, $\kappa_{\mathrm{cal}}^{(l)}(T)$, can be calculated by the following relation,

$$\kappa_{cal}^{l}(T) = \kappa_{Ag}(T) \times \left(1 - \frac{ab}{wt}\right). \tag{5}$$

In Fig. 17, the calculated $\kappa_{\text{cal}}^{(l)}(T)$ was also presented. $\kappa(T)$ of the l-sample was well reproduced by $\kappa_{\text{cal}}^{(l)}(T)$ for $(ab/wt) = 0.45$, except for around the peak structure at low temperature. The (ab/wt) value is the same as the filling factor of the superconductor, f_{sc}, which is defined by the volume ratio of Bi-2223 filaments to the whole tape and is roughly estimated to be 0.36, because Ag-ratio $(V_{\text{Ag}}/V_{\text{Bi}-2223})$ was 1.8. The slight difference in f_{sc} suggests the small amount of Ag, about 15% of total Ag parts, does not contribute to the thermal transport. This might originate from the complicated heat path in the core region. $\kappa_{\text{cal}}^{(w)}(T)$ of the tape along the w-direction and $\kappa_{\text{cal}}^{(t)}(T)$ along the t-direction, respectively, are given by

$$\kappa_{\text{cal}}^{(w)}(T) = \kappa_{Ag}(T) \times \left(1 - \frac{b}{t}\right), \tag{6}$$

and

$$\kappa_{cal}^{(t)}(T) = \kappa_{Ag}(T) \times \left(1 - \frac{a}{w}\right). \tag{7}$$

The similar fitting procedure was adopted for $\kappa(T)$ of the w- and t-samples, and then $a/w = 0.96$ and $b/t = 0.47$ were obtained independently. $(a/w) = 0.96$ multiplied by $(b/t) = 0.47$ equals to 0.45. The resultant value is consistent with the ab/wt value determined by the analysis for the l-sample, which supports the validity of our model. Consequently, we can conclude that the measured $\kappa(T)$ of the 'stacked' w- and t-samples correspond to those of the tape along w- and t-directions, because the thickness of the solder layer between the tapes is negligibly thin and the contact thermal resistance is ignored. The mechanism of the thermal transports for the w-direction is almost the same as that for the l-direction. The value of $a/w = 0.96$ means that the applied heat flows through only about 4% of the total Ag region for the t-direction. The volume ratio of the edges of the tape to the whole tape, $V_{\text{edge}}/V_{\text{tape}}$, is roughly estimated to be 5%. Therefore, $(a/w) = 0.96$ is a plausible value and we can conclude for the t-direction that the routes of the applied heat are both edges of the tape. The similar analysis was also shown in the Ref. 27.

The temperature dependence of the thermal resistances $R_r(T)$ and $R_z(T)$ of the parallelepiped samples cut from a double-pancake coil along the radial (r) and the thickness (z) directions, respectively, was reported.[28] The double-pancake coil was wound with DI-BSCCO tape sheathed with silver alloy. Both $R_r(T)$ and $R_z(T)$ increased monotonically with decreasing temperature. The thermal transport in the coil was analyzed by use of the parallel and series heat current circuit of DI-BSCCO tapes and insulators. The obtained value of the thermal contact resistance is not a universal one, which depends on the condition of the interface between the materials. Using the results, the temperature rise in the coil can be estimated, when a hot-spot creates. These analyses might play an important role in the design of the cooling system.

7. Summary

We have introduced the thermal conductivity $\kappa(T)$ of Bi-2223 wires sheathed with Ag-based alloys. The thermal conductivity of sheathed Ag-based alloys was also summarized. For DI-BSCCO tapes and the stacked or sandwiched DI-BSCCO tapes, the anisotropic thermal conductivities were measured and analyzed using the heat flow model. The $\kappa(T)$ data is useful and valuable for calculating the heat intrusion and thermal stability for the application of the superconducting coils and power current leads.

Acknowledgments

The work of the development of Ag–Au alloy sheath material was performed with the Hitachi Cable Ltd. I would like to thank T. Sasaoka for his valuable discussion. The work of the thermal properties of DI-BSCCO was performed with the SEI, Ltd. I would like to thank Y. Yamada and J. Fujikami of SEI for their valuable discussions.

References

1. J. Shimoyama, T. Morimoto, H. Kitaguchi, and H. Kumakura *et al.*, *Jpn. J. Appl. Phys.* **31**, L163 (1992).

2. Y. Yamada, B. Obst, and R. Flükiger, *Supercond. Sci. Technol.* **4**, 165 (1991).
3. S. Kobayashi, K. Yamazaki, T. Kato, K. Ohkura, E. Ueno, K. Fujino, J. Fujikami, N. Ayai, M. Kikuchi, K. Hayashi, K. Sato, and R. Hata, *Physica C* **426–431**, 1132 (2005).
4. S. Yamade, N. Ayai, J. Fujikami, S. Kobayashi, E. Ueno, K. Yamazaki, M. Kikuchi, T. Kato, K. Hayashi, K. Sato, H. Kitaguchi, and J. Shimoyama, *Physica C* **463–465**, 821 (2007).
5. T. Sasaoka, K. Nomura, J. Sato, S. Kuma, H. Fujishiro, M. Ikebe, and K. Noto, *Appl. Phys. Lett.* **64**, 1304 (1994).
6. H. Fujishiro, M. Ikebe, K. Noto, T. Sasaoka, and K. Nomura, *Cryogenics* **33**, 1086 (1993).
7. K. P. Weiss, M. Schwarz, A. Lampe, R. Heller, W. H. Fietz, A. Nyilas, S. I. Schlachter, and W. Goldacker, *IEEE Trans. Appl. Supercond.* **17**, 3079 (2007).
8. H. Fujishiro, M. Ikebe, T. Naito, K. Noto, S. Kohayashi and S. Yoshizawa, *Jpn. J. Appl. Phys.* **33**, 4965 (1994).
9. H. Fujishiro, M. Ikebe, K. Nakasato, and T. Naito, *Czech. J. Phys.* **46**, Suppl. S5, 2761 (1996).
10. T. Naito, H. Fujishiro, and Y. Yamada, *IEEE Trans. Appl. Supercond.* **19**, 3034 (2009).
11. H. Fujishiro, M. Ikebe, K. Noto, M. Matsukawa, T. Sasaoka, K. Nomura, J. Sato, and S. Kuma, *IEEE Trans. Mag.* **30**, 1645 (1994).
12. R. S. Crisp and J. Rungis, *Phil. Mag.* **22**, 217 (1970).
13. J. L. White, *Trans. AIME* **215**, 178 (1959).
14. J. Markali and P. Thoresen, *Acta Chem. Scand.* **15**, 31 (1961).
15. H. S. Park, B. K. Ji, J. H. Lim, J. Joo, S-B. Jung, W. Nah, J. Yoo, J. Ko, and H. Kim, *IEEE Trans. Appl. Supercond.* **11**, 3277 (2001).
16. S. H. Jang, J. H. Lim, J. H. Kim, J. J. Joo, W. Nah, J. S. Volf, H. K. Liu, and M. Apperley, *IEEE Trans. Appl. Supercond.* **13**, 2956 (2003).
17. M. Putti, C. Ferdeghini and G. Grasso, *Int. J. Mod. Phys.* **B14**, 2908 (2000).
18. S. D. Peacor, J. L. Cohn and C. Uher, *Phys. Rev.* **B43**, 8721 (1991).
19. M. Ikebe, H. Fujishiro, T. Naito, and K. Noto, *J. Phys. Soc. Jpn.* **63**, 3107 (1994).
20. H. E. Boyer and T. L. Gall. In *American Society for Metals.* Ohio: Metals Park, 1–53 (1985).
21. H. Fujishiro, T. Naito, M. Ikebe, and K. Noto, *J. Cryo. Super. Soc. Jpn.* **28**, 533 (1993) (in Japanese).
22. T. Naito, H. Fujishiro, and Y. Yamada, *Cryogenics* **49**, 429 (2009).
23. T. Naito, H. Fujishiro, and Y. Yamada, *J. Phys. Conf. Series* **234**, 022025 (2010).
24. Y. S. Touloukian, R. W. Powell, C. Y. Ho, and P. G. Klemens. In *Thermophysical Properties of Matter.* New York-Washington: IFI/Plenum, 588 (1970).

25. C. Y. Ho, M. W. Ackerman, K. Y. Wu, S. G. Oh, and T. N. Havill, *J. Phys. Chem. Ref. Data* **7**, 1091 (1978).
26. T. Naito, H. Fujishiro, and J. Fujikami, *IEEE Trans. Appl. Supercond.* **21**, 2828 (2011).
27. T. Naito, H. Fujishiro, G. Osabe, and J. Fujikami, *Physica C* **484**, 159 (2013).
28. T. Naito, H. Fujishiro, and Y. Yamada, *Physica C* **470**, 1397 (2010).

Chapter 1.8

Thermal Stability of Bi-2223 Wires

Vitaly Vysotsky*, Alexander Rakhmanov† and Nikolai Zmitrenko‡

*Russian Scientific Cable R&D Institute, Moscow, Russia

†Institute Theoretical and Applied Electrodynamics, Moscow, Russia

‡Keldysh Institute of Applied Mathematics RAS, Moscow, Russia

The review novel approaches are presented to describe quench/heat developments in HTS Bi-2223-based devices. These approaches permit determination of thermal stability level for both: Uniform and non-uniform heating of HTS devices, particularly during overheat conditions. Thermal quench/thermal runaway current and characteristic time of heating development can be evaluated. At non-uniform heating blow-up, regimes with temperature localization could happen that should be taken into account by HTS cables designers. Review of studies of cooling of HTS wires by boiled liquid nitrogen is also presented.

1. Introduction

Superconducting devices made of Bi-2223 superconductors were the very first ones able to work at temperatures up to temperature of liquid nitrogen. High operating temperature in comparison with liquid helium temperature demanded novel approaches to understand and to describe stability and quench development in Bi-2223 based superconducting devices. Such approaches were developed first for quasi-uniform case and described in Refs. 1–7. The universal scalable analytical model has been developed[1] permits determination of thermal quench ("thermal runaway")[8] current I_q, which is the maximum stable current at certain cooling conditions of HTS devices. With good cooling this current may exceed what we call critical current with low cooling — I_q can be less than critical current.[5–7] The theory developed was very well confirmed by many experiments and permits to find out conditions for stable or unstable regime for Bi-2223 based superconducting devices with high accuracy.[1–7]

Another problem is connected with overload regimes. The power electro-technical devices (power cables, transformers, etc.) made with HTS wires should have one general feature — they must withstand fault currents dozen times more than their operating currents (if we do not consider special current limiting devices). A concern is whether the fault current will over-heat the HTS conductor and degrade the superconducting properties or burn-out the HTS tape altogether.

We studied the problem of overload regimes in HTS devices both theoretically and experimentally. With numerical experiments, we showed that in long HTS objects the blow-up regimes with strong heat localization take place.[9–12] In adiabatic conditions, always the temperature rise happens fast ("thermal runaway").[8] In conditions with cooling, two possible modes are possible: Stable and unstable, like it was found for the quasi-uniform heating by the analytical model[1,6,7] mentioned above. There is the sharp border between stable and unstable regimes in relation to the current density change. At the unstable regime the heat localization takes place in case of the disturbance presented and instability development time is rather small in comparison with the uniform heating.[12]

With the experimental study of overload regimes in superconducting tapes cooled by boiling liquid nitrogen it was found that there is sharp switch between cooling regimes from convectional cooling to nucleate boiling (NB) and cooling.[13–16] It is known that at NB regime the heat transfer coefficient from a surface can be by orders of value more in comparison with the convection cooling.[17,18] In this case Bi-2223 wire can survive much higher overload current due to much better cooling. The numerical model developed permitted to describe this phenomenon with good coincidence between calculated and experimental data.[13–16]

In this chapter, we present the review of studies mentioned on stability and quench/heat development of Bi-2223 wires and devices.

2. Novel Approaches to Describe Quench/Heat Development

The analytical model of thermal quench/heating development in HTS devices was first presented in Ref. 1 and is described in details in Refs. 6 and 7. It used the standard power law for voltage–current characteristic of HTS superconductors: $E = E_o[I/I_c(T)]^n$. Here, I is a transport current, E is electric field in a superconductor, n is the slope of the $E - I$ curve, I_c is the parameter usually called "critical current", while in our approach it is just a *parameter* that means current corresponding to the electric field E_0. By solution of usual heat balance equation it was shown[1] that near the thermal quench current (TQC) I_q, analytical expressions could be found for two cases. If $I < I_q$, the temperature stabilize at some level T_q–T_f. If $I > I_q$, the temperature rises with strong acceleration after time t_q.[1–7] Two major regimes do exist: Stable and unstable shown in Fig. 1. The following major expressions were found:
For the TQC I_q:

$$\frac{I_q}{I_c(T_0)} = \frac{n}{n+1}\left[\frac{hP(T_c - T_0)}{nE_0 I_c(T_0)}\right]^{1/(n+1)} \tag{1}$$

Characteristic times:

$$t_h = \frac{CA}{Ph} \qquad t_f = t_h\sqrt{\frac{2I_q}{|I - I_q|(n+1)}} \tag{2}$$

Here, T_0 and T_c are ambient temperature and critical temperature of a superconductor, T_q is a characteristic temperature at which fast temperature rise starts at time t_q, t_f is a time necessary to heat a sample to equilibrium temperature $T_q - T_f$ at $I < I_q$.[1] t_h is a characteristic thermal time, where C is a volumetric specific heat, A and P

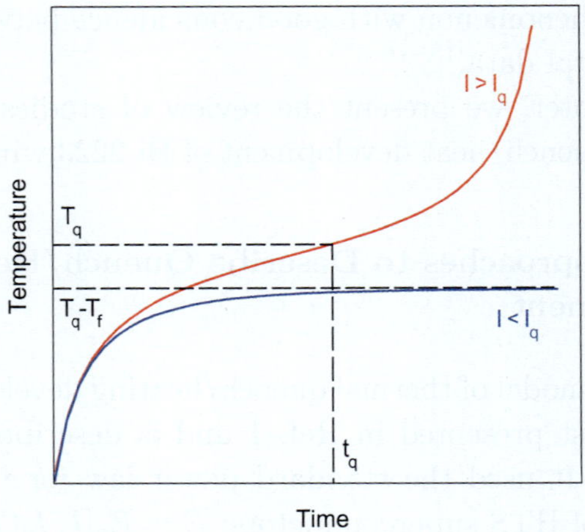

Fig. 1. Temperature traces for two general modes: Unstable (when thermal run-away happens) $I > Iq$ and stable at $I < Iq$.

are conductor's cross section and effective cooling perimeter, respectively, and h is a heat removal coefficient. More details about the model are presented in Refs. 1, 6, and 7.

It was also shown that expression (1) for $I > I_q$ are universal and permits the scaling for the widest variety of superconducting devices. The expression above were extensively verified by experiments as it is shown in Fig. 2 as one of examples.[1–7]

From this model confirmed by the experimental studies, several important conclusions about stability of HTS devices can be made.

The major is the difference in quench/heat development behavior between LTS and HTS devices. The high specific heat at usual operating temperatures for HTS devices leads to the very strong difference of characteristic times of the heat processes development t_h. Specific heat is about 2,000 times more for HTS at LN_2 temperatures. Heat removal coefficient h is about 15 times (LN_2 cooling) more for HTS, but much less for indirect cooling by cryo-coolers. In any case, the specific heat development time τ_h is at least 200 times and more long at LN_2 temperature. In principle, it makes easier protection of HTS devices as one could have more time for detecting voltage or resistive zone and for activation of protection devices. Switch from the superconducting to the real normal zone with full resistivity in

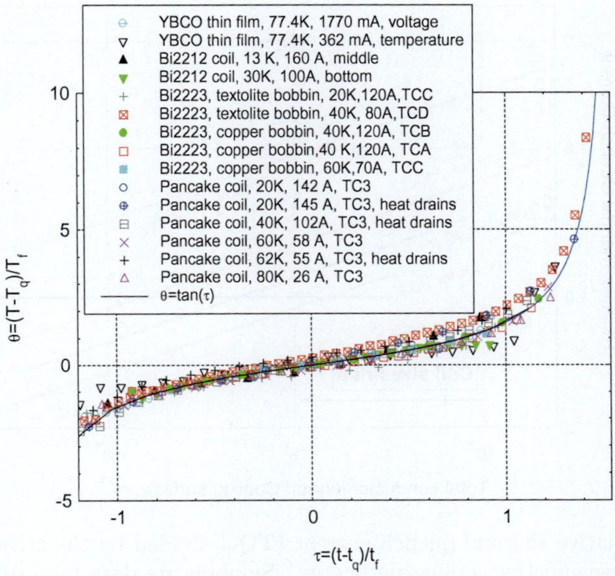

Fig. 2. Example of scaling.[6,7] Dimensionless temperature θ versus dimensionless time τ for experiments with different HTS objects.

LTS devices occurs very quickly, during $\sim 10^{-5}-10^{-3}$ s.[17,18] In HTS devices this process could be quite slow (even up to $\sim 10^{3}-10^{4}$ s)[2–4], like it is shown in Fig. 1. Thus it is difficult to determine real normal zone in HTS at high temperatures and one could speak about *propagation of the resistive zone* only, *not normal zone propagation*!

The introduced specific TQC I_q or thermal runaway current has resembled the critical current for LTS devices.[1,6,7] Important thing is that depending on size of the device and cooling conditions, TQC could be more than HTS critical current I_c at certain temperature or less than it. The major consequence from the implementation of this theory is the necessity of changing design criteria for HTS devices. Usually, design criteria for any superconducting device includes: Maximum allowable operating current I_{oper} and the time t_{prot} when protection devices should be fully activated if a quench happens. The time mentioned is determined by the maximum allowed heating temperature T_{max}.

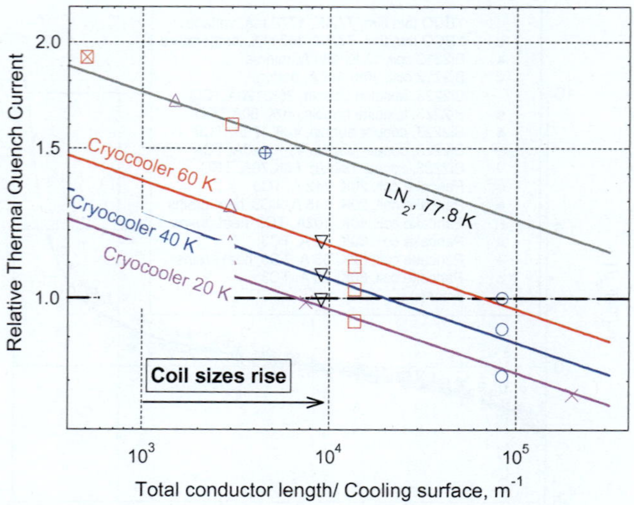

Fig. 3. Relative thermal quench current (TQC divided by the critical current) *versus* inverse effective cooling perimeter.[4] Symbols are data from different HTS coils from Refs. 3, 19–22, solid lines- calculations by (1) for $n = 12$.

LTS devices work if operating current is less than the critical current $I_{oper} < I_c$ and this is the "critical current design criterion". HTS devices may work if current is more than I_c, but may not work even at currents less than I_c, especially in large devices. This is illustrated in Fig. 3, where the TQC divided by the critical current ("relative TQC") is shown *vs.* characteristic parameter of total conductor length divided by the cooling perimeter.[7] Because heat release is in the volume of a device while heat removal is from the surface, their ratio reduces with sizes. In this case thermal quench or thermal runaway current may be less than the critical current. Thermal runaway current design criteria should be used[8], i.e., I_{oper} *must be less* than I_q.

The reasonable criteria for the heating temperature should be also changed. From the maximum heating temperature T_{max} used for LTS devices[17,18] to the temperature T_q, at which the slope of $T(t)$ is drastically changing for HTS devices (see Fig. 1).

The change of the protection time criteria is from the time necessary to heat a device up to T_{max} (LTS devices) to the time t_q, when the slope of $T(t)$ curve is changing (for HTS devices), i.e., t_{prot} must be less than t_q.

Both these criteria for t_{prot} and T_{max} should be used because beyond T_q and t_q, the temperature rise is so fast that it could barely be controlled.

It is necessary to note that the low index n, while reducing the "critical current" or increasing "temperature of current sharing", may lead to better stability if good enough cooling is provided. It may be important for large CICC cables, where the reduction of n was observed.[23] The theory developed for HTS superconducting objects with the low index n may also be used for large CICC cables.

The approaches presented and discussed in Refs. 1–7 were developed for quasi-uniform heating but as we will see below all major conclusions are valid for non-uniform heating too.

3. Non-Uniform Cases and HTS Device at Overloading Conditions

Next step in implementation of new approaches to quench/heat development in HTS devices is analysis of non-uniform heating for long or large objects, like HTS power cables.

Non-uniform cases are important if the characteristic heat length $l_h = \sqrt{Ak/Ph}$ is less than the characteristic size of HTS-devices. Here k is heat conductivity. Once the characteristic heat length is large enough (good heat conductivity and low cooling, which are in most cryo-cooler cooling cases), one may consider a winding as quasi-uniform and use the formulas (1)–(2) above and other expressions from Refs. 1, 6, and 7. For windings, the heat length may be presented[6,7] as $l_h = \sqrt{Vol \cdot k \cdot G^{-1}}$, where Vol is the total volume of the winding and G is the effective cooling per surface unit. So, the heat length depends on a winding's volume and cooling. In Ref. 4, the typical values of the characteristic heat length were analyzed depending on characteristic winding size (cubic root of the winding volume) for different temperatures at cryo-cooler cooling and liquid nitrogen (LN_2) cooling. It was shown that l_h are rather large (~ 1 m and more) in the case of cryo-cooler cooling and even large windings may be considered as quasi-uniform with all parameters averaged along the winding.[1,3]

In case of LN_2 cooling, heat length l_h becomes low (1–10 cm) and strong non-uniformity may appear. We observed the non-uniformity at LN_2 cooling in the four-pancake magnet described in Refs. 2–4.

As we mentioned above, HTS cables being installed to a power grid may undergo overloading regimes when current *forcibly* becomes much more than its operating value and critical current. In this case, there is no way to talk about normal zone and its propagation. HTS cables are obviously "long HTS objects" cooled by liquid nitrogen and, therefore, they are rather vulnerable to the non-uniformity appearance.

Due to strong non-linearity with temperature of the heat release and material parameters of HTS devices, some specific phenomena could happen: *Blow-up regimes with the heat localization.*[9–12]

The study of these phenomena (beside the experiments) could be done only by numerical analysis. We performed numerical studies of this phenomenon with several computer-based experiments.[9–12] The numerical solution of the standard thermal balance equation[1,9,12] was performed with parameters that were as much as possible close to the reality. Depending on parameter combinations (current density, cooling, initial disturbance, etc.) different heating modes may appear.[11,12]

In adiabatic cases eventually, a fast temperature rise happens with heat localization (while may be after rather long time). This is illustrated in Fig. 4.

In the presence of cooling, two modes of the heat development do exist, *absolutely similar to the analytical model* for quasi-uniform case-stable and unstable. This is illustrated in Figs. 5 and 6. In the stable mode, the initial disturbances disappear. In the unstable mode, the very fast, actually catastrophic temperature rise happens with the heat localization (see Fig. 6). The time till the temperature runaway starts in the unstable regime becomes rather short with a current density rise. Like in the analytical model, this time can be considered as the safety parameter.

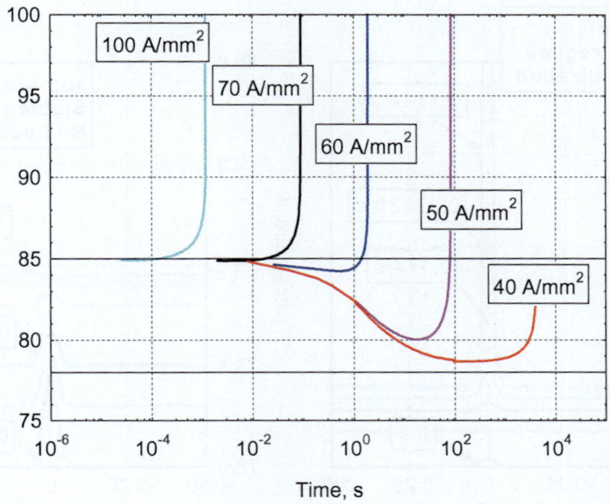

Fig. 4. Instability developments in the adiabatic cases for different current density in an HTS tape. Slow decay eventually changes to the fast rise of the temperature.[11]

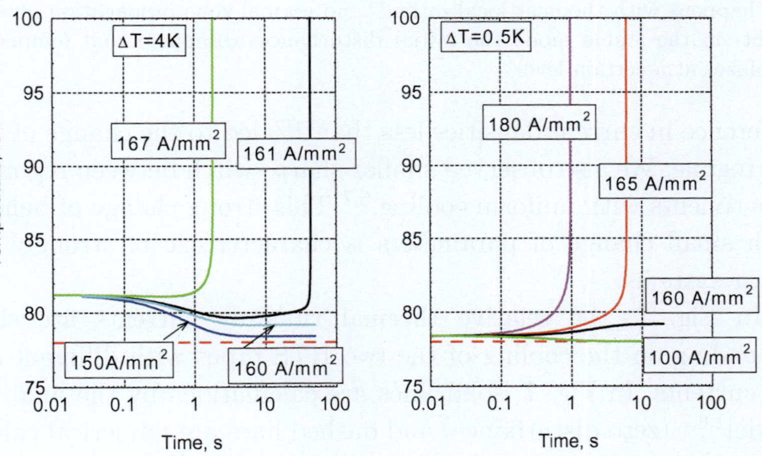

Fig. 5. The examples of the instability developments with time in the cases with cooling. Different initial disturbances levels ΔT are shown. Small change of the transport current leads to the change of regimes.

It is necessary to note that when switch from stable to unstable mode happens, transport current changes very little. Say, in Fig. 5 left, one can see that current density 160 A/mm^2 corresponds to the stable mode; while at 161 A/mm^2, the instability happens. The

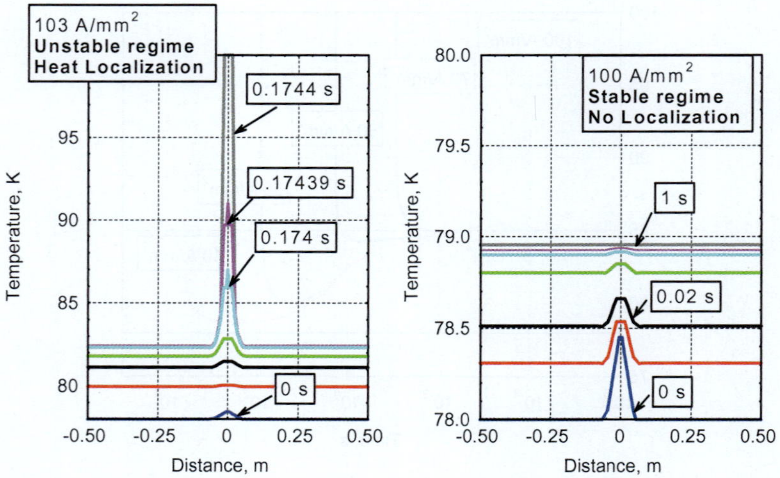

Fig. 6. The examples of the heating developments in the presence of the initial disturbance in the cases with cooling. Different initial disturbance levels are shown. Left- in the unstable mode the fast, actually a catastrophic temperature rise happens with the heat localization[11], no normal zone propagation observed. Right- in the stable mode the initial disturbances disappear and temperature stabilizes at a certain level.

difference in current densities less than 1% led to the change of heating regime. We also observed similar sharp switch between regimes in experiments with uniform cooling.[6,7] This strong change of behavior with small change of parameters is characteristic of strongly nonlinear systems.

In Fig. 7, the relative thermal runaway currents are shown depending on the cooling of the two HTS tapes with different critical currents. In Fig. 7, solid lines are calculations by the analytical model[1,6,7] (zero disturbance) and dashed lines are numerical calculations by the upgraded model[12] and by the model[11] with the temperature disturbances in the center of the sample ($\Delta T \sim 0.5$–4 K). In the model[11] the heat release was approximated for the simplicity by the power function, while in the upgraded model[12] we used the real heat release function. One can see the practical coincidence of calculations with different methods. Symbols shown in Fig. 7 are the experimental data. It was shown in Ref. 1 that the thermal runaway current I_q is proportional to the heat removal coefficient h like $I_q \sim h^{1/n+1}$ (see

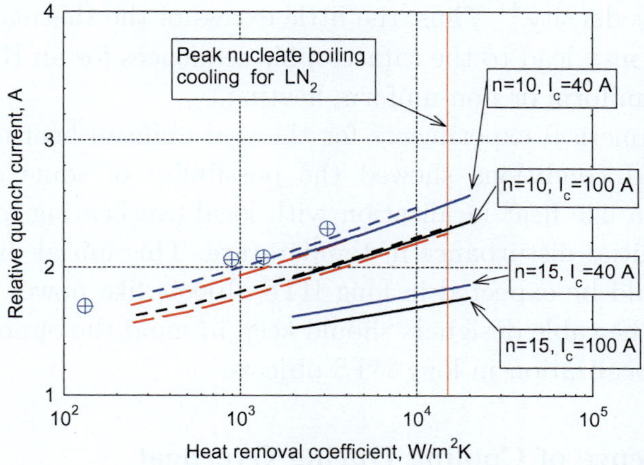

Fig. 7. Relative thermal runaway currents (transport current divided by the critical current) *versus* the heat removal coefficient. Solid lines- analytical calculations by theory[1,4]. The upgraded model calculations[11] (short dashed lines), the experimental data[12] (symbols) and the data from calculations by the model[11] (long dashed lines) are also shown.

expression (1) above). That means very weak dependence on cooling at the high index n. It also permits more or less precise evaluation of the thermal runaway current for different HTS devices.

One can see that the *behavior* of the long HTS devices *with non-uniform heating* is quite close to the *behavior of HTS devices with uniform* (or quasi-uniform) heating. The models for both cases work and provide close results that well coincide with the experiments when determining limiting operating current or heating development time. In both models, we did not use any "superconducting" words: Like "normal zone", "normal zone propagation" or "minimum propagating energy". Anyway, reasonable predictions of HTS devices' behavior could be done and models are quite clear from the physical point of view.

For both the uniform and non-uniform heating, sharp switching between stable and non-stable mode does exist. Less than 1% of the transport current change may lead from the stability to the fast temperature rise. The time of the temperature rise reduces drastically with the current rise, and inversely proportional to the square of the current density.[11] Thus, the little excess of the thermal runaway current I_q may lead to the catastrophic sequences for an HTS device for both, uniform or non-uniform heating.

The numerical experiments for the non-uniform heating of HTS at overload conditions showed the possibility of some dangerous phenomena like heat localization with local overheating in presence of some initial disturbance in temperature. This initial disturbance always could be expected in long HTS objects like power cables. In any case, the cable designers should keep in mind the opportunity of the heat localization in long HTS objects.

4. Influence of Cooling During Overheat

In the Secs. 2 and 3 above, we discussed stability and quench-thermal runaway behavior of Bi-2223 based HTS devices when they are working at currents near TQC. Also possibility of heat localization and blow-up regimes were shown by numerical experiments. In this section, we will discuss the peculiarities of quench/heat development in HTS wires at boiling liquid nitrogen, when transport current is much more than the critical current of an HTS tape, i.e., at overload conditions.

Our goal was to learn how to evaluate transient processes of heat and voltage developments in HTS tapes in boiling liquid nitrogen at currents more than critical one, but less than thermal runaway current. Some similar studies were performed with LTS (for example),[24,25] and HTS (for example).[26,27]

The study of overload behavior of different HTS tapes cooled by liquid nitrogen is presented in Refs. 13–16. In these works, we observed sharp peaks in the voltage during rectangular current pulse.[13,14] In the Refs. 15 and 16, we explained these peaks by analysis of transient regimes in liquid nitrogen. Even in homogeneous nitrogen, it can withstand a certain overheating before switching to a nucleate boiling. The sharp peak in the voltage occurs after temperature exceeds certain level before a nucleate boiling (NB) starts.

We tested Bi-2223 based samples without and with insulating cover at current overloading conditions. Details of experiments are

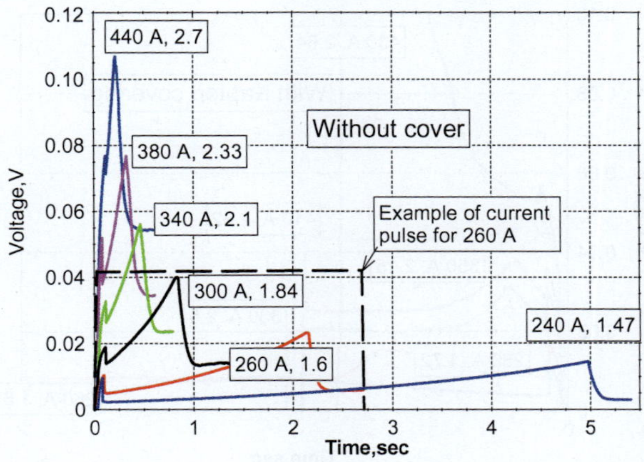

Fig. 8. Typical voltage recordings for tests *versus* time for different current and (I/Ic) for Bi-2223 HTS non-covered.[15] Example of current pulse is shown for 260 A shot.

presented in Refs. 11–15. Typical voltage recordings are shown in Figs. 8 and 9 for Bi-2223 brass laminated tape. Kinks in the beginnings of curves are due to inductive voltage noise at fast current ramp. In boxes, the current magnitudes and a ratio of I/I_c are shown. Typical curves at overloads (when transport current is constant during a pulse) are as follows: First fast rise of the voltage or temperature, then fast drop when cooling switches from convection to nucleate boiling.[13,14] Then voltage or temperature becomes stable until current pulse stops (see Fig. 8 with an example of 260 A current pulse). These are stable modes after cooling switched. If current is high enough, unstable mode occurs. As an example, the $V(t)$ curve for $I > I_q$ is shown for $I \sim 430$A and $I/I_c \sim 2.64$ in Fig. 9. This is an unstable case.

To describe the observed phenomenon, the numerical model was developed.[15,16] The suggestion in the model was overheating of the surface of an HTS tape in comparison with the bath temperature before switching from convection cooling to nucleate boiling and cooling. To simulate heat/voltage development of HTS tape at overloads for heat-transfer coefficient (α_{boil}), we used the expression:

$$\alpha_{boil} = C_{boil} \Delta T^n. \tag{3}$$

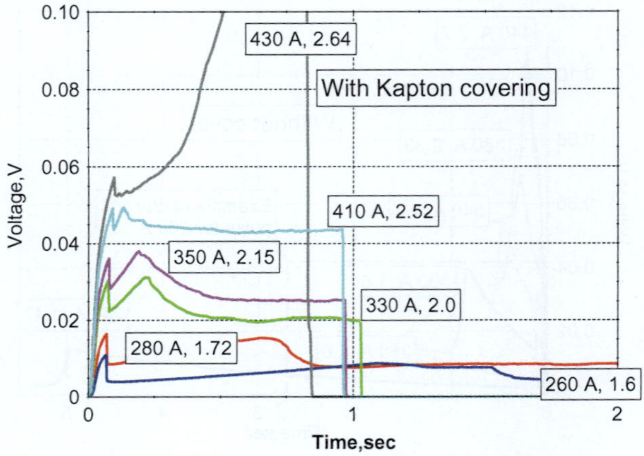

Fig. 9. Typical voltage recordings for tests *versus* time for different current and (I/Ic) for Bi-2223 HTS covered by Kapton^{TM}.[15] At 430 A, the unstable mode occurs.

Both coefficients C_{boil} and n were determined from our measurements. The difference between overheating temperatures and boiling bath temperature ΔT_{oh} were determined from experiments as well. In Fig. 10, the dependencies of maximum voltages on current are shown for Bi-2223-based HTS. In Fig. 11, the dependencies of times before switching to NB *versus* current is shown for ReBCO tapes. One can see very good coincidence of experimental and calculated data. Therefore, the model[15,16] can predict switch from convection to NB and cooling and may be used for evaluation of HTS devices' behavior at overloads by current.

When sample is covered by a Kapton^{TM}, switch from convection to NB and cooling happens faster. Overheating temperatures when NB activates are about same for certain tests, and it is less when sample is covered by Kapton^{TM}. The suggested model returned data well coinciding with experiments and can be used for evaluation of HTS devices' behavior at overloads. It was determined by experiments and by calculations that I_q/I_c for ReBCO HTS tapes is approximately 1.5 and does not depend on covering, due to high resistivity of a substrate in such tapes. For Bi-2223 HTS tapes, I_q/I_c ratio is practically directly proportional to the coefficient of heat transfer, and for Bi-2223 tapes, tested is 2.6–2.8.[15,16]

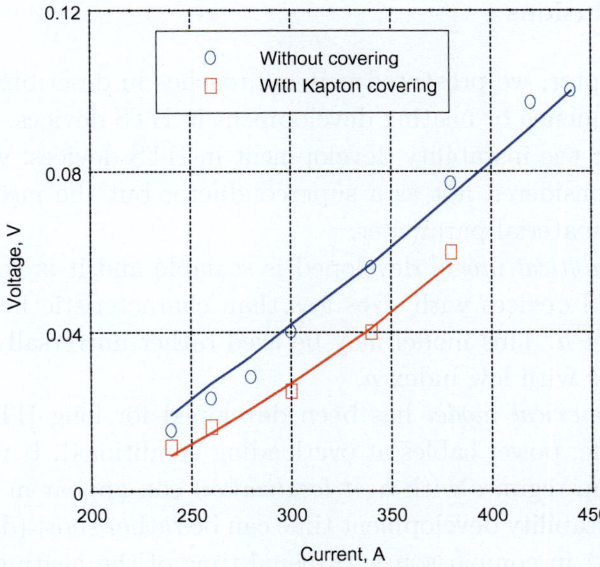

Fig. 10. Comparison of maximum voltages for Bi-2223 HTS tapes with and without cover.[15] Symbols- experiment, solid lines- calculations by the model.

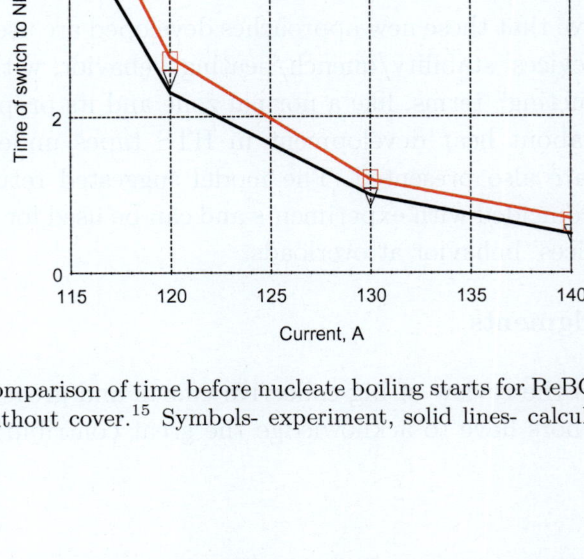

Fig. 11. Comparison of time before nucleate boiling starts for ReBCO HTS tapes with and without cover.[15] Symbols- experiment, solid lines- calculations by the model.

5. Conclusions

In this chapter, we presented new approaches in describing the stability and quench or heating development in HTS devices. The models consider the instability development in HTS devices, while HTS device is considered not as a superconductor but the medium with non-linear material parameter.

The analytical model developed is scalable and it is good to use for the HTS devices with sizes less than characteristic heat length: $l_h = \sqrt{Ak/Ph}$. This model may be used rather universally, even for LTS-devices with low index n.

The numerical model has been developed for long HTS objects (for example, power cables at overloading conditions). It was shown that blow-up regimes with *heat localization* can appear in long HTS objects. Instability development time can be rather short (due to heat localization) in comparison with usual time of the heating development in HTS devices.

The threshold current I_q weakly depends on cooling characteristics and the stability of HTS devices. This current separates stable and unstable modes. The increase of the transport current in an HTS device by less than 1% may lead to the switching from the stable to unstable mode.

Important safety parameter is the heating development time that quickly decays with the transport current rise which inversely proportional to the square of the current.

We believe that these new approaches developed are useful to analyze HTS-devices' stability/quench/heating behavior, without using "superconducting" terms, like a normal zone and its propagation.

Results about heat development in HTS tapes under overload conditions are also presented. The model suggested returned data which well coincided with experiments and can be used for evaluation of HTS devices' behavior at overloads.

Acknowledgments

This work is the review of big collective efforts and great teamwork and the authors have to acknowledge the great contribution to this work of their co-authors: G. H. Karapetyan, V. V. Repnikov, E. A. Lobanov, S. S. Fetisov, Yu. A. Ilyin, V. V. Zubko, and S. M. Ryabov.

References

1. A. L. Rakhmanov, V. S. Vysotsky, and Yu. A. Ilyin *et al.*, Universal scaling law for quench development in HTSC devices, *Cryogenics* **40**(1), 19–27 (2000).
2. V. S. Vysotsky, Yu. A. Ilyin, and T. Kiss *et al.*, Thermal quench study in HTSC pancake coil, *Cryogenics* **40**(1), 9–17 (2000).
3. Yu. A. Ilyin and V. S. Vysotsky *et al.*, Stability and quench development study in small HTSC magnet, *Cryogenics* **41**(9), 665–674 (2001).
4. V. S. Vysotsky, Yu. A. Ilyin, and A. L. Rakhmanov, Stability and quench development in HTS magnets: Influence of cooling and material parameters, *Adv. in Cryog. Eng.* **47**, 481–488 (2002).
5. V. S. Vysotsky, Yu. A. Ilyin, A. L. Rakhmanov, and M. Takeo, Quench development analysis in HTSC coils by use of the universal scaling law, *IEEE Trans. Appl. Supercond.* **11**(1), 1824 –1827 (2001).
6. V. S. Vysotsky, V. E. Sytnikov, A. L. Rakhmanov, and Yu. A. Ilyin, Analysis of stability and quench in HTS devices — new approaches, *Fusion Engineering and Design* **81**, 2417–2424 (2006).
7. V. S. Vysotsky, A. L. Rakhmanov, and Yu. A. Ilyin, Novel approaches to describe stability and quench of HTS devices, In *Superconductivity Research Developments*, J. R. Tobin (ed.). New York: Nova Science Publishers, Inc, 221–237 (2007).
8. A. Ishiyama and H. Asai, *IEEE Trans. Appl. Supercond.* **10**(1), 1834–1837 (2000).
9. A. L. Rakhmanov, V. S. Vysotsky, and N. V. Zmitrenko, Quench development in long HTS objects- the possibility of "blow-up" regimes and a heat localization, *IEEE Trans. Appl. Supercond.* **13**(2), 1942–1945 (2003).
10. V. S. Vysotsky, V. E. Sytnikov, A. L. Rakhmanov, and N. V. Zmitrenko, Quench development analysis in long HTS objects, *Proceedings of EUCAS-2003*; *IOP Conference Series* **181**, 580–583 (2004).
11. V. S. Vysotsky, V. E. Sytnikov, and V. V. Repnikov *et al.*, Heating development analysis in long HTS objects with cooling, *IEEE Trans. Appl. Supercond.* **15**(2), 1655–1658 (2005).
12. V. S. Vysotsky, G. H. Karapetyan, and V. E. Sytnikov *et al.*, Heating development analysis in long HTS objects- updated results, *J. Phys.: Conf. Ser.* **43,** 877–880 (2006).
13. V. S. Vysotsky, S. S. Fetisov, and V. E. Sytnikov, Peculiarities on voltage–current characteristics of HTS tapes at overloading conditions cooled by liquid nitrogen, *J. Phys.: Conf. Ser.* **97**, 012015 (2008).
14. S. S. Fetisov, V. S. Vysotsky, and V. E. Sytnikov, Test of HTS tapes cooled by liquid nitrogen at overloading conditions, *IEEE Trans. Appl. Supercond.* **19**(3), 2411–2414 (2009).

15. S. S. Fetisov, V. S. Vysotsky, and V. V. Zubko, HTS tapes cooled by liquid nitrogen at current overloads, *IEEE Trans. Appl. Supercond.* **21**(3), 1323–1327 (2011).
16. V. V. Zubko, S. M. Ryabov, S. S. Fetisov, and V. S. Vysotsky, Heat transfer simulation to liquid nitrogen from HTS tapes at the overload currents, *Paper Wed-Af-P2.3-12 presented at ICEC-25 conference*, July (2014).
17. Y. Iwasa, *Case Studies in Superconducting Magnets.* New York: Plenum Press (1994).
18. M. N. Wilson, *Superconducting Magnets.* Oxford, UK Clarendon Press (1983).
19. S. S. Torii, S. Akita, and K. Ueda, Transport current properties of double-pancake coils wound by Ag-sheated Bi-2223 tapes, *IEEE Trans. Appl. Supercond.* **9**(2), 944–947 (1999).
20. H. Kumakura, H. Kitaguchi, K. Togano, and T. Hasegawa. *Cryogenics* **38**, 163–167 (1998).
21. H. Kumakura, H. Kitaguchi, K. Togano, and H. Wada, Stability of a Bi-2223 refrigerator cooled magnet, *Cryogenics* **38**, 639–842.
22. O. Shevchenko, H. J. G. Krooshoop, and B. ten Haken *et al.*, An optimized BSCCO/Ag resonator coil for utility use, *IEEE Trans. Appl. Supercond.* **10**(1), 849–852 (2000).
23. N. N. Martovetsky, P-L. Bruzzone, and B. Stepanov *et al.*, Effect of conduit material on CICC performance under high cycling load, *IEEE Trans. Appl. Supercond.* **15**(2), 1367–1370 (2005).
24. H. Boschman, *On The Resistive Transition of Composite Superconductors.* PhD thesis, Twente University, Netherlands (1990).
25. G. B. J. Mulder and L. J. M. van de Klundert, Thermally and magnetically controlled superconducting rectifiers, *IEEE Trans. Magn.* **MAG-25**, 1819–1823 (1989).
26. F. Chovanec and P. Usak, Instabilities above critical current region in Bi-2223/Ag superconducting coils cooled by liquid nitrogen, *Cryogenics* **42**, 543–549 (2002).
27. W. Bailey, E. A. Young, Y. Yang, and C. Beduz, Boiling heat transfer to a liquid nitrogen pool from Ag sheathed BiPb2223 tapes carrying over-current, *Supercond. Sci. Technol.* **19**, 276–282 (2006).

Chapter 1.9

Thin Films of Bi-2223

Akiyoshi Matsumoto
National Institute for Materials Science,
1-2-1 Sengen, Tsukuba, Ibaraki 305-0047, Japan

To investigate the potential of high T_c Bi-2223 superconductors, as well as the growth mechanism of Bi-2223 phases, Bi,Pb-2223 thin films were fabricated by a combination of sputtering and annealing methods. The resulting films had J_c values almost one order of magnitude greater than those reported for wires. The growth mechanism by which a Bi-2223 single phase is obtained is highly complex since this material contains six elements, and many impurity phases may be present. Studies of thin films such as those detailed herein assist in understanding this growth mechanism.

1. Introduction

This chapter focuses on $(Bi,Pb)_2Sr_2Ca_2Cu_3O_x$ (Bi,Pb-2223) thin films. To obtain high performance Bi,Pb-2223 wires, it is vital to first perform basic research, such as with thin films, so as to determine the full potential of Bi,Pb-2223 superconductors, as well as the growth mechanism of the Bi,Pb-2223 phase.

The superconductor Bi,Pb-2223 has shown promise as a means of reducing energy consumption in various practical applications, such as power cables, magnets and motors. As discussed in previous chapters, the critical current density (J_c) exhibited by the long tapes associated with such devices is approximately 6×10^4 A/cm^2 at 77 K when in a self-field. To date, very long tapes (over 2 km) having a critical current (I_c) over 200 A have been fabricated,[1] and similar critical current values have been realized in many other applications. However, to allow for high performance applications, it is necessary to increase the J_c values that may be obtained, and thin-film studies performed soon after the discovery of Bi-2223 (Pb was not contained in this study) were able to achieve a J_c of 3.8×10^5 A/cm^2 at 77 K in a self-field.[2] To clarify the difference between thin films or single crystals and recently developed tapes, it is important to fully investigate the potential of Bi,Pb-2223.

Since its discovery,[3] the Bi-Sr-Ca-Cu-O (BSCCO) system has attracted considerable attention due to its high transition temperature (T_c). Initially, many researchers attempted to fabricate high temperature Bi-2223 thin films using various fabrication processes, such as pulsed laser deposition (PLD),[4–5] radio-frequency (RF) and DC sputtering,[6–10] molecular beam epitaxy,[11–13] and chemical vapor deposition (CVD),[14,15] and succeeded in fabricating stable Bi-2223 thin film superconductors with high T_c values. These initial efforts focused on the separation of Bi-2223 and Bi-2212, and the fabrication of high-quality layered structures with high T_c. Few studies have investigated the J_c properties of Bi-2223 thin films.

Typically, commercial Bi,Pb-2223 tapes contain Pb because the addition of Pb to the bulk material promotes the formation of Bi-2223.[16,17] However, reports on Pb-doped BSCCO films[18,19] indicate that much of the original Pb is lost from the sample during the requisite high-temperature annealing. Pb-doped Bi-2223 thin films have been prepared by RF magnetron sputtering,[20] DC magnetron sputtering,[21–24] PLD,[25] and CVD.[26,27] The RF sputtering method and the associated post-annealing processes are particularly useful in the formation of Bi-2223. Herein we discuss the fabrication of Bi,Pb-2223 thin films with very high J_c values ($\sim 10^6$ A/cm^2 at 77 K in a self-field) and zero-resistance T_c (typically around 100 K), obtained by a refined three step (deposition + annealing + post-annealing) process. Results obtained from the assessment of the temperature and magnetic field dependencies of the transport properties in such films are also provided.

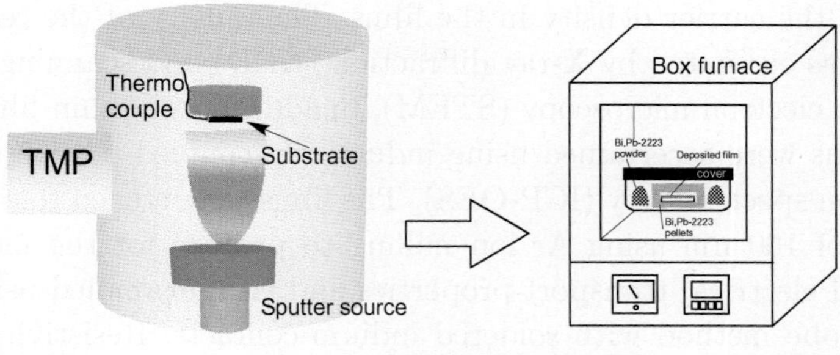

Fig. 1. Schematic representation of the fabrication process. As-deposited films were fabricated by a conventional sputtering method as shown in the figure on the left. The substrate temperatures were controlled by a thermocouple attached nearby. Following deposition, films were heat-treated in a box furnace while surrounded by Bi,Pb-2223 pellets at various temperatures and holding times.

2. The Fabrication of Bi,Pb-2223 HTS (High Tc Superconductor) Thin Films

Bi-2223 thin films were fabricated on $SrTiO_3$ (STO) (100) substrates by RF magnetron sputtering, as shown in left side image of Fig. 1. The sputtering temperature was 973 K at an O_2 pressure of 50 Pa. The sample holder was maintained at 973 K using a resistive heater and a thermocouple located near the substrate. Since the Bi-2223 pellets for sputtering target contained six elements, controlling the thin film composition was generally quite challenging, and so it was crucial to fabricate the target by sputtering. Target pellets were fabricated by heat treatment at 1108 K for 10 h and subsequent pressing and sintering at 1,113 K for 10 h. Our group has produced thin films using many different targets with various nominal compositions but, in this study, a target with the nominal composition of $Bi_{2.0}Pb_{0.5}Sr_{2.0}Ca_{3.0}Cu_{3.0}O_x$ was employed because the composition of the as-deposited film was very close to the stoichiometric composition. To improve their superconducting properties, as-deposited films were annealed in the presence of Bi,Pb-2223 pellets at 1,113 K for 100 h in an atmosphere of Bi, Pb and O, as shown in the right side image of Fig. 1. The annealing process produced Bi,Pb-2223 thin films approximately 300 nm in thickness. These films were subjected to post-annealing at 723 K for 10 h under a flow of oxygen to control the carrier density in the films. The quality of the resulting films was evaluated by X-ray diffraction (XRD) and scanning transmission electron microscopy (STEM). In addition, the thin-film compositions were ascertained using inductively coupled plasma optical emission spectrometry (ICP-OES). The films were etched to a bridge width of 100 nm using Ar ion milling to prepare for the measurement of electrical transport properties and then examined using the four-probe method with soldered indium contacts. Resistivity measurements were also performed while applying various temperatures, magnetic fields and angles between the film surface and the external magnetic field.

2.1. *Superconducting properties of Bi,Pb-2223 thin films*

Figure 2 presents XRD patterns obtained from the as-deposited and annealed films, both plotted on the same scale to allow comparison. The XRD pattern for the as-deposited film exhibits intense Bi-2223 $00l$ peaks, indicating that the main phase of the film consisted of

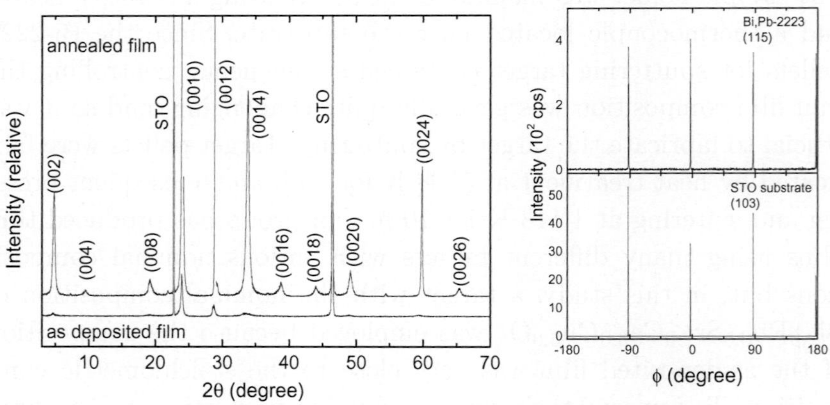

Fig. 2. XRD patterns for as-deposited and annealed films. The majority of the peaks generated by the as-deposited film are 001 peaks from the Bi-2223 phase. Following the annealing process, the peaks are stronger and sharper. The image on the right shows ϕ-scans for the Bi,Pb-2223 (115) peak and the STO substrate (103) peak. These peaks appear at essentially the same positions, suggesting that the Bi,Pb-2223 grains grew in an epitaxial manner.

Bi-2223 and that the grains were aligned parallel to the *c*-axis of the STO (100) substrate. The lattice parameters of the STO substrate and the Bi-2223 were 3.905 and 5.420 A, respectively. The lattice mismatch between the STO (110) and Bi-2223 was less than 2%, suggesting that this substrate was suitable for epitaxial growth of the Bi-2223 phase. Quadrupole peaks were observed for both the as-deposited and annealed films in ϕ-scan data obtained from the Bi-2223 and the STO substrate. These data provide evidence that the Bi-2223 films had an in-plane texture. Furthermore, the STO and Bi-2223 peak positions were coincident, demonstrating that the Bi-2223 grew epitaxially on the STO substrate. The peaks generated by the Bi,Pb-2223 thin film following annealing were much more intense and much sharper than those for the as-deposited film. Thus, the annealed film evidently possessed a much higher degree of crystallinity than the as-deposited film.

Data from ICP-OES analysis did not show the presence of Pb in the as-deposited films, even when the target contained a large amount of Pb. However, this same analysis identified Pb in the annealed films, demonstrating that Pb can be introduced into a thin film while still obtaining a high degree of crystallinity.

From the plots of resistivity as a function of temperature presented in Fig. 3, it is evident that the as-grown film had a $T_{\mathrm{c}}^{\mathrm{zero}}$ of 72 K, together with a high temperature transition, ΔT. These values are attributed to the low degree of crystallinity in this film, as well as the variation in the composition of Bi-2223 grains. In contrast, a T_c^{zero} value of 105 K and a more abrupt transition were obtained for the Bi,Pb-2223 thin film after annealing. This observation confirms that a combination of sputtering deposition and annealing is an effective means of obtaining high quality of Bi,Pb-2223 thin films, as has been previously reported.[4–6]

Figure 4 presents $J_c - B$ plots for various Bi-2223 films and tapes, obtained at 77 K. From these data, it can be seen that the previous thin film data obtained by our group were approximately equal to the results reported for the material DI-BSCCO, developed by Sumitomo Electric Industries (SEI), Ltd. for use in commercial Bi,Pb-2223 tapes.[1] The values for both materials are much lower than those

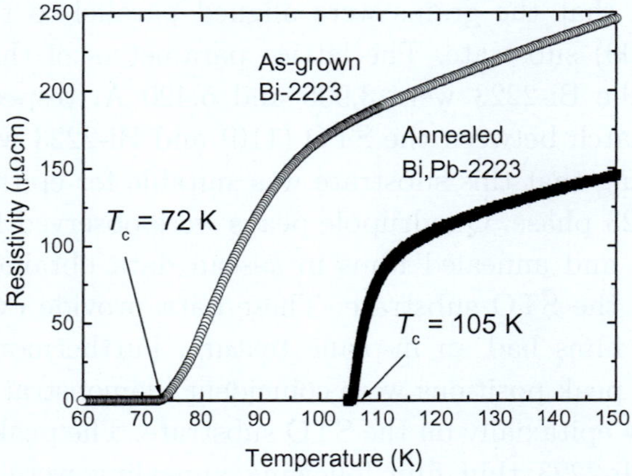

Fig. 3. Temperature dependence of resistivity for an as-deposited Bi-2223 film fabricated by sputtering and an annealed Bi,Pb-2223 film. The as-deposited film exhibits a T_c of 72 K, whereas this value is drastically increased, to over 100 K, by annealing.

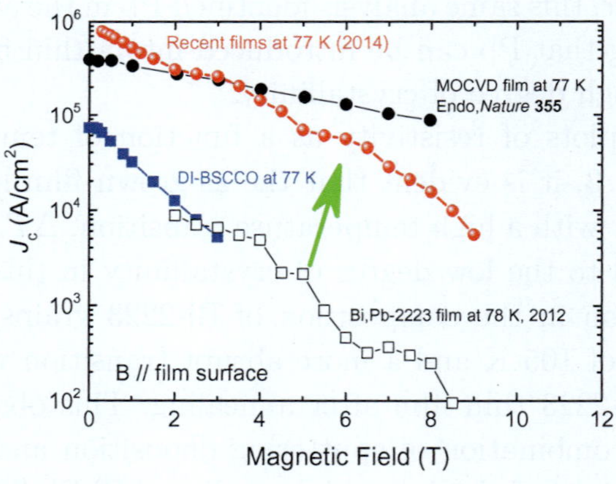

Fig. 4. Transport J_c values versus applied magnetic field strength at 77 K. Data is shown for the present Bi,Pb-2223 thin film after optimization of target composition and annealing conditions, a previously researched Bi,Pb-2223 film[28] as well as for DI-BSCCO[1] and MOCVD films[2] for comparison.

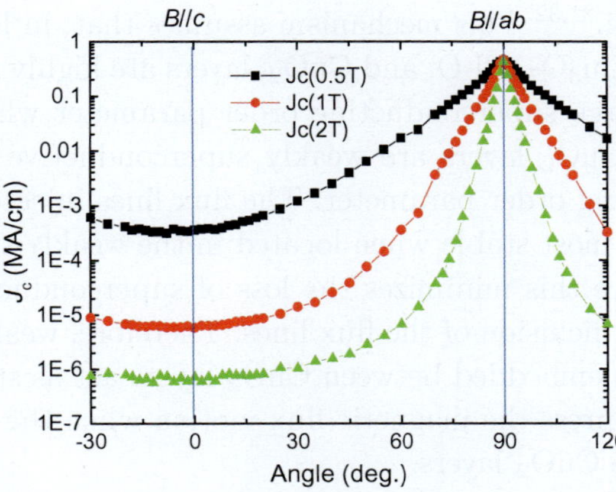

Fig. 5. J_c values at 77.3 K in a magnetic field applied parallel and perpendicular to the ab plane, $J_c\ //(H, T)$, in a Bi,Pb-2223 thin film after annealing.

reported by Endo[2] for films generated by CVD. One possible reason for these low values is low carrier density due to the lack of oxygen flow in the post-annealing process. By applying post-annealing, our group successfully fabricated Bi,Pb-2223 thin films exhibiting very high J_c values of 0.9 MA/cm^2 at 77 K in a self-field.

To understand the J_c behavior of the present Bi,Pb-2223 thin films, J_c values were measured as a function of the angle between the applied magnetic field and the film surface (the *ab*-plane). Figure 5 summarizes the J_c data as a function of θ at 77.3 K under magnetic fields of 0.5, 1 and 2 T. The J_c values were relatively high when the applied field was parallel to the *ab*-plane; however, they decreased markedly at an angle of 5° as the magnetic field component in the *c*-direction increased. Such an angular dependence has also been reported for Yttrium Barium Copper Oxide (YBCO), for which J_c is generally higher in fields parallel to the *ab*-plane than in fields along the *c*-direction.[30] However, the direction of the magnetic field evidently had a much greater effect on the J_c value of the Bi-2223 than has been observed for YBCO. This result suggests that the superconducting state of this film is very sensitive to the magnetic field orientation. The present experimental results can be explained by the intrinsic pinning mechanism proposed by Tachiki and Takahashi.[31,32] This mechanism assumes that, in layered oxides such as $YBa_2Cu_3O_7$, Bi-O, and CuO_2, layers are highly superconductive with a large superconducting order parameter while the layers between the CuO_2 layers are weakly superconductive with a small superconducting order parameter. The flux lines injected parallel to the layers are most stable when located in the weakly superconducting layers since this minimizes the loss of superconducting cohesive energy due to inclusion of the flux lines. Therefore, weakly superconducting layers embedded between CuO_2 layers act as strong pinning centers to suppress the magnetic flux motion when the flux lines are parallel to the CuO_2 layers.

2.2. *Growth mechanism of Bi,Pb-2223*

Bi-2223 is typically formed from the reaction between Bi-2212 and CaCuPbO, a process that has been confirmed in the bulk material. It is very important to fully understand this formation mechanism and research based on thin films represents an ideal means of doing so. It is also important to obtain an ideal microstructure in this material so as to obtain optimal performance, and this can be readily accomplished during the sputtering process by alternately depositing Bi-2212 and CaCuPbO through exchanging the target. This technique was demonstrated by our group, using the same process applied to obtain the Bi-2223 layer to fabricate a multilayered precursor film, followed by heat treatment.

Figure 6 shows images of the multilayered precursor, obtained by cross sectional high angle annular dark field STEM (HAADF-STEM). Here, the dark and bright regions indicate the Bi-2212 and CaCuPbO layers and a total of six layers can be seen. The different layer thicknesses for the two phases result from the requirement to obtain a final composition in which Bi:Sr:Ca:Cu = 2:2:2:3. The image on the right was obtained at a higher magnification near the center of the left side image. The white region in this image represents an amorphous area, with Bi-2212 layers both above and below. From the results of EDX analysis for each element, the concentration of Bi was very low in the white region. These results suggest that the sputtering

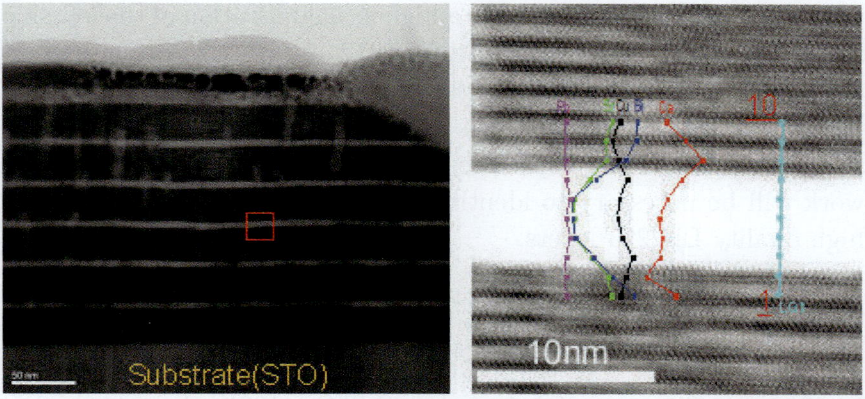

Fig. 6. HAADF-STEM images of a multilayered precursor with 6 × 6 layers of Bi-2212 and CaCuPbO. The bright and dark regions represent CaCuPbO and Bi-2212, respectively.

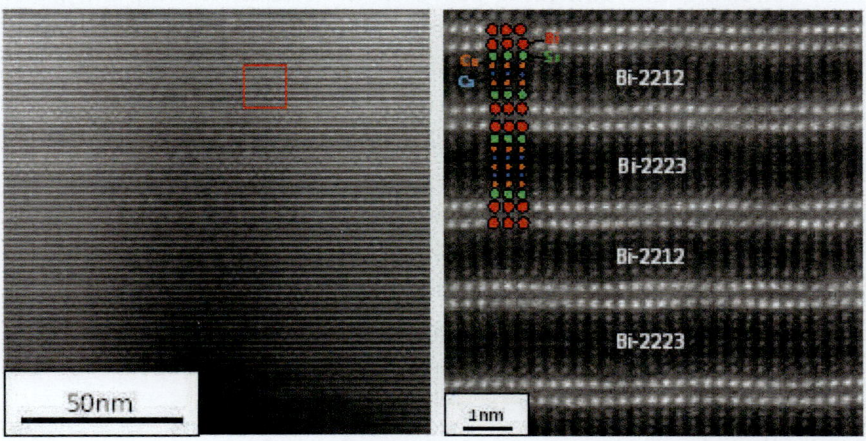

Fig. 7. HAADF-STEM images of an annealed film of the multilayered precursor with 6 × 6 layers of Bi-2212 and CaCuPbO. Bi-2212 and Bi-2223 microstructures are seen. The right side shows a more highly magnified image, in which the rows of bright dots indicate the Bi-atom columns.

method was very effective at generating a multilayered precursor of the Bi-2212 + CaCuPbO system. Figure 7 presents cross sectional HAADF-STEM images of a Bi,Pb-2223 thin film following annealing of the multilayered precursor film in Fig. 6 at 840° C for 100 h. Here the right side image again shows a higher magnification of the left side image. Atomic scale layers of both Bi-2212 and Bi-2223 can be seen in this image, confirming that the reaction of Bi-2212 and CaCuPbO to give Bi,Pb-2223 proceeded during the heat treatment step. However, since some Bi-2212 are still present in the finished product, future work will be necessary to identify the conditions required to obtain high quality Bi-2223 layers.

3. Conclusion

Bi-2223 thin films were fabricated on $SrTiO_3$ (100) substrates by RF sputtering to allow basic research concerning the potential applications of this material. The Bi,Pb-2223 phase was obtained by annealing as-deposited films while surrounded by Bi,Pb-2223 pellets. Following this annealing process, the transition temperature of the Bi,Pb-2223 was increased to above 100 K, a value almost equal to that of bulk Bi,Pb-2223. The highest J_c value of 0.9×10^6 A/cm^2 at 77 K and 0 T was obtained from the Bi,Pb-2223 thin film, and this value is much higher than those exhibited by tape materials. To better understand the growth mechanism of Bi-2223, a multilayered precursor film composed of alternating Bi-2212 and CaCuPbO layers was also fabricated. After annealing, a Bi,Pb-2223 phase was obtained in this film. In future, it will be vitally important to continue to perform such basic research on these thin films to obtain further information concerning high performance Bi-2223.

References

1. H. Kitaguchi , K. Takahashi, H. Kumakura, T. Hayashi, K. Fujino, N. Ayai, and K. Sato, *Supercond. Sci. Technol.* **22**, 045005 (2009).
2. K. Endo, H. Yamasaki, S. Misawa, S. Yoshida, and K. Kajimura, *Nature* **355**, 327 (1992).
3. H. Maeda, Y. Tanaka, M. Fukutomi, and T. Asano, *Jpn. J. Appl. Phys.* **27**, L209 (1988).
4. N. Asano, K. Shinohara, N. Matsunami, and Y. Takai, *Supercond. Sci. Technol.* **12**, 203 (1899).
5. E. Kume, H. Fujino, X. Zhao, and S. Akai, *Physica C* **412–414**, 1354 (2004).

6. K. Harada, S. Kishida, T. Matsuoka, T. Maruyama, H. Tokutaka, and K. Fujimura, *Jpn. J. Appl. Phys.* **35**, 4297 (1996).
7. M. Holiastou, V. Psyharis, D. Niarchos, P. Halibach, U. Frey, and H. Adrian, *Supercond. Sci. Technol.* **10**, 712 (1997).
8. K. Matsumoto, M. Ogura, K. Katsurahara, S. Kishida, and H. Tokutaka, *Supercond. Sci. Technol.* **15**, 1757 (2002).
9. M. Ohkubo, J. Geerk, G. Linker, and O. Mayer, *Appl. Phys. Lett.* **67**, 2403 (1995).
10. A. Alem, A. E. Al-Salami, K. A. Ziq, and H. J. Adrim, *Supercond. Nov. Magn.* **24**, 159 (2011).
11. Y. Nakayama, I. Tsukada, and K. Uchinokura, *J. Appl. Phys.* **70**, 4371 (1991).
12. I. Tsukada, Y. Nakayama, I. Terasaki, A. Maeda, and K. Uchinokura, *Supercond. Sci. Technol.* **4**, S116 (1991).
13. A. Tsukamoto, K. Imagawa, M. Hiratani, K. Kanehori, and K. Takagi, *Jpn. J. Appl. Phys.* **30**, L830 (1991).
14. A. D. Berry, D. K. Gaskill, R. T. Holm, E. J. Cukauskas, R. Kaplan, and R. L. Hney, *Appl. Phys. Lett.* **52**, 1743 (1998).
15. J. M. Zhang, J. Zhao, H. O. Marcy, L. M. Tonge, B. W. Wessels, T. J. Marks, and C. R. Kannewurf, *Appl. Phys. Lett.* **54**, 1166 (1988).
16. M. Takano, J. Takada, K. Oda, H. Kitaguchi, Y. Miura, Y. Ikeda, Y. Tomii, and H. Mazaki, *Jpn. J. Appl. Phys.* **27**, L1041 (1988).
17. U. Balachandran, D. Shi, D. I. Dos Santo, S. W. Graham, M. A. Patel, B. Tani, K. Vandervoort, H. Claus, and R. B. Poeppel, *Physica C* **156**, 649 (1988).
18. A. Tanaka, T. Machi, N. Kamehara, and K. Niwa, *Appl. Phys. Lett.* **54**, 1362 (1989).
19. R. Escudero, E. Chavira, and D. Rios-Jara, *Appl. Phys. Lett.* **54**, 1576 (1989).
20. S. K. Dew, N. R. Osborne, P. J. Mulhern, and R. R. Parsons, *Appl. Phys. Lett.* **54**, 1929 (1989).
21. Y. Hakuraku, S. Higo, and T. Ogushi, *Appl. Phys. Lett.* **55**, 1569 (1989).
22. Y. Hakuraku, S. Koba, S. Higo, A. Nakao, and T. Ogushi, *Supercond. Sci. Technol.* **9**, 23 (1996).
23. K. Kuroda, K. Kojima, M. Tanioku, K. Yokoyama, H. Kuroki, and K. Hamanaka, *Jpn. J. Appl. Phys.* **29**, 1066 (1990).
24. H. Masumoto, T. Goto, and T. Hirai, *Appl. Phys. Lett.* **55**, 498 (1989).
25. H. Tabata, T. Kawai, M. Kanai, O. Murata, and S. Kawai, *Jpn. J. Appl. Phys.* **28**, L430 (1989).
26. J. M. Zhang, H. O. Marcy, L. M. Tonge, B. W. Wessels, T. J. Marks, and C. R. Kannewurf, *Appl. Phys. Lett.* **55**, 1906 (1989).
27. N. Takahashi, D. Kanematsu, A. Koukitu, H. Seki, and Y. Kamioka, *Jpn. J. Appl. Phys.* **32**, L1648 (1993).
28. A. Matsumoto, H. Kitaguchi, T. Doi, T. Kajihara, and S. Hata, *IEEE Trans. Appl. Supercond.* **23**, 7500504 (2013).
29. A. Matsumoto and H. Kitaguchi, *Supercond. Sci. Technol.* **27**, 015002 (2014).

30. H. Raffy, S. Labdi, O. Laborde, and P. Monceau, *Supercond. Sci. Technol.* **4**, S100 (1991).
31. K. Watanabe, S. Awaji, N. Kobayashi, H. Yamane, T. Hirai, and T. Muto, *J. Appl. Phys. Lett.* **69**, 1543 (1991).
32. M. Tachiki and S. Takahashi, *Solid State Commun.* **70**, 291 (1989).
33. M. Tachiki and S. Takahashi, *Solid State Commun.* **72**, 1083 (1989).

PART 2
Fabrication

Chapter 2.1

Development and Manufacture of Bi-2223 Wires

Shin-ichi Kobayashi

Superconductivity & Energy Technology Department,
Sumitomo Electric Industries, Ltd.,
1-1-3, Shimaya, Konohana-ku, Osaka, Japan 554-0024

This chapter reviews Ag-sheathed (Bi, Pb)$_2$Sr$_2$Ca$_2$Cu$_3$O$_x$ (Bi-2223) wire made by the powder-in-tube technique (PIT). The currently leading high-temperature superconductors (HTS) wire technology for practical use is Bi-2223 wire, made by the controlled over-pressure (CT-OP) sintering process. The CT-OP process uses pressures up to 30 MPa during heat treatment. The technique densifies the Bi-2223 filaments and enhances the uniformity of the electrical and mechanical performance in the Bi-2223 wire. Today, Bi-2223 wires are used in most HTS applications, such as power cables, many kinds of magnets, and motors for ship propulsion and electric vehicles.

1. Introduction

Since the discovery of bismuth-based high-temperature superconductors in 1988 in Japan by Maeda *et al.*,[1] a great deal of development at works have been done to produce superconducting wires with industrial applications. The family of bismuth-based high-temperature superconductors include several superconducting phases, such as $Bi_2Sr_2CuO_x$ (Bi-2201)[2] $Bi_2Sr_2CaCu_2O_x$ (Bi-2212), and $Bi_2Sr_2Ca_2$ Cu_3O_x (Bi-2223).[3] Conduction in all three superconducting phases occurs preferentially along the CuO_2 planes. Critical temperatures of these superconducting phases are 20, 80 and 110 K, respectively. The use of superconductivity in supplying electrical power and in magnets depends on superconducting wires with high critical current densities (J_c) at temperatures where cryogenic losses are acceptable.

Bi-2223 wire has been manufactured commercially by several companies.[4–8] Manufacturer's goals for industrial Bi-2223 wires are to achieve the desired high electrical and mechanical performance while producing long lengths that are robust, and to do so efficiently and on a large enough scale to meet the expected market demand. Although much work has been done to refine many different wire fabrication techniques,[4–12] most current efforts focus on the powder-in-tube (PIT) technique. One of the important advantages of the PIT technique is that it is easily adaptable to a multi-filamentary approach by bundling the wires. The importance of a multi-filamentary structure is that it makes "react-and-wind" wire possible. The other important aspect of the PIT technique is that during the deformation process it is possible to introduce texture into the platelet-shaped grains. As the material is squeezed after a rolling process, the platelets become aligned. Appropriate texture is a very important part of achieving high electrical performance in Bi-2223 wires.

We have found strong correlations between J_c and Bi-2223 phase homogeneities[13] and between J_c and grain alignments.[14,15] The highest J_c's are obtained from wires with the highest volume fraction of the Bi-2223 phase and the smallest misalignment angle. The typical J_c at 77.3 K and self-field for commercial Bi-2223 wires ranges from 50 to 60 kA/cm.[2] This is far below the highest local J_c of 300 kA/cm^2 estimated by magneto optical (MO), imaging[16] suggesting that several factors in the microstructure of Bi-2223 wires limit current flow. To further enhance J_c, the connectivity between Bi-2223 grains must be improved by further reducing porosity and by increasing the number of strong links. This can be achieved with small grain misorientation, clean grain boundaries, and reducing Bi-2212 content.[17,18] The controlled over-pressure (CT-OP) sintering process can reduce porosity, deformation cracks and Bi-2212 content and increases J_c.[8] Controlling the oxygen content using a post-annealing process can also increase J_c at a low temperature and under a magnetic field.[19]

This can result in increasing the critical current (I_c) value and the yield of manufacturing runs, which has a direct impact on applications by reducing the price as measured in \$/kAm. This chapter addresses Ag-sheathed Bi-2223 wire produced by the PIT technique and the CT-OP process. Transport properties and mechanical characteristics of the conductors are also discussed.

2. Processing Technology

2.1. *Metal-sheathing and rolling process*

Bi-2223 wires are fabricated from a mixture of Bi-2212 and other phases by the PIT technique, as shown in Fig. 1. In the PIT technique, the Bi-2223 precursor powder, which is a mixture of Bi-2212 and other oxides with starting compositions in the range of Bi : Pb : Sr : Ca : Cu = 1.7–1.8 : 0.3–0.4 : 1.8–2.0 : 1.9–2.1 : 3.0–3.1[18,20–23] is packed into a silver (Ag) tube or an Ag alloy tube and sealed to form a billet. The billet is then drawn and bundled to form a multifilament composite and transformed into a wire shape by being drawn through a series of dies with gradually decreasing diameters. Finally, the wire is usually rolled into a flat wire.

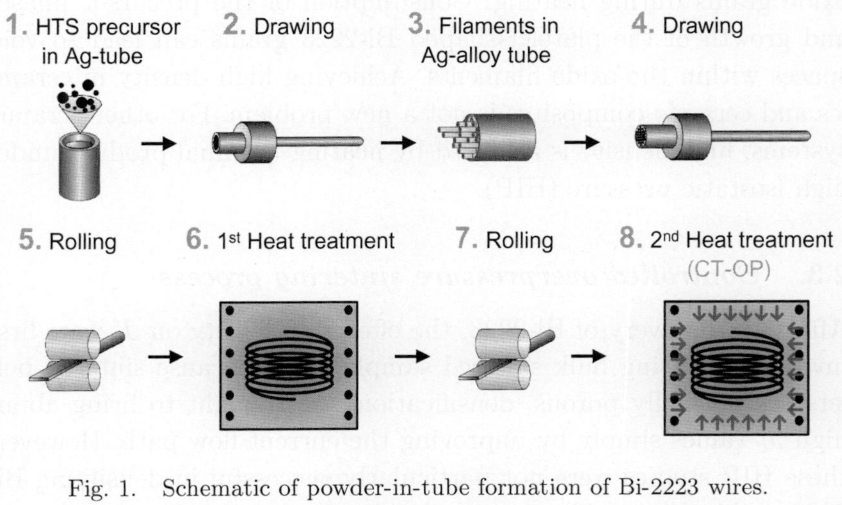

Fig. 1. Schematic of powder-in-tube formation of Bi-2223 wires.

2.2. *Thermo-mechanical process*

Thermo-mechanical processing is carried out using two heat treatment steps and an intermediate rolling step. During the first heat treatment step, Bi-2212 and other phases are converted into Bi-2223 under atmospheric pressure. Bi-2223 formation is known to be a complicated reaction. When Bi-2223 forms from a mixture of Bi-2212 and non-superconducting phases, the random growth of the platelet-shaped grains causes the Bi-2223 wire to expand. After the first heat treatment step, an intermediate rolling step is carried out, to align the Bi-2223 grains parallel to the tape surface and to densify the oxide filaments. Then the Bi-2223 grains are reconnected during the second heat treatment step. It is ideal for increasing I_c to sinter the grains to a high density. However, micro-cracks formed in the oxide filaments during the intermediate rolling step cannot be removed completely in the second heat treatment step, Jiang *et al.*[24] measured the oxide filament density at various points in the thermo-mechanical processing and reported that even though the intermediate rolling step densifies the oxide filaments, the filament density decreases in the final heat treatment. After the final heat treatment step, the Bi-2223 wire still contains about 25% porosity and cracks, thought to be caused by gas evolution and the growth of nonaligned oxide grains during heating. Consumption of the precursor phases and growth of the platelet-shaped Bi-2223 grains can lead to void spaces within the oxide filaments. Achieving high density in ceramics and ceramic composites is not a new problem. For other ceramic systems, high density is achieved by heating the final product under high isostatic pressure (HIP).

2.3. *Controlled overpressure sintering process*

After the discovery of Bi-2223, the effects of density on J_c were first investigated using bulk sintered samples.[25,26] Because sintered pellets are normally porous, densification was thought to bring about high J_c values simply by improving the current flow path. However, these HIP studies were not particularly successful in densifying Bi-2223 and retaining its superconducting properties. In conventional

Bi-2223 wire heat treatment, the wire first expels and then consumes oxygen during the Bi-2223 phase conversion. Nevertheless, these systems could not control oxygen partial pressure (pO_2) during heat treatment.

HIP studies have borne fruit since the early 2000's, and it is possible for new HIP systems to control pO_2 during heat treatment. One of the successful HIP systems is a hot-wall, laboratory-scale flowing-gas overpressure (OP) system that was built at the University of Wisconsin-Madison.[27,28] Yuan *et al.*[27] reported that the OP process eliminates pores and repairs cracks, resulting in the densification of oxide filaments and an increase in the J_c of a short Bi-2223 wire.

At the same time, Sumitomo Electric Industries (SEI) designed a controlled overpressure (CT-OP) sintering process using a cold-wall HIP.[8,29] This system can be operated up to 30 MPa and 900°C. Figure 2 shows a manufacturing-scale system with an outside diameter of 1.5 m and a height of 2 m. The CT-OP system uses a zirconia cell to measure the O_2 content of the outlet gas and to maintain the desired total pressure (p_{total}) and pO_2 by using solenoid valves during heat treatment. This system does not affect the expulsion and consumption of O_2 by Bi-2223 or the oxidation of the hot furnace wall.

3. Bi-2223 Wire Products

Bi-2223 wire is now optimized to meet the specific requirements of every application. SEI provides five Bi-2223 wire products (DI-BSCCO) using the CT-OP process.

Type H is a standard bare wire with the ability to handle a high current density. Figure 3 shows a photomicrograph of a cross section of Type H. This wire, measuring approximately $0.23 \times 4.3\,\text{mm}^2$, comprises 121 filaments of Bi-2223 embedded in a metal matrix of oxide-dispersion strengthened silver alloy. Type G has reduced thermal conduction due to the use of silver–gold alloy. This is the wire for a current lead.

The type HT series offers a much higher level of mechanical durability. For applications where high tensile stresses and tensile strain

Fig. 2. CT-OP furnace, which can be operated up to 30 MPa and 900°C.

Table 1. DI-BSCCO wire specifications.

	Type H	Type HT-CA	Type HT-SS	Type HT-NX	Type G
Width (mm)	4.3	4.5	4.5	4.5	4.3
Thickness (mm)	0.23	0.36	0.30	0.32	0.23
I_c (A)	200	200	200	200	200
Critical Wire Tension at RT (N)	80	280	230	410	50
Critical Bending Diameter at RT (mm)	70	60	60	40	110

Fig. 3. Cross-sectional photomicrograph of Type H wire.

Fig. 4. Cross-sectional photomicrograph of Type HT-CA wire showing 50 μm thin strips of copper alloy on the top and bottom of the tape.

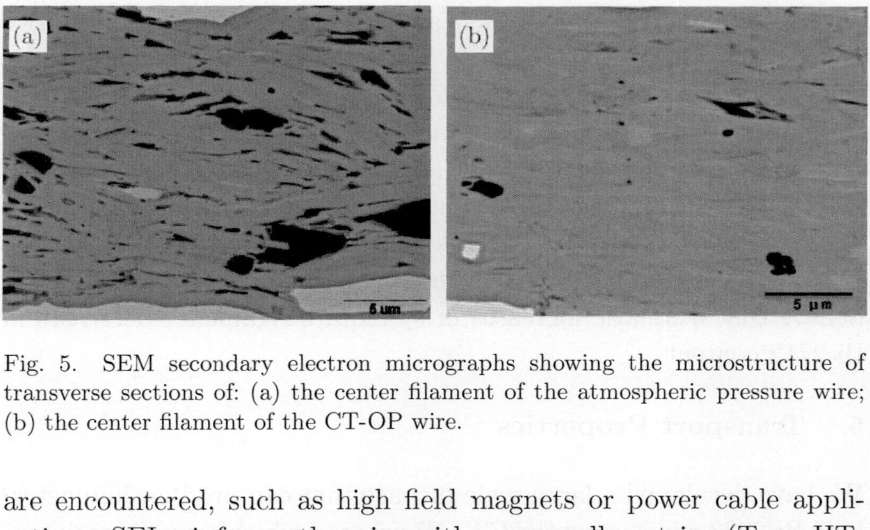

Fig. 5. SEM secondary electron micrographs showing the microstructure of transverse sections of: (a) the center filament of the atmospheric pressure wire; (b) the center filament of the CT-OP wire.

are encountered, such as high field magnets or power cable applications, SEI reinforces the wire with copper alloy strips (Type HT-CA), stainless steel strips (Type HT-SS) and nickel alloy strips (Type HT-NX). Type HT-SS and Type HT-NX are designed for use in applications where the wire must be mechanically strong, such as high field magnets and motors. Type HT-CA is designed for use in applications where the wire must be mechanically strong and have low splice resistance, such as cable applications. Figure 4 shows a photomicrograph of a cross-section of Type HT-CA.

4. Microstructure

Figure 5 shows SEM images of transverse sections of atmospheric pressure and CT-OP processed wires. The bulk of the gray regions is Bi-2223. The light gray streaks are Bi-2212. The dark gray and black particles are alkaline earth cuprates (AEC), CuO and $(Pb,Bi)_3Sr_2Ca_2CuO_x$ (3221). The dark gray regions indicate pores and cracks. The CT-OP wire seems to consist of well-connected Bi-2223 grains, which are particularly evident in the atmospheric pressure wire. The Bi-2223 grains in the atmospheric pressure filament are separated by pores and cracks.

Hellstrom *et al.*[27,28] showed similar microstructures for over-pressure processed wire. They reported that removing the current obstructions of porosity, cracks, and residual 2212 should improve grain connectivity and result in better electrical connectivity for superconducting current in Bi-2223.

The microstructure of these tapes shows large, plate-like grains, but imperfect alignment. There are still many Bi-2212 grains and secondary phases. The Bi-2212 grains appear to separate the Bi-2223 grains. These Bi-2212 grains may reduce I_c by obstructing the current paths required to maintain long-range current continuity. We believe that dramatic increases in J_c require significant decreases in Bi-2212 content.

5. Transport Properties

We have developed a large scale furnace that can produce kilometer-long Bi-2223 wires using the CT-OP process, which suppresses potential bubbles during the wire sintering step. When the long Bi-2223 wires are sintered at atmospheric pressure, bubbles occur infrequently.[31] Bubbles cause degradation of I_c by expanding and damaging the filament. The CT-OP process achieves uniform long wire performance. Figure 6 shows the I_c and n-value distribution in the longitudinal direction of long wires made with the CT-OP process, indicating that uniform performance is achieved over all 1300 m. The I_c of long wires was measured every 4 m. The average n-value was 18 defined in the region of 10^{-6} to 10^{-5} V/cm.

5.1. *Anti-ballooning performance*

Densification of the oxide filaments prevents liquid nitrogen penetration during long term exposure to liquid nitrogen. Liquid nitrogen

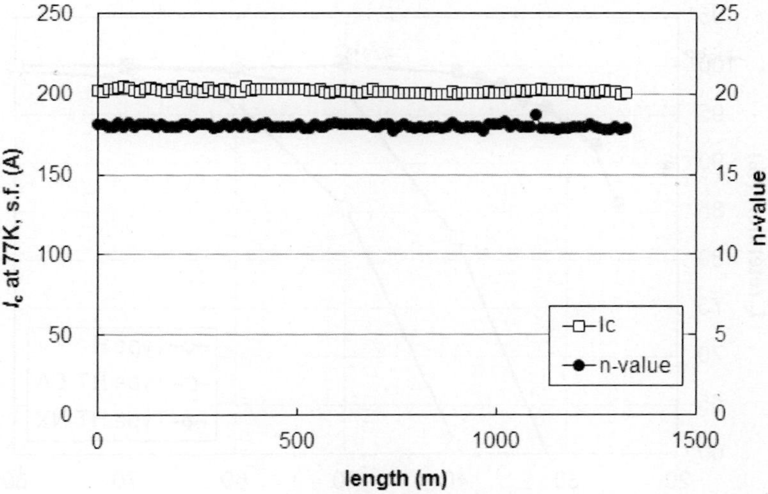

Fig. 6. Continuously measured I_c and n-values of a 1300 m long Bi2223 wire made with the CT-OP process. The data were obtained by the stepwise four-terminal measurement of transport current at 77 K and self-field.

can seep into the porous regions of filaments through defects in the outer sheath of the wire, so that when the Bi-2223 wire warms up, the trapped nitrogen expands and causes cryogenic delamination. Ballooning causes considerable degradation of I_c because the filament is destroyed.[30] Ballooning is a sign of pores within the filaments. The CT-OP process has been found to heal cracks and pores in filaments, so it is expected that the wires will not show ballooning. It has been confirmed to date such wires have not ballooned after being exposed to pressurized liquid nitrogen.

6. Mechanical Properties

In applications such as motors, magnets and generators, the HTS wires must be resistant to mechanical stress and strain. An HTS wire is subjected to large stresses and strain when it is wound into a coil and the magnet is cooled and energized. The tensile strain tolerance of about 0.1% to 0.35% before 95% I_c degradation limits the functionality of Bi-2223 wires.[32] Although the critical tensile stress of the CT-OP processed wires has increased, the critical strain has

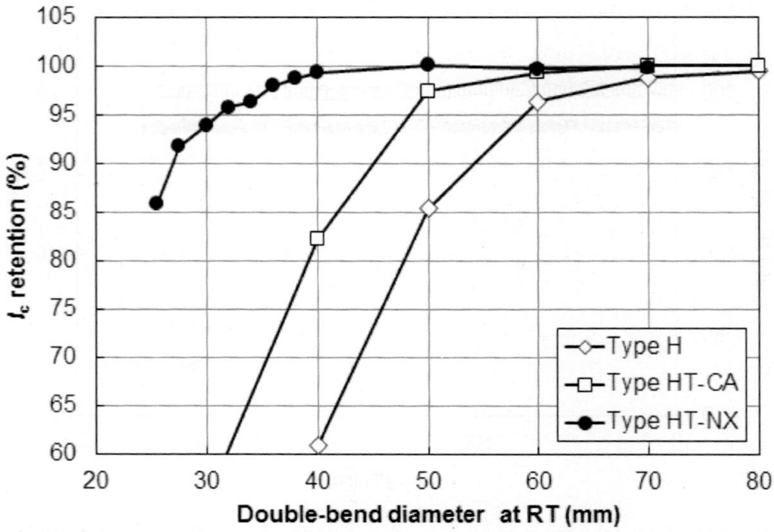

Fig. 7. Double-bend diameter dependence of normalized I_c for various wires at room temperature.

not improved.[8] Therefore, the tensile strain tolerance of the I_c of Bi-2223 wires is improved by laminating the wires with reinforcing materials such as stainless steel. High strength materials laminated on both sides of the Bi-2223 wire clearly produce enhanced resistance to mechanical stress and strain, and improved I_c retention. The mechanical properties and their influence on the I_c of DI-BSCCO were reported by Osamura *et al.*[33]

Figures 7 and 8 show the double-bend diameter and tensile strength dependence of normalized I_c for Bi-2223 wires laminated with various materials. Clearly, lamination with nickel alloy improved both mechanical properties. The double-bend diameter at room temperature reached 35 mm for Type HT-NX, and its tensile strength at 77.3 K is over 430 MPa.

Especially for rotating electrical machinery such as motors and generators, designers frequently require fatigue data as they design for long lifetime reliability. Shin *et al.* showed the results of axial fatigue test on non-laminated Bi-2223 wire. The data indicated that at stress levels up to 115 MPa there is 95% I_c retention up to 10^6

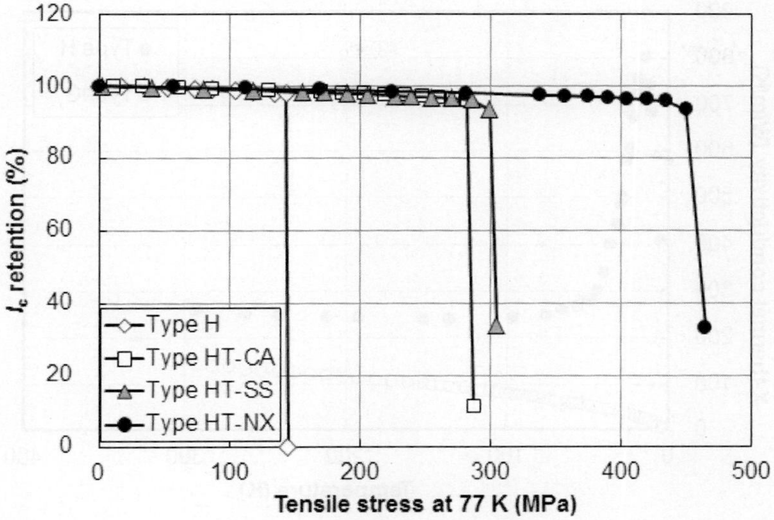

Fig. 8. Tensile strength dependence of normalized I_c for various wires at 77.3 K.

cycles. However, for laminated Bi-2223 wires, the same I_c retention for the same 10^6 cycles was demonstrated at stress levels up to 320 MPa.[34] Hojo *et al.* tested for tensile fatigue along the longitudinal axis of stainless steel laminated DI-BSCCO, showing that at stress levels up to 315 MPa there is 98% I_c retention up to 10^6 cycles.[35]

7. Thermal Properties

An important parameter for the design of HTS current leads is the thermal conductivity of the HTS conductors. For use as the current leads of superconducting magnets, the thermal conductivity of the current leads must be low enough to reduce the heat transfer from current leads to magnets. In the case of Bi-2223 tapes, Bi-2223 oxide filaments are encased in an Ag–Au alloy matrix, in order to provide high current density with low thermal conductivity. Fujishiro *et al.* reported that the thermal conductivity of Ag–Au alloy tapes decreased drastically at low temperatures with increasing Au content after the heat treatment necessary to realize

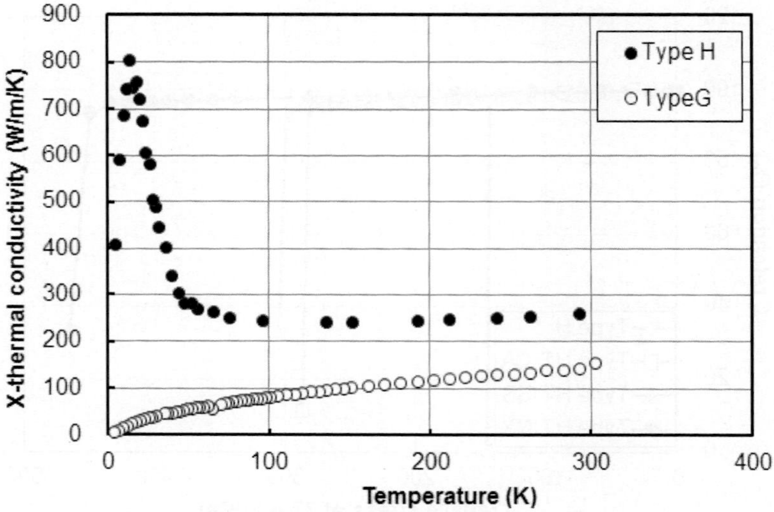

Fig. 9. Temperature dependence of thermal conductivity for Type H and Type G.

the superconducting wire.[36] Naito *et al.* measured the temperature dependence of the thermal conductivity of DI-BSCCO sheathed with Ag and Ag–5.4wt%Au alloy.[37] At around 77K, thermal conductivity values for the wires were about 260 W/m/K and 80 W/m/K, respectively. The thermal conductivity of the Ag–5.4wt%Au alloy tape has a very low value that decreases monotonically at temperatures from 250 K to 10 K.

References

1. H. Maeda, Y. Tanaka, M. Fukutomi and T. Asano, *Jpn. J. Appl. Phys.* **27**, L209 (1988).
2. J. Akimitsu, A. Yamazaki, H. Sawa and H. Fujiki, *Jpn. J. Appl. Phys.* **26**, L2080 (1987).
3. M. Takano, J. Takada, K. Oda, H. Kitaguchi, Y. Miura, Y. Ikeda, Y. Tomii and H. Hazaki, *Jpn. J. Appl. Phys.* **27**, L1041 (1988).
4. J. Kellers and L. Masur, *Physica C.* **372–376**, 1040 (2002).
5. T. Arndt, A. Aubele, H. Krauth, M. Munz and B. Sailer, *IEEE Trans. Appl. Supercond.* **15**, 2503 (2005).
6. H. P. Yi, X. H. Song, L. Liu, R. Liu, J. Zong, J. S. Zhang, A. K. M. Alamgir, Q. Liu, Z. Han and Y. K. Zheng, *IEEE Trans. Appl. Supercond.* **15**, 2507 (2005).

7. J. Kellers, M. Backer, C. Buhrer, J. Muller, A. Rath, S. Remke and J. Wiezoreck, *IEEE Trans. Appl. Supercond.* **15**, 2522 (2005).
8. S. Kobayashi, T. Kato, K. Yamazaki, K. Ohkura, K. Fujino, J. Fujikami, E. Ueno, N. Ayai, M. Kikuchi, K. Hayashi, K. Sato and R. Hata, *IEEE Trans. Appl. Supercond.* **15**, 2534 (2005).
9. S. X. Dou and H. K. Liu, *Supercond. Sci. Technol.* **6**, 297 (1993).
10. J. Tenbrink, M.Wilhelm, K. Heine and H. Krauth, *IEEE Trans. Magn.* MAG-**27**, 1239 (1991).
11. K. Heine, J. Tenbrink and M. Thoner, *Appl. Phys. Lett.* **55**, 2441 (1989).
12. G. N. Riley, Jr. and W. L. Carter, *Am. Ceram. Soc. Bull.* **72**, 91 (1993).
13. S. Kobayashi, T. Kaneko, N. Ayai, K. Hayashi, H. Takei and R. Hata, *Physica C* **357–360**, 1115 (2001).
14. T. Kaneko, S. Kobayashi, K. Hayashi and K. Sato, *Advances in Supercond* **4**, 907 (1997).
15. S. Kobayashi, S. Yamade, M. Kikuchi, K. Yamazaki, J. Fujikami, N. Ayai, K. Hayashi, K. Sato, R. Hata and H. Kitaguti, *Supercond. Sci. Technol.* **21**, 54009 (2008).
16. K. Patnaik, D. M. Feldmann, Y. Polyanskii, X. Y. Cai, E. E. Hellstrom, D. C. Larbalestier and Y. Hung, *IEEE. Tran. Appl. Supercond.* **13**, 2930 (2003).
17. B. Hensel, G. Grasso and R. Flukiger, *Phys. Rev. B* **51**, 15456 (1995).
18. Q. Li, G. N. Riley Jr., R.D. Parrella, S. Fleshler, M. W. Rupich, W. L. Carter, J. O. Willis, J. Y. Coulter, J. F. Bingert, V. K. Sikka, J. A. Parrell and D. C. Larbalestier, *IEEE Trans. Appl. Supercond.* **7**, 2026 (1997).
19. T. Nakashima, S. Kobayashi, T. Kagiyama, K. Yamazaki, M. Kikuchi, S. Yamade, K. Hayashi, K. Sato, J. Shimoyama, H. Kitaguchi and H. Kumakura, *Physica C* **471**, 1086 (2011).
20. K. Sato, T. Hikata, H. Mukai, M. Ueyama, N. Shibuta, T. Kato, T. Masuda, M. Nagata, K. Iwata and T. Mitsui, *IEEE Trans Magn.*, MAG **27**, 1231 (1991).
21. H. Wilhelm, H. W. Neumuller, and G. Reis, *Physica C* **185–189**, 2399 (1991).
22. R. Flukiger, Y. Huang, F. Marti, M. Dhalle, E. Giannini, R. Passerini, E. Bellingeri, G. Grasso and J.-C. Grivel, *IEEE Trans. Appl. Supercond.* **9**, 2430 (1999).
23. M. O. Rikel, A. Wolf, S. Arsac, M. Zimmer and J. Bock, *IEEE Trans. Appl. Supercond.* **15**, 2499 (2005).
24. J. Jiang, X. Y. Cai, A. A. Polyanskii, L. A. Schwartzkopf, D. C. Larbalestier, R. D. Parrella, Q. Li, M. W. Rupich and G. N. Riley, Jr., *Supercond. Sci. Technol.* **14**, 548 (2001).
25. H. Seino, K. Ishizaki and M. Takata, *Jpn. J. Appl. Phys.* **28**, L78 (1989).
26. S. X. Dou, H. K. Liu, M. H. Apperley, K. H. Song, C. C. Sorrell, K. E. Easterling, J. Nisaka and S. J. Guo, *Physica C* **167**, 525 (1990).
27. Y. Yuan, J. Jiang, X. Y. Cai, D. C. Larbalestier, E. E. Hellstrom, Y. Huang and R. Parrella, *Appl. Phys. Lett.* **84**, 2127 (2004).
28. E. E. Hellstrom, Y. Yuan, J. Jiang, X. Y. Cai, D. C. Larbalestier and Y. Huang, *Supercond. Sci. Technol.* **18**, S325 (2005).

29. T. Kato, S. Kobayashi, K. Yamazaki, K. Ohkura, M. Ueyama, N. Ayai, J. Fujikami, E. Ueno, M. Kikuchi, K. Hayashi, and K. Sato, *Physica C* **412–414**, 1066 (2004).
30. L. Masur, D. Parker, M. Tanner, E. Podtburg, D. Buczek, J. Scudiere, P. Caracino, P. Corsaro and M. Nassi, *IEEE Trans. Appl. Supercond.* **11**, 3256 (2001).
31. H. S. Ha, S. S. Oh, D. W. Ha, K. J. Song, Y. K. Kwon and K. S. Ryu, *Physica C* **372–376**, 956 (2002).
32. M. Sugano, K. Osamura and M. Hojo, *Supercond. Sci. Technol.* **16**, 571 (2003).
33. K. Osamura, S. Machiya, H. Suzuki, S. Ochiai, H. Adachi, N. Ayai, K. Hayashi and K. Sato, *Supercond. Sci. Technol.* **21**, 54010 (2008).
34. H.-S. Shin, J. R. C. Dizon, K.-H. Kim, S.-S. Oh and D.-H. Ha, *Cryogenics* **46**, 378 (2006).
35. M. Hojo, K. Osawa, T. Adachi, Y. Inoue, K. Osamura, S. Ochiai, N. Ayai and K. Hayashi, *Physica C* **470**, 1373 (2010).
36. H. Fujishiro, M. Ikebe, K. Noto, T. Sasaoka and K. Nomura, *Cryogenics* **33**, 1086 (1993).
37. T. Naito, H. Fujishiro and Y. Yamada, *Cryogenics* **49**, 429 (2009).

Chapter 2.2

Microstructure Evolution of Bi-2223 Wires

Takayoshi Nakashima
Sumitomo Electric Industries, Ltd.,
1-1-3 Shimaya Konohana-ku Osaka, Japan 554-0022

In terms of the ideal microstructure for Bi-2223 wires with high I_c performance, Bi-2223 volume fraction along with distributions of impurity phases and Bi-2223 grain alignment were reviewed. This chapter summarized the characteristics of the present available wires, which had over 95% of Bi-2223 volume fraction and the degree of texturing ~6°. The results showed that there were many issues to address for further improvement of microstructure, though the wire had high I_c, enough for the current application use.

1. Introduction

High critical current (I_c) is one of the most attractive characteristics for applications of $(Bi,Pb)_2Sr_2Ca_2Cu_3O_{10+x}$ (Bi-2223) wires. The ideal microstructure of Bi-2223 wires for high I_c or high critical current density (J_c) should satisfy the following points.

(1) High superconducting transition temperature (T_c) matrix:

Any effective pinning centers have not been found for Bi-2223 wires. Therefore, T_c of Bi-2223 matrix is the predominant factor mainly on intergrain J_c properties, and partly on intragrain J_c properties.

(2) High Bi-2223 volume fraction:

Each Bi-2223 grains are required to be well connected in order to ensure current paths. High Bi-2223 volume fraction can increase the chance of good connectivity. The distribution of impurity phases blocking the paths should be controlled.

(3) Well-aligned Bi-2223 grains:

Bi-2223 has a *d*-wave symmetry in an order parameter, leading to an anisotropy in the various superconducting properties. The large mis-orientation angles between neighboring Bi-2223 grains spoil intergrain J_c property, thus they should be eliminated.

T_c is a composition-dependent factor; however, it is also variable in dome-shaped fashion in regard to non-stoichiometric oxygen contents. Unfortunately, the accurate composition dependence of T_c has not been clarified because quite a few reports have optimized oxygen contents. In this chapter, the latter two issues (2), and (3) are discussed.

2. Formation of Bi-2223 and the Other Phases

The standard Bi-2223 wires are fabricated by powder-in-tube (PIT) method. PIT method often brings about homogeneity in microstructure of the filaments because full mixing of the inner powder is impossible in the middle processes. Besides, Bi-Pb-Sr-Ca-Cu-O multi-element system has difficulty in increasing Bi-2223 volume fraction. Basically, Bi-2223 wires are composed of six or seven elements, Bi-Pb-Sr-Ca-Cu-O or -Ag except for an additive element in Ag alloy. In terms of thermodynamic degree of freedom ($F = C - P + 1$ in the general expression for condensed matter; F is degree of freedom, C and P are numbers of compositions and phases, respectively), up to seven phases ($P \leq 7$ for $C = 6$ in Bi-Pb-Sr-Ca-Cu-O) can coexist in the filaments at equilibrium state.[1] Main phases appearing in the filaments are summarized in Table 1. Bi-2223 wires generally have three kinds of superconducting phases, Bi-2223, Bi-2212, and Bi-2201. The non-superconducting phase is roughly classified into four

Table 1. Classified phases seen in usual Bi-2223 wires.

	Formula	Abbrev.
superconducting phase	$(Bi,Pb)_2Sr_2CuO_y$	Bi2201
	$(Bi,Pb)_2Sr_2CaCu_2O_y$	Bi2212
	$(Bi,Pb)_2Sr_2CaCu_3O_y$	Bi2223
Non-superconducting phase		
Pb-rich phase	$(Pb,Bi)_3Sr_2Ca_2CuO_y$	Pb3221
	Ca_2PbO_4	CPO
Alkaline Earth Cuprate (AEC)	$(Sr,Ca)CuO_2$	1:1
	$(Sr,Ca)_2CuO_3$	2:1
	$(Sr,Ca)_3Cu_5O_8$	14:24
Other cuprate	CuO	—
Non-cuprate	$Bi_3Sr_4Ca_3O_x$	Bi3430

groups, "Alkaline earth cuprate", "Pb-rich phase", "Other cuprate (especially CuO), and "Non-cuprate (especially $Bi_3Sr_4Ca_3O_x$)". In many cases, we can observe more species in the same filaments than the theoretically predicted maximum at equilibrium state.

This discrepancy appears to be attributed to homogeneity in the actual filaments.

The species, amounts, distributions of the phases are dependent on the initial compositions and thermo-mechanical processes. In this chapter, the results in all the figures are observed with our own wires[2] for systematic study. The wire is composed of 121 filaments and fabricated according to standard PIT processing, first rolling- first heat treatment; second rolling- second heat treatment. The final size of the wire is ~4.0 mm in width and ~0.23 mm in thickness.

2.1. *Bi-2223, Bi-2212, and Bi-2201*

Figure 1 shows scanning electron microscopy (SEM) image of a cross section of the typical wire after the second heat treatment. The fine Bi-2212 was observed in Bi-2223 matrix. Bi-2201 is too few to be seen in Fig. 1, however, the existence can be confirmed in X-ray diffraction (XRD) patterns. Bi-2212 (T_c ~85 K) and Bi-2201 (T_c ~20 K) are

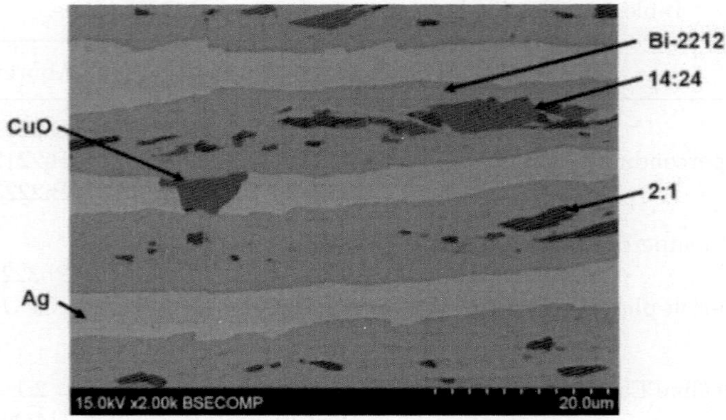

Fig. 1. SEM image of the typical Bi-2223 wire after the second heat treatment. The polished cross section in the longitudinal direction was observed. The matrix in each filaments are Bi-2223. The white phases, dark grey phases, and light grey phases in the filaments are Bi-2212, 2:1, and 14:24 phases, respectively. The grey phase in Ag is CuO.

low T_c phases. For application at over 20 K, we can consider them to be non-superconducting phases. The amount of Bi-2201 is negligible or small compared with that of Bi-2212. Therefore, the ratio of Bi-2223 to sum of Bi-2223 and Bi-2212 is defined as"Bi-2223 volume fraction", which is calculated with XRD peak intensity originating from each phases. It is known that J_cs of Bi-2223 wires are positively related to the Bi-2223 volume fractions over certain values.[3,4]

Bi-2223 is inherently hard to form. Bi-2212 and liquid phases are required to form and increase Bi-2223 quickly.[5] This is why the precursors including Bi-2212 are usually employed. The mechanism of Bi-2223 formation has not been completely established yet. The behavior and the resultant microstructure are similar to peritectic reaction. PbO is added to the starting powder because the additive promotes the generation of Bi-2223[6] and stabilizes the formed structure in which Pb ions partially substitute for Bi sites. Ag also benefits Bi-2223 formation through a decrease of temperature for generation of liquid phases.[7] This accounts for the fact that Bi-2223 grains begin to form at the filaments/Ag interfaces in an early stage of heat treatment.

Bi-2223 volume fraction takes at least tens of hours in heat treatment to reach over 95% even when the starting compositions and the sintering parameters are optimized. In the actual wire processing, rolling and heat treatment are repeated alternately for densification and Bi-2223 grain alignment. Such repetitive thermo-mechanical processing dramatically shortens the overall sintering time. Bi-2212 phases tend to remain in Bi-2223 grains in the formation behavior like peritectic reaction, thus they lose the chance of the reaction with the other phases. In the intermediate rolling after heat treatment, the Bi-2223 grains are broken and the inner Bi-2212 phases are exposed to the other residual phases, hence Bi-2223 formation is considered to be promoted in the subsequent heat treatment.

2.2. *AEC*

AEC has a wide variety of materials in addition to those listed in Table 1. The solid solution between Sr and Ca is allowed in each AEC species, which makes these phases more stable and the removal of them difficult. Among all the AEC phases, 1:1, 2:1, and 14:21 phases are particularly focused on by many researchers because they almost always appear in the fabricated filaments. In Fig. 1, 2:1 and 14:21 phases can be seen. As discussed in Sec. 3, Bi-2223 grains forming around the filaments/Ag interfaces push out AEC particles. As a result, there are few AEC particles around the filaments/Ag interfaces where Bi-2223 grains are well-aligned and better part of current are believed to flow. We can take advantage of this formation manner in order to reduce AEC particles. The well-aligned Bi-2223 layers are formed prior to condensation of AEC particles, hence fabricating the thinner filaments, the thickness of which is equivalent to that of the Bi-2223 layers, and gives no room for gathering of AEC particles. In fact, AEC phases sometimes disappear in the thinner parts around the edges of the filaments.

2.3. *CuO*

Cu in the filaments diffuses into Ag sheathes during heat treatment and forms CuO at the filaments/Ag interfaces or inside of Ag

Fig. 2. Wire surface after the first heat treatment. Black particles are CuO phases.

sheathes (Fig. 1). The CuO particles at the interfaces have a potentially harmful effect on the well-aligned Bi-2223 layers. Nonaka *et al.*[8] has reported that heat treatment at relatively lower temperature increases CuO particles. Ionescu *et al.*[9] has found that CuO is more stable in air than in reduced oxygen atmosphere during heat treatment.

It is noteworthy that CuO also precipitates on the outer surfaces of the wires. Figure 2 shows the surfaces of the wire after the first heat treatment. CuO particles are large enough to be visible to the naked eye as black dots. Higher temperature promotes the diffusion of Cu, and more CuO particles can be seen inside and/or outside of Ag sheathes. The CuO particles on the wire surfaces appear after the first heat treatment, however, they hardly increase in the second heat treatment. This indicates that Bi-2223 grains at the filament/Ag interfaces, which are formed in the first heat treatment, hinder Cu in the filaments from diffusing into Ag sheathes. Although the amounts and sizes are dependent on the heat treatment conditions, the filaments inevitably lose fair amounts of Cu. Nevertheless, the starting compositions near the stoichiometric mixture ratio, Bi(Pb):Sr:Ca:Cu = 1.9–2.1:1.7–2.1: 1.7–2.1 :2.9–3.1, are favored because these lead to good I_c performance.

The excess Cu in Ag sheathes seems to harden the wires and coarse the wire surfaces after the first heat treatment. The influence on subsequent rolling process has yet to be reported.

2.4. *Pb-rich phases*

Pb ions are highly related to both the intragrain and intergrain J_c properties. Accordingly, the control of Pb-rich phases is a hot issue. Since Pb ions in Pb-rich phases have high valence — tetravalent, more Pb-rich phases form in oxidation atmosphere ($P_{O2} > 1\,kPa$). While, in reduced oxygen atmosphere, Pb ions are incorporated into the superconducting phases in divalent state along with a decrease of Pb-rich phases. It is possible to adjust the movement of Pb ions between the Pb-rich phases and the superconducting phases by annealing in the controlled oxygen partial pressure even after these phases are completely formed.

Figures 3(a) and 3(b) show SEM images of Pb-rich phases precipitated from Bi-2223 grains after the second heat treatment. At the final stage in the treatment, the wires were sintered at 783°C

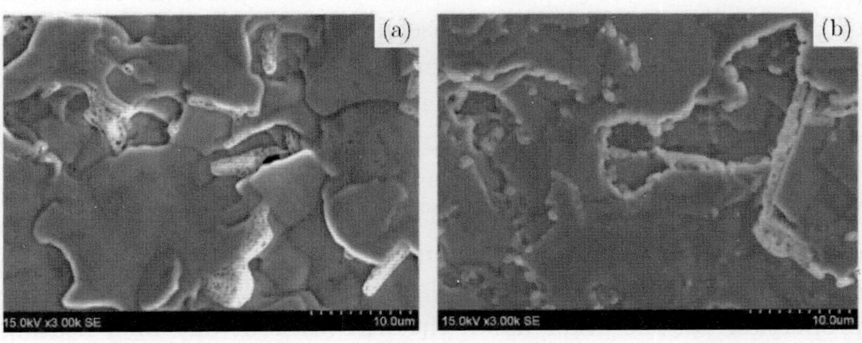

Fig. 3. SEM images of the filament surfaces after the second heat treatment. The wires were sintered at 783°C (a) and 700°C (b) in the final heating stages. Ag sheathes were removed by etching with $NH_3(aq)/H_2O_2(aq)$ solution.

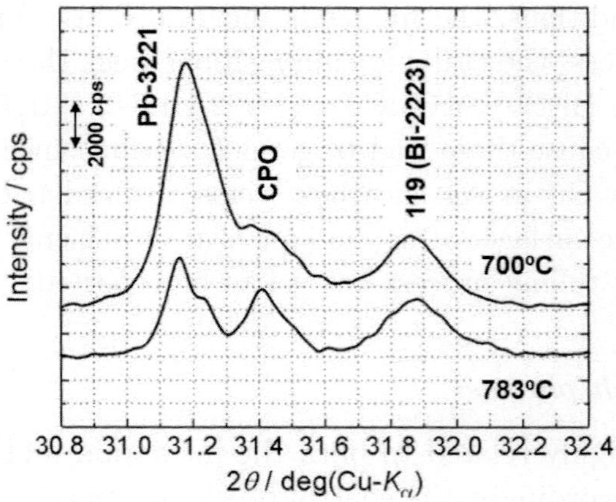

Fig. 4. X-ray diffraction patterns of the wires after the second heat treatment. The wires were sintered at 783°C (a) and 700°C (b) in the final heating stages. The peaks of Bi-2223 (119) reflection, Pb-3221, and CPO are indicated.

[Fig. 3(a)] and 700°C [Fig. 3(b)] in flowing 8% O_2/N_2 gas for 24 h, respectively. The Pb-rich phases precipitate at the edges of Bi-2223 grains. They were increased and surrounded Bi-2223 grains after sintering at 700°C. The XRD patterns of the corresponding wires are shown in Fig. 4. Pb-3221 and CPO were confirmed for both the wires. The heat treatment at 700°C increased only Pb-3221 phases. Shimoyama *et al.*[10] has reported on the effect of Pb-3221 phase formation on T_c with our wires. In the report, the formation of Pb-3221 phase increased T_c (onset), and the maximum T_c (onset) ~115.2 K was achieved by post-annealing at 650–700°C in air. Their results also showed that the enhancement of T_c (onset) accompanied broader transition width, suggesting poor connectivity between Bi-2223 grains. Actually, the wire sintered at 700°C showed ~30 A lower I_c than that intered at 783°C when their non-stoichiometric oxygen contents were full. This indicates that the Pb-3221 particles surrounding Bi-2223 grain boundaries degraded intergrainnular J_c properties. On the other hand, the wire without any final heating stages for formation of Pb-rich phases showed ~10 A lower I_c than that sintered at 700°C. There seems to be an acceptable amount of the formed Pb-rich phases for improvement of I_c.

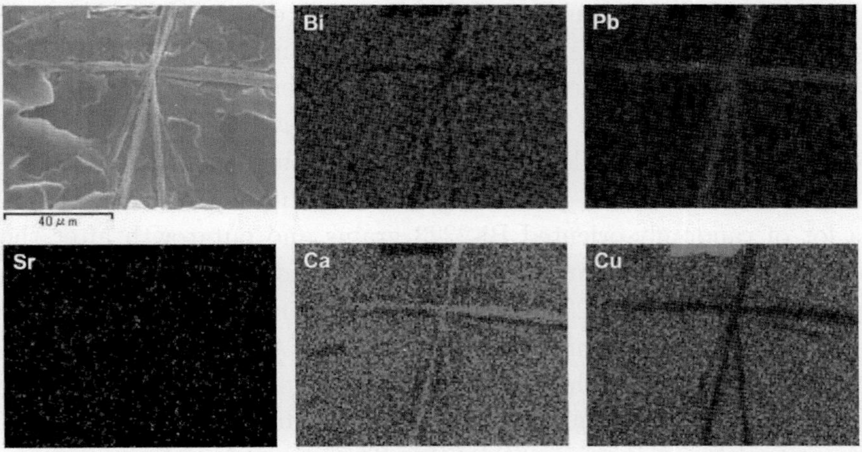

Fig. 5. SEM image and EDX mappings of the etched filament surface after the first heat treatment at 845°C. Ag sheathes were removed by etching with $NH_3(aq)/H_2O_2(aq)$ solution.

The heat treatment at temperatures around the upper limit for the Bi-2223 formation can generate CPO particles. Figures 5(a) and 5(b) show SEM image and the corresponding element mappings analyzed with an energy dispersive X-ray spectroscopy (EDX) of CPO particles. The particles appeared on the surface of filaments after the first heat treatment at 845°C. They are rod-shaped materials with length of 100 μm. There were no CPO particles on the surface before furnace cooling, indicating that the CPO particles were formed directly from the Bi-2223 grains. Moreover, the sintering temperature dependence of the CPO formation suggested that the compositions of the formed Bi-2223 grains were determined by the temperature. Unlike Pb-3221, it is difficult to control the formation of CPO phase independently. This is why the influence of CPO formation on I_c has not been revealed well, although CPO particles are crossing several Bi-2223 grains.

3. Bi-2223 Grain Alignment

The transport current significantly drops when the mis-orientation angle is over 4° in high temperature superconductors. Especially for

Bi-2223, the grain boundaries with the angle of over 8° hardly serve as current paths. The texturing of Bi-2223 wire has put the upmost importance on improvement of *c*-axis alignment.

Since the grains of Bi-based materials have plate-like morphology, they are aligned along their *c*-axis by rolling or pressing processing. If the mechanical processing is not optimized, we can confirm a lot of randomly-oriented Bi-2223 grains and outgrowth after the subsequent heat treatment. In the case of the proper mechanically-processed wires, Bi-2223 grains near the filaments/Ag interfaces spontaneously align their *ab*-planes in parallel to Ag surfaces during the first heat treatment. The drive for the alignment has not been located.

The behaviors of microstructure during the first heat treatment are summarized in Figs. 6(a)–6(h). The wires were heated up to 835°C and sintered at the temperature in flowing $8\%O_2/N_2$ gas for 0, 5, and 10 h and then quenched in air. The wire furnace cooled after sintering for 10 h was also prepared. Figures 6(a)–6(h) show SEM images of the cross sections in the longitudinal direction and the surfaces of the filaments. When the temperature reached 835°C [Figs. 6(a) and 6(e)], the filaments were composed of small, $\sim 3\,\mu$m, Bi-2212 grains as a major phase. The fine, $\sim 2\,\mu$m, AEC particles were dispersed homogeneously in the filaments. After sintering at 835°C for 5 h [Figs. 6(b) and 6(f)], the Bi-2223 grains with the size of $\sim 10\,\mu$m were generated along the filaments/Ag interfaces.

AEC particles were kept away from the interfaces and grew up to $5\,\mu$m. We can see some liquid phases inside the filaments and the surfaces of Bi-2223 grains. After sintering at 835°C for 10 h [Figs. 6(c) and 6(g)], the thickness of the Bi-2223 layers along the interfaces became $\sim 3\,\mu$m and the liquid phases still remained. Many randomly-oriented Bi-2223 grains were generated around the centers of the filaments. They appeared to push away the AEC particles. Finally, after the furnace cooling [Figs. 6(d) and 6(h)], the AEC particles remained around the centers of the filaments. Some Bi-2212 grains can be seen over the whole filaments. In contrast, the liquid phases completely disappeared. As seen in Figs. 6(b) and 6(c), the Bi-2223 grain alignment from the interfaces failed to continue up to the centers of the filaments. Note that the AEC particles and the liquid phases were

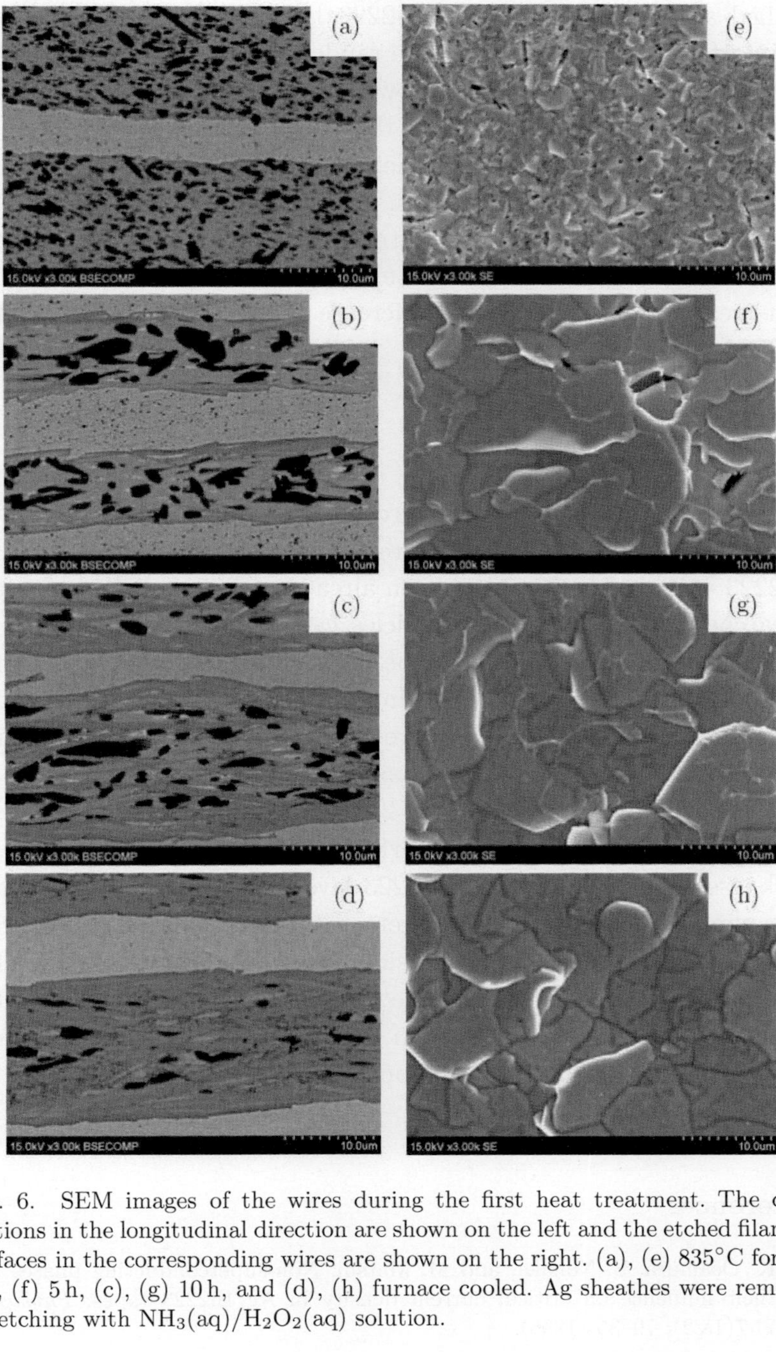

Fig. 6. SEM images of the wires during the first heat treatment. The cross sections in the longitudinal direction are shown on the left and the etched filament surfaces in the corresponding wires are shown on the right. (a), (e) 835°C for 0 h, (b), (f) 5 h, (c), (g) 10 h, and (d), (h) furnace cooled. Ag sheathes were removed by etching with $NH_3(aq)/H_2O_2(aq)$ solution.

pushed away by the growing Bi-2223 layers and did not necessarily interfere with the growth. The randomly-oriented Bi-2223 grains were more likely to block the growth front. As for their generation, it is possible that the AEC and the liquid phases pushed away by the growth front accompanied the composition change in the unreacted regions around the centers, resulting in the random Bi-2223 nucleation. If the explanation is true, the adjustment of the sintering parameters depending on the ever-changing compositions during heat treatment will potentially increase the thickness of the aligned Bi-2223 layers.

4. Conclusion

The formation behaviors of microstructure in Bi-2223 wires were overviewed using our present wires. The methods for over 95% of Bi-2223 volume fraction have been already well established by many researchers. AEC and CuO particles can be reduced, however, the diffusion of Cu into Ag should be taken into account. T_c enhancement and degradation of Bi-2223 intergranular coupling by formation of Pb-rich phases are competitive effects, leading to a permissible amount of the formed Pb-rich phases for I_c improvement.

The observed wires show I_c of 200 A with the degree of texturing $\sim 6°$when all the processing is performed. Despite such high I_c, the thickness of the well-aligned Bi-2223 layers was only 2–3 μm. This is approximately 1/4–1/3 of the average filament thickness. The random Bi-2223 nucleation was considered to limit the thickness, thus the first heat treatment should be optimized from that viewpoint.

We believe that the study on a composition-dependent T_c, which was not discussed here, has a potential to be a mother lode for higher I_c.

References

1. K. Osamura, S. Nonaka, and M. Matsui, Non-superconducting phases and their influence on critical current density in Ag/Bi2223 tapes, *Physica C* **257**(1&2), 79–85 (1996).

2. K. Sato, S. Kobayashi, and T. Nakashima, Present status and future perspective of bismuth-based high-temperature superconducting wires realizing application systems, *Jpn. J. Appl. Phys.* **51**(1), 010006 (2012).
3. K. Sato, T. Hikata, H. Mukai, T. Masuda, M. Ueyama, H. Hitotsunagi, T. Matsui, and M. Kawashima, Electromagnetic properties and structures of BiPbSrCaCuO superconducting wires, *Adv. Supercond.* **2**, 335–340 (1990).
4. T. Kaneko, S. Kobayashi, K. Hayashi, and K. Sato, Consideration of J_c limiting factors in Bi-2223 Ag-sheathed tapes, *Adv. Supercond.* **9**(2), 907–910 (1997).
5. S. X. Dou and H. K. Liu, Ag-sheathed Bi(Pb)SrCaCuO superconducting tapes, *Supercond. Sci. Technol.* **6**(5), 297–314 (1993).
6. T. Hatano, K. Aota, S. Ikeda, K. Nakamura, and K. Ogawa, Growth of the 2223 phase in leaded Bi-Sr-Ca-Cu-O system, *Jpn. J. Appl. Phys.* **27**(2), L2055–L2058 (1988).
7. S. X. Dou, H. K. Liu, J. Wang, M. H. Apperley, C. C. Sorrell, S. J. Guo, B. Loberg, and K. E. Easterling, Microstructure and flux pinning in superconducting Bi-Pb-Sr-Ca-Cu-O wires, *Physica C* **172**(1&2), 63–70 (1990).
8. S. Nonaka and K. Osamura, Influence of calcining temperature on critical current density of Ag/Bi2223 tapes, *Physica C* **281**(2&3), 205–210 (1997).
9. M. Ionescu, J. McKinnon, S. Zhou, and S. X. Dou, Phase formation and evolution of (Bi,Pb)2223 inside Ag tapes in air and reduced oxygen partial pressure using synchrotron X-rays, *Physica C* **400**(1&2), 81–88 (2003).
10. J. Shimoyama, T. Kato, S. Kobayashi, K. Yamazaki, K. Hayashi, and K. Sato, Possible breakthrough for higher T_c discovered in Bi(Pb)2223 practical tapes, *Jpn. J. Appl. Phys.* **44**(50), L1525–1528 (2005).

Chapter 2.3

Lamination of Metals onto Bi-2223/Ag Wires

Alexander Otto

Solid Material Solutions, LLC,
P. O. Box 437 Groton MA, USA 01450,
and 55 Middlesex St. #205 N.
Chelmsford MA, USA

The technology, engineering science and motivation for laminating metal strips onto Bi-2223/Ag tapes are reviewed. The approach allows for the production of Bi-2223/Ag tape with high critical current (I_c) that can be laminated with many types of thin metal strips in order to improve axial stress, strain, bend, and surface indent tolerance as well as mitigation of cryogen penetration and thermal instability. Advances to attain target mechanical properties are enabled by an analytical model that identifies required modifications to internal stress, strain, and failure modes. The model also enables the selection of materials, architectures and process parameters for producing very robust tape. In a recent advance, laminated super alloy strips increased Bi-2223 tape axial stress tolerance from 120 MPa to 540 MPa, and axial strain tolerance from 0.25–0.4%, with a 30% reduction in whole wire current density from the added strip areas, but with no reduction in I_c.

1. Introduction

Bi-2223 conductors must meet the specific mechanical property and stability requirements of diverse applications. However, development of a single Bi-2223/Ag wire that meets these diverse mechanical property and stability requirements has proven to be prohibitively challenging.

This is because the Bi-2223 filament–silver matrix combination was developed in order to attain high oxide and wire current densities (J_c and J_e). Specifically, it was found that sheet rolling drawn round wire to a high aspect tape was the most effective, low cost, and scalable method, in combination with heat treatments, to increase J_c.

Silver is the only material that is both non-reactive with Bi-2223 at heat-treat conditions, and that allows the required rapid oxygen diffusion. Because of oxygen's high diffusivity, alloyed metals like Mg or Al can also be oxidized *in situ* to form oxide dispersion strengthened (ODS)-silver with elevated yield stress and strain. However, Bi-2223-compatible ODS forms of silver are still not adequate for providing the tape with the mechanical properties required in most commercial applications, as illustrated in Fig. 1(a).

This is because the biggest problems with silver are (1) Its low elastic low modulus, a property that cannot easily be changed, and (2) The need for silver to remain ductile enough, even in ODS form, to hold together through tape forming rolling that is applied *after* a first heat treatment that initiates formation of high J_c Bi-2223, and that also converts the silver alloy matrix to its less ductile ODS form.

In addition to ODS silver's limitations, the tape architecture seen in Fig. 1(b) and brittle nature of Bi-2223 further challenge the development of robust tape in two ways: (1) The outer Bi-2223 filaments are typically only about 10 micrometers from the tape surface, where the silver offers inadequate protection from local indent loading during some handling and environmental conditions, (2) Variations in the Bi-2223 filaments provide for lower energy crack propagation sites that in combination with the low modulus and flow stress of silver enable cracks to propagate from filament to filament at relatively low strain levels.

Because of these factors, there is a need to improve select mechanical properties of Bi-2223 tape for different applications, without impairing its superconducting properties. Among several methods investigated, solder lamination of metal strips to the Bi-2223/Ag tape has proven to be best suited for this purpose because of its lower cost,

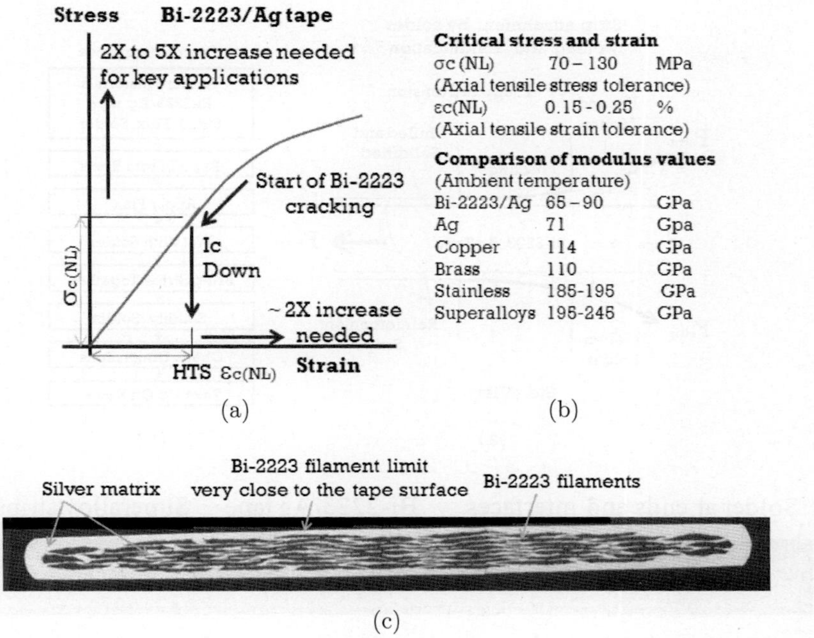

Fig. 1. Tensile stress-strain-I_c schematic for Bi-2223/Ag tape (a) and stress-strain-modulus properties (b) illustrate the low composite and silver modulus values that result in low stress tolerance for the type of Bi-2223/Ag tape shown in cross section (c).

greater flexibility in material and process parameter selection, and applicability to any Bi-2223/Ag tape.

2. Method for Producing Lamination Reinforced Bi-2223 Wire

The lamination reinforcement method illustrated in Figs. 2(a) and 2(b) consists of the in line solder wetting of thin metal strips and Bi-2223/Ag tape, bringing them in contact while they are immersed in molten solder, and solidifying the solder while tooling holds them all in place and removes excess solder to form a uniform, rectangular cross-sectioned tape. This method has been developed by both Sumitomo Electric Industries (SEI) and AMSC into industrial processes with the capability to laminate many types of metals.[1–13]

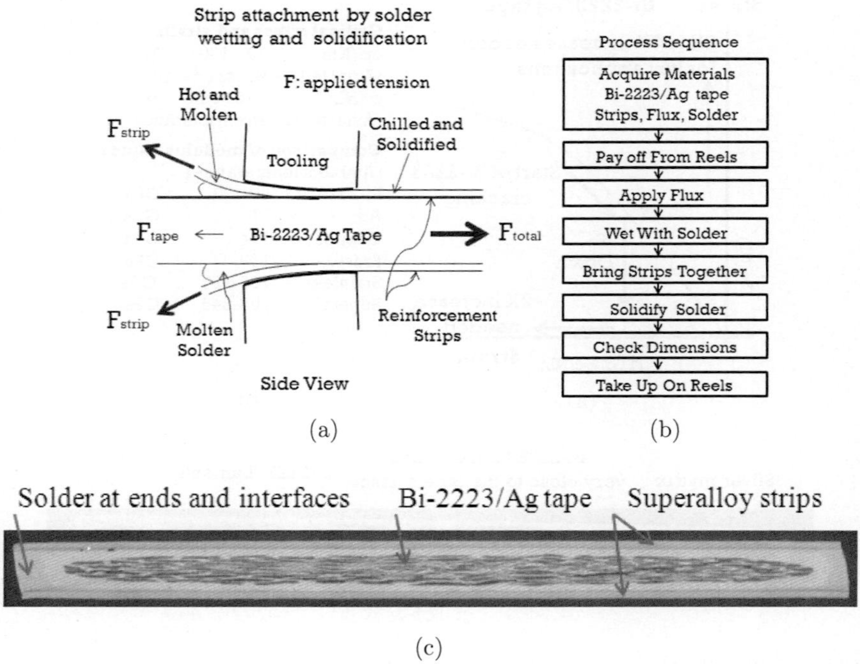

Fig. 2. Illustration of the lamination method that is applied to modify the mechanical properties of Bi-2223 tapes by the solder attachment of thin metal strips in (a) using the process sequence in (b) to produce, for example, a super alloy strip laminated tape (c) The sum of the back tensions at the pay off and frictional drag add up to total force $F_{\text{total}} = 2 \times F_{\text{strip}} + F_{2223} + F_{\text{friction}}$ on the laminated tape at the take up (Photo provided by SEI).

The first development and use of solder laminated Bi-2223/Ag tape was in the mid 1990s for the production of hermetically sealed, and strengthened tape that was used in the first in-grid HTS power cable demonstration.[3,4] Since then, Bi-2223/Ag quality has been improved, notably by SEI, and lamination's hermetic benefit is now less important than its mechanical property and thermal stabilization benefits.[5–13]

2.1. *Wetting*

The strip and Bi-2223/Ag tape surfaces must be in conditions that allow for rapid and effective wetting of the solder, and formation of high strength bonds upon solidification. Strips used include copper, brass, stainless steel, super alloys and copper sheathed molybdenum (ref). Although copper and brass strips can be wetted by most solders, the surfaces of stainless steel and super alloy strips are chemically treated in order to make them acceptably wetted by the solder.

2.2. *Flux*

A suitable flux that is applied just before soldering helps to ensure that all surfaces are wetted everywhere rapidly enough to allow high speed processing, since even storage of the strip and the Bi-2223/Ag tape leads to some local deterioration in wetting. The flux is applied in the thinnest layer that still sufficiently enhances high speed wetting.

2.3. *Solder*

The solder is selected to provide a high degree of adhesion and adequate strength in order to prevent interfacial peeling and solder fracture when large stresses and strains are applied to the tape. In order to provide the most acceptable product, lead free compositions are selected from among high tin commercial formulations that also contain silver and copper for strengthening and metallurgical stability.

2.4. *Bi-2223/Ag tape*

A thin chemical parting agent material is applied to Bi-2223/Ag tape surfaces in order to prevent sintering together of contacting silver surfaces at the temperature used in the final Bi-2223 forming heat treatment. After heat treatment is completed, this material must be removed in order to allow adequate solder wetting of the silver surfaces during lamination. Silver sulfide, formed by reaction with airborne hydrogen sulfide, must also be removed if it is not mitigated with commercial getter materials and sealed Bi-2223/Ag tape storage.

2.5. *Stress and strain evolution to the desired condition by lamination*

Tensile stress and strain tolerance are among the most vital properties improved by lamination. In order to avoid I_c loss, an externally generated axial strain must remain inside the strain window of the tape, defined as the range from axial compression through to axial tensile strain in which I_c is not degraded. Four mechanisms are used by lamination to improve stress and strain tolerance as depicted in Fig. 3.

1. A high modulus material can be applied without any back tension to the Bi-2223/Ag tape, increasing the tape's modulus and as a result getting to the same critical strain as before, which requires a proportionally higher stress level as shown by **(1)** in Fig. 3.
2. Secondly, adding back tension only to the strips elastically stretches them. After attachment and removal of the applied tension, the strips rebound until equilibrium in internal forces is reached, with the Bi-2223/Ag tape in axial compression compared

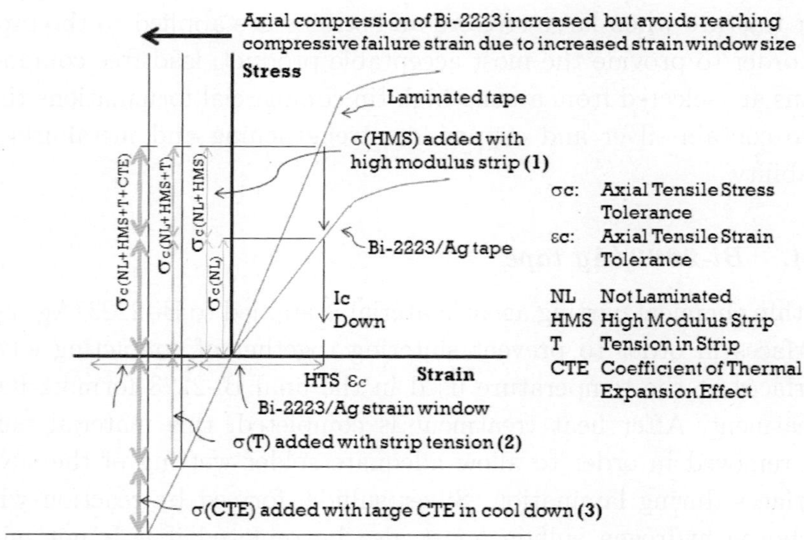

Fig. 3. Illustration of the way tensile stress and strain tolerance can be increased by four mechanisms in a laminated Bi-2223/Ag tape.

to where it was before. This increases the axial strain required to reach the Bi-2223's pre-laminated critical strain as shown by **(2)** in Fig. 3.

3. Thirdly, using a strip with a higher axial coefficient of thermal expansion (CTE) than the Bi-2223/Ag tape will, upon cooling down from solder solidification to cryogenic temperatures, increase the axially compressed state of the Bi-2223 and add to the tensile critical strain as shown by **(3)** in Fig. 3.
4. Fourthly, the higher modulus strips impair the ability of strain to localize for tensile or shear cracks to propagate until larger strains are imparted on the composite, thereby increasing its strain tolerance window as shown by **(4)** in Fig. 3.

3. Lamination Strengthened Bi-2223 Wire Design

Based on these mechanisms, a theoretical framework has been developed and from that a model established and validated that is used to design laminated Bi-2223 tapes with target stress and strain properties.[5]

3.1. *Back tension*

The laminated tape can be made with different back tensions in the constituents, subject to limitations imposed by the materials and methods of assembly. Once the tensions are removed, the mechanically coupled composite relaxes axially to equilibrium where the constituents may be under considerable residual axial stress and strain.

The residual axial Bi-2223 strain effect of the pre-coupling axial tensile force, F_i, is estimated for any number of constituents with elastic modulus E_i and cross sectional area A_i by applying Eq. (1) to each constituent as it relaxes from its initial to final state within the constraints imposed by mechanical coupling among all constituents.

$$\mathrm{d}\varepsilon_i = \mathrm{d}F_i/A_iE_i. \tag{1}$$

Solving *via* integration from the initial to final force and strain states yields Eq. (2), where the "*f*" and "*o*" superscripts denote coupled-relaxed and pre-coupling forces in each constituent.

$$A_iE_i(\varepsilon_i^f - \varepsilon_i^o) = A_iE_i\Delta\varepsilon_i = F_i^f - F_i^o. \quad (2)$$

Since the constituents are coupled, the relaxation strain, $\Delta\varepsilon_i$, is identical for all constituents. However the final axial strain state of the Bi-2223 provides the link to I_c retention, and therefore Eq. (3) provides the relaxation strain of interest, where ε_w^f and ε_w^o are the Bi-2223/Ag final strain after relaxation and initial back tension strain.

$$\Delta\varepsilon_I = \varepsilon_w^f - \varepsilon_w^o. \quad (3)$$

Substituting Eq. (3) into Eq. (2) for all constituents then yields Eq. (4), where ε_w^o is given by Eq. (5)

$$A_iE_i(\varepsilon_w^f - \varepsilon_w^o) = F_i^f - F_i^o. \quad (4)$$

$$\varepsilon_w^o = F_w^o/(A_wE_w). \quad (5)$$

At equilibrium after relaxation,

$$\Sigma F_i = 0. \quad (6)$$

Combining Eqs. (4), (5), and (6) then yields Eq. (7) for the post-relaxation axial strain of the Bi-2223 due to pre-tension in the strips.

$$\varepsilon_w^f = F_w^o/(A_wE_w) - \Sigma F_i^o/\Sigma A_iE_i. \quad (7)$$

3.2. *Thermal mismatch*

The laminated tape is mechanically coupled at a process temperature from materials with differing CTEs. The laminated tape is then cooled to the I_c and strain test temperature. The tape CTE C_c in Eq. (8) is determined from the cross-section area fractions of the constituents, Af_i, their thermal expansion coefficients, C_i, and elastic modulus by applying the constraint of mechanical coupling and Eq. (6).

$$C_c = \Sigma(Af_iE_iC_i)/\Sigma(A_iE_i). \quad (8)$$

The CTE-related shift in axial strain state of Bi-2223 is then calculated by Eq. (9) from the difference between the test temperature and the mechanical coupling temperature, ΔT. If $C_c > C_w$ then axial compressive strain on the BSCCO is increased, where C_w is the Bi-2223/Ag CTE.

$$\Delta\varepsilon_{CTE} = (C_c - C_w)\Delta T. \quad (9)$$

3.3. *Pre-stress and thermal effects combined*

The strain window is shifted by the residual strain in the Bi-2223 tape from constituent pre-tensions and CTE effects. In the elastic region these axial strains are additive. As a result, the pre-reinforced strain window midpoint for the Bi-2223/Ag tape, $\varepsilon^{o}_{\text{wmp}}$, can be adjusted by the pre-tension and CTE-induced strains to estimate the laminated tape strain window midpoint. However the shift in window position is opposite in sign and equal in magnitude to the residual strain additions provided by Eqs. (7) and (9), yielding Eq. (10) for calculating the new strain window midpoint position relative to the pre-reinforced Bi-2223/Ag wire strain window midpoint.

$$\varepsilon^{f}_{\text{wmp}} = \varepsilon^{o}_{\text{wmp}} - \Delta\varepsilon_{\text{CTE}} - \varepsilon^{f}_{w}. \tag{10}$$

Substitution of Eqs. (7) and (9) yields Eq. (11) for $\varepsilon^{c}_{\text{wmp}}$

$$\varepsilon^{f}_{\text{wmp}} = \varepsilon^{o}_{\text{wmp}} - (C_c - C_w)\Delta T - [F^{o}_{w}/(A_w E_w) - \Sigma F^{o}_{i}/\Sigma A_i E_i]. \tag{11}$$

4. Comparison of Model Calculations to Measurements

A detailed study has been completed in which the strain window midpoint and size, as well as failure stress levels, were experimentally determined for a variety of laminated strip materials, as well as some plated materials. The design model was validated on the basis of data from this study, in preparation for its use in designing laminated Bi-2223 tape products.

4.1. *I_c retention strain ranges*

Measured strain window midpoints are compared to the calculated values from Eq. (11) in Fig. 4. The measured data is linear with respect to the calculated data, with a slope of 1.11 and an R^2 correlation coefficient of 0.91, validating the model and demonstrating the relatively well-behaved strain window midpoint changes, realized from widely varied added material attributes and process conditions.

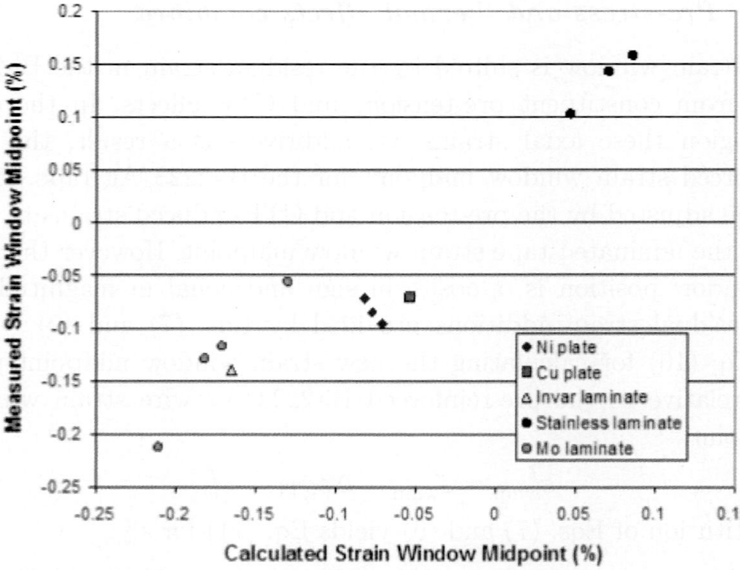

Fig. 4. Data from a study of strain window size and midpoint location, as well as compressive and tensile critical strains with a variety of added materials and conditions, illustrates good correlation between measured and calculated midpoint shifts due to different material properties. The measured Bi-2223/Ag tape midpoint was at −0.06%.[5]

From the fit line, the approximate strain window midpoint is given by Eq. (12).

$$\varepsilon_{\text{wmp}}^{f*} = 1.11\varepsilon_{\text{wmp}}^{c} + 0.03. \quad (12)$$

The deviation from the straight line values of the Cu-clad Mo laminated data and the plated data, while generally still following the linear dependence, suggests that the exact physical dimensions or averaged properties of some of the constituents are off. Some of this variation is also due to the limitations of the employed compression side test.[2]

4.2. *Strain window size*

As shown in Fig. 5, axial strain window size for I_c retention varied by two-fold from Bi-2223/Ag tape at the low end to molybdenum laminated tape at the high end. All added materials increased strain

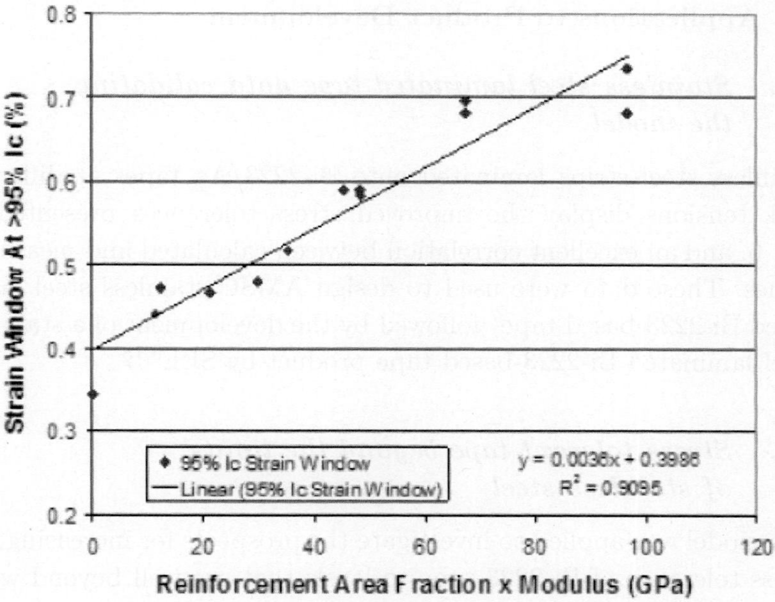

Fig. 5. Data from the study described in Sec. 4 illustrate a large benefit of lamination with high modulus strips. The nearly doubled strain window greatly expands opportunities for utilizing Bi-2223 tape in mechanically demanding environments.[5]

window size, with an ~ linear correlation of strain window size to the product of added materials' area fraction and its modulus of elasticity.

Therefore, axial strain window size for reinforced Bi-2223/Ag tape can be estimated from Eq. (13) for different laminated materials even though the mechanisms giving rise to this correlation are not yet fully identified.

$$\Delta\varepsilon_{w95\%Ic} = 0.0036\ Af_r E_r + 0.40 \tag{13}$$

The dependence observed in Fig. 5 provides very large benefit for improving mechanical properties, because by addition of higher modulus strips and increasing lamination tension in the strips, the axial strain tolerance can be improved greatly, while keeping the axial compressive strain tolerance fixed at a reasonable level, or even increasing slightly.

5. Applications to Product Development

5.1. *Stainless steel laminated tape data validating the model*

Stainless steel strips laminated onto Bi-2223/Ag tapes at different back tensions display the improved stress tolerances presented in Fig. 6, and an excellent correlation between calculated and measured values. These data were used to design AMSC stainless steel laminated Bi-2223-based tape[7] followed by the development of a stainless steel laminated Bi-2223-based tape product by SEI.[9,13]

5.2. *Stress tolerant tape beyond the limits of stainless steel*

The model was applied to investigate the prospects for increasing the stress tolerance of Bi-2223 tape to levels that are well beyond what

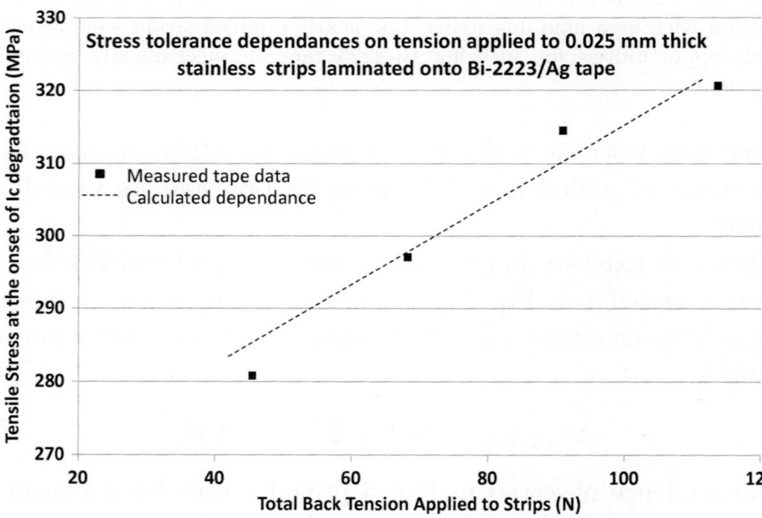

Fig. 6. Measured and calculated stress tolerances of Bi-2223/Ag tape laminated with stainless steel strips at increasing back tension levels, illustrates the predictive correlation that is possible with the described analysis method, provided materials' properties are sufficiently known, and the lamination process is adequately controlled.[7]

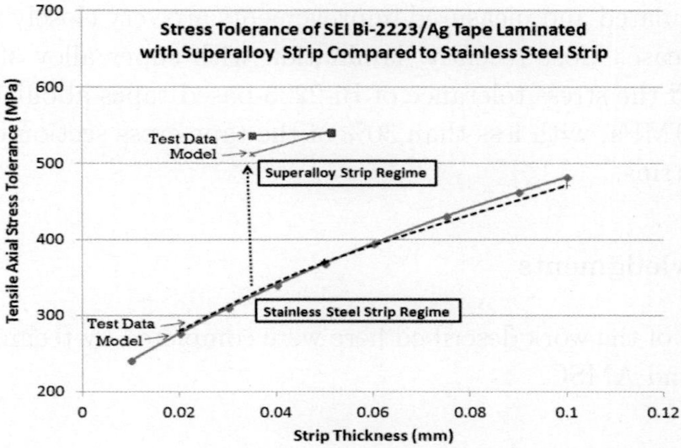

Fig. 7. Measured and calculated stress tolerances of Bi-2223/Ag tape laminated with stainless steel and super alloy strips illustrating the large benefit of super alloy compared to stainless steel strips, and the predictive correlation between calculated and measured data.
Source: Data from SEI, analysis by Otto.[14]

can be attained with stainless steel of the same thickness, with a goal of reaching 500 MPa tensile stress tolerance by adding less than 30% by area of laminated strip. A series of alloys were identified that could be manufactured into the required strip form with elastic modulus values in the 220 GPa to 240 GPa regime, with yield strains as high as 1% and with reasonably high CTEs in the 12 – 14 ppm K range.[14] The results from laminated tapes produced with these strips in Fig. 7 illustrate the new high strain tolerance regimes possible with these strip materials and the utility of the analysis for developing laminated HTS products.

6. Conclusions and Outlook

This article has focused on the lamination of Bi-2223/Ag tape with metal strips that improve its mechanical and thermal stabilization properties. A product and process design model for improving Bi-2223-tape based on mechanical properties has been validated by the results obtained with stainless steel and super alloy strip lamination.

The calculated and measured improvements are very closely matched in each case. Most recently, lamination with super alloy strips has increased the stress tolerance of Bi-2223-based tapes about four-fold to ~ 500 MPa, with less than 30% of the tape cross section occupied by the strips.

Acknowledgments

Portions of the work described here were completed by technical staff at SEI and AMSC.

Variable List

A	Area on short transverse cross section
A_f	Area fraction on short transverse cross section
C	Coefficient of thermal expansion (CTE)
C_c	Reinforced composite axial CTE
C_w	Bi-2223/Ag axial CTE
E	Young's modulus of elasticity in the axial direction
ε	Strain
$\Delta\varepsilon$	Strain difference
F	Axial force
ΔT	Temperature difference (final temperature–initial temperature)

References

1. B. B Gamble, G. N. Riley, J. D. Scudiere, M. D. Manlief, D. M. Buczek, and G. L. Snitchler, *USA Patent 5*, 801,124 (Filed 1996).
2. B. Ten Haken , A. Godeke, H. J. Schuver, and H. H. J. Ten Kate, *Adv. Cry. Eng.*, 651 (1996).
3. D. M. Buczek, L. J. Masur, P. K. Miles, F. Sivo, D. Marlowe, E. R. Podtburg, D. R. Parker, S. Fleshle, J. D. Scudiere, P. Metra, M. Nassi, M. Rahman, and D. W. Von Dolen, *IEEE Trans. Appl. Superconduct.* **7**, 1140 (1997).
4. L. J. Masur, D. R. Parker, M. Tanner, E. R. Podtburg, D. M. Buczek, J. D. Scudiere, P. Caracino, S. Spreafico, P. Corsaro, and M. Nassi, *IEEE Trans. Appl. Superconduct.* **11**(1), 3256 (2001).
5. A. Otto, E. Harley, and R. Mason, *Supercond. Sci. Technol.* **18**, S308 (2005).

6. D. T. Ryan, L. Li, X. Huang, J. W. Bray, E. T. Laskeris, K. Sivasbramaniam, A. D. Gadre, J. M. Fogarty, E. J. Harley, A. Otto, and A. D. Ouden, *IEEE Trans. Appl. Superconduct.* **15**, 3684 (2005).
7. A. Otto, E. R. Podtburg, R. Mason, and P. Antaya, *IEEE Trans. Appl. Superconduct.* **17**, 3071 (2006).
8. S. Ochiai, H. Rokkaku, K. Morishita, J. K. Shin, S. Iwamoto, H. Okuda, M. Hojo, K. Osamura, M. Sato, A. Otto, E. J. Harley, and A. Malozemoff, *Supercond. Sci. Technol.* **20**(3), 202 (2007).
9. S. Ochiai, H. Rokkaku, J. K. Shin, S. Iwamoto, H. Okuda, K. Osamura, M. Sato, A. Otto, and A. P. Malozemoff, *Supercond. Sci. Technol.* **21**, 075009 (2008).
10. S. Machiya, K. Osamura, H. Suzuki, T. Kato, N. Ayai, K. Hayashi, and K. Sato, *Supercond. Sci. Technol.* **21**, 054007 (2008).
11. K. Osamura, S. Machiya, H. Suzuki, S. Ochiai, H. Adachi, N. Ayai, K. Hayashi and K. Sato, *Supercond. Sci. Technol.* **21**, 054010 (2008).
12. K. Yamazaki, T. Kagiyama, M. Kikuchi, S. Yamade, T. Nakashima, S. Kobayashi, G. Osabe, J. Fujikami, K. Hayashi, and K. Sato, *Supercond. Sci. Technol.* **25**, 054015 (2012).
13. K. Osamura, S. Machiya, S. Ochiai, G. Osabe, K. Yamazaki, and J. Fujikami, *Supercond. Sci. Technol.* **26**, 045012 (2013).
14. A. Otto and K. Yamazaki, Results from SEI development program, *presented at the ICMC Conference*, 20 June, 2013.

Chapter 2.4

Quality Assurance of Bi-2223 Wires

Goro Osabe

Sumitomo Electric Industries, Ltd
Superconductivity Technology
Department 1-1-3, Shimaya,
Konohana-ku, Osaka, 554-0024, Japan
osabe-goro@sei.co.jp

Activity of quality assurance of commercial Bi-2223 wire, Bi, Pb(2223) wire as described in this chapter, is introduced, and the specifications of the spliced wire are proposed. Statistical data and experimental data related to wire specifications are discussed based on the probability theory. Detecting the defects for protection against the quench is a highly important task for Bi, Pb(2223) wire quality assurance. The method of electrical inspection such as critical current I_c and n-value is explained. For inspecting the uniformity of the wire, the hall probe measurement method has been developed. Uniformity of the critical current of Bi, Pb (2223) is discussed.

1. Introduction

Since the discovery of Bi, Pb(2223) superconductor, Sumitomo Electric Industries, Ltd. has proceeded with research and development. Now, Bi, Pb(2223) superconductor wire has successfully completed the verification stage to demonstrate its practicality for use in superconducting magnets, motors and power cables. If a superconducting wire does not have the competitiveness of an industrial material, it will be difficult to survive on the market. Recently, requests for high strength, uniformity of the wire properties and cost of the wire are becoming more severe for considering real practical use of superconducting application. From this view point, a method of quality control and philosophy of quality assurance becomes important. In this chapter, experimental and statistical data related to the specification of several kinds of Bi, Pb(2223) wire and the development of the spliced technique for considering the commercial long tape with the kilo-meter over-length of Bi, Pb(2223) wire are introduced.

2. Specifications of Bi, Pb(2223) Wire

The powder-in-tube method is used for manufacturing Bi, Pb(2223) wires.[1,2] The process is constituted of many processes such as powder making, drawing, rolling, and several wire heat treatments. After final heat treatment, the Bi, Pb(2223) wire with the length of up to 2000 m has electrical inspection carried out, such as critical current I_c and n-value at 77.3K, and geometrical inspection at RT such as width, thickness, and length. After these inspections, Bi, Pb(2223) wire is wrapped and shipped to the customer together with a data sheet of the inspection results. Table 1 shows the specifications of Bi, Pb(2223) wire commercialized as DI-BSCCO manufactured by Sumitomo Electric Industries, Ltd.

To assure the Bi, Pb(2223) wire quality, basically only by inspection can all wire length be guaranteed the characteristics over the entire shipping length. This kind of inspection is so-called the in-line inspection, and the manufacturer can ensure the specifications with the customer. On the other hand, some of the physical properties, such as mechanical properties, are important but difficult to measure over the entire length. In this case, off-line sampling inspections are carried out as typical data for shipping wire. Off-line sampling inspections such as critical wire tension, critical tensile strength, critical wire strain and critical bending diameter are listed on the specifications as typical values. Critical value of the mechanical properties was determined as retaining 95% I_c.

Table 1. The specifications of Bi, Pb(2223) wire commercialized as DI-BSCCO manufactured by Sumitomo Electric wire.[3]

	Bi, Pb(2223)					
Use		Magnet	Cable	High Field Magnet	Current Lead	AC Application
Ag Sheath Material	Ag	Ag	Ag	Ag	Ag-Au 5.4wi%	Ag
Twist	—	—	—	—	—	yes
Lamination						
Material	—	Stainless steel	Cu alloy	Ni alloy	—	Cu alloy
Thickness (μm)	—	20	50	30	—	50
Critical Current						
I_c@sf(A)	170–200	170–200	170–200	170–200	170–200	60–70
Size						
Width (mm)	4.3 ± 0.2	4.5 ± 0.1	4.5 ± 0.1	4.5 ± 0.2	4.3 ± 0.2	2.8 ± 0.1
Thickness (mm)	0.23 ± 0.01	0.29 ± 0.02	0.34 ± 0.02	0.31 ± 0.03	0.23 ± 0.01	0.32 ± 0.02
Mechanical Properties						
Critical Wire Tension@RT (N)*	80	230	280	410	80	150
Critical Tensile Strength@77K (MPa)*	130	270	250	400	130	270
Critical Tensile Strain@77K (%)*	0.2	0.4	0.3	0.5	0.2	0.4
Critical Double Bending Diameter@RT (mm)	80	60	60	40	80	40

*95% I_c retention, typical value

3. Thickness and Width of Bi, Pb(2223) Wires

Width and thickness of shipping wires are important requirements for wire quality assurance. Such geometrical inspection systems are equipped with a reel-to-reel winding machine. A high accuracy width sensor is used for the width inspection and a contact-type thickness roller is used for thickness inspection. Controlling the inspection temperature is important to increase the accuracy. Furthermore, in addition to the average value of the width and thickness of the total shipping length, the fluctuation value of the local value of thickness and width are informative for detecting the quality abnormality.

Figures 1 and 2 show the distribution data of the width and thickness of Bi, Pb(2223) wire and Cu alloy 50 μm thick laminated wire. The number of the calculation data of the Bi, Pb(2223) wire is 576 wires and that of laminated wire is 397 wires. From probability theory, the continuous random variable such as thickness and width has the Gaussian distribution. Average and standard deviation of the thickness of the Bi, Pb(2223) wire are 0.23 mm and 0.002 mm, respectively. COV of the thickness is 1.0%. Average and standard deviation of the width of the Bi, Pb(2223) wire are 4.3 mm and 0.07 mm, respectively. COV of the width is less than 1.6%. The distribution of the thickness is smaller than that of the width because thickness of the Bi, Pb(2223) wire is the controlled value at the rolling process of phase inversion method (PIT) method. On the other hand, average

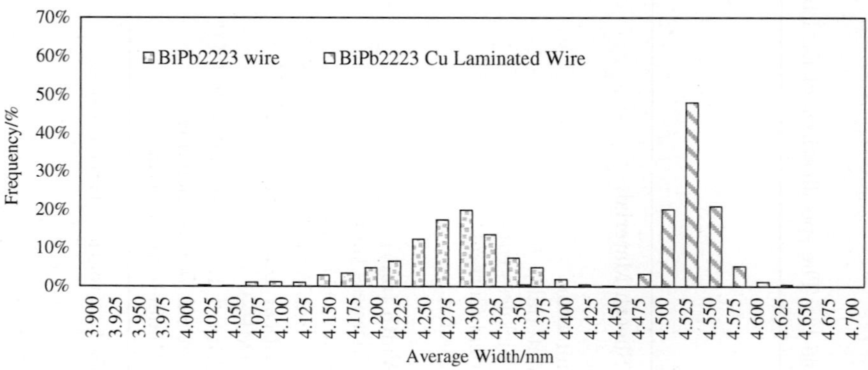

Fig. 1. The distribution of the average wire width of the Bi, Pb(2223) and Bi, Pb(2223) laminated wire with the thickness of 0.05 mm Cu alloy.

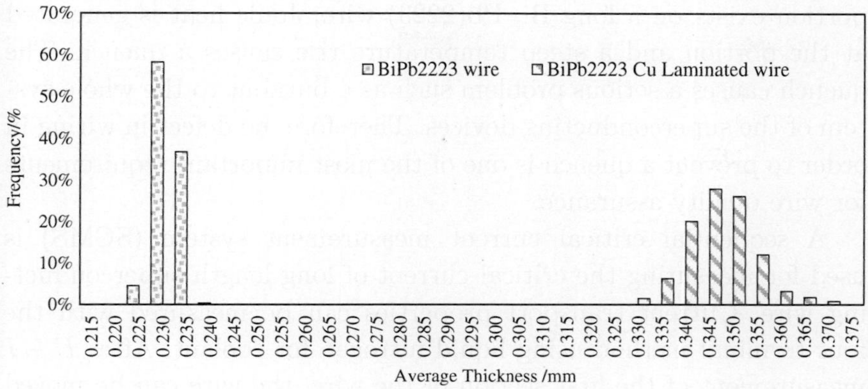

Fig. 2. The distribution of the average wire thickness of the Bi, Pb(2223) and Bi, Pb(2223) laminated wire with the thickness of 0.05 mm Cu alloy.

and standard deviation of the thickness of the laminated wire are 0.34 mm and 0.007 mm, respectively. COV of the thickness is 2.0%. Average and standard deviation of the width of the laminated wire are 4.5 mm and 0.027 mm, respectively. COV of the width is 0.6%. The distribution of the width of the Cu laminated wire depends on the distribution of the width of the lamination material itself. After lamination, distribution of the Bi, Pb(2223) bare wire is hidden and it is bigger than that of the Bi, Pb(2223) bare wire because solder thickness causes the wider distribution during the lamination process and it is important for reducing and making uniform thickness of the solder for wire application.

The specification limits usually are required to be inside the $\pm 3\sigma$. It means, the values less than three standard deviations away from the mean accounts for 99.73%. The manufacturing process achieves a higher level of control with low σ, as it may set higher sigma. The specifications of Bi, Pb(2223) and laminated wires are considered with the average value of the entire wire length with $\pm 3\sigma$ as listed in Table 1.

4. Critical Current I_c and n-Value

Critical current I_c and n-value at 77.3 K are inspected completely throughout the length of the Bi, Pb(2223) wire. If one defective portion exists on a long Bi, Pb(2223) wire, Joule heat is generated at the portion and a steep temperature rise causes a quench. The quench causes a serious problem such as a burnout to the whole system of the superconducting devices. Therefore, no defect in wiring in order to prevent a quench is one of the most important requirements for wire quality assurance.

A sequential critical current measurement system (SCMS) is used for measuring the critical current of long length superconducting wire. Current transport properties can be measured with the four-terminal method using liquid nitrogen for coolant. After $E-I$ measurement of the first section of the wire, the wire can be moved forward using a reel-to-reel winding system and is measured at the next section. This step continues automatically from the beginning to the end of the total length. Using this inspection system, one can automatically detect the defect for total length or an arbitrary interval of the wire without cutting out the wire. The value of critical currents I_c is defined by the electric field of 10^{-6} V/cm, usually. The relationship between the voltage and the current of the superconductor approximately satisfies the power-law model written as

$$\mathrm{V} = V_c \left(\frac{I}{I_c}\right)^n ,$$

where the exponent of I/I_c is the n index value (n-value) and V_c is the voltage at the critical current I_c. The n-value is usually defined by fitting the curve with power-law, with the electric field range from 10^{-6} V/cm to 10^{-7} V/cm. Minimum interval I_c or n-value of all sections of the shipping wire should meet the specifications. Figure 3 shows the typical data of the critical current I_c and n-value at 77.3K inspected completely throughout the length of the Bi, Pb(2223) wire.

To protect from the quench and to detect the defect is a highly important task for Bi, Pb(2223) wire quality assurance. Recently, a defect of HTS wire can be measured using a hall sensor array.[4] The advantages of the measurement are high resolution and high inspection speed to compare with the SCMS.

Sumitomo Electric has been developing the inspection system including the function of the defect detection measurement system

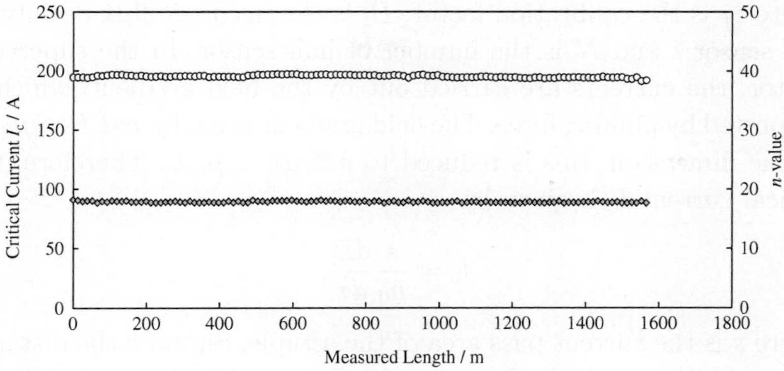

Fig. 3. Typical I_c and n-value data at 77.3 K throughout the entire length of the Bi, Pb(2223) wire. Average critical current and n-value are 195A and 17.8, respectively. (i.e. I_cx $L = 306$ kAm, $I_c = 448$A/cm).

(DDMS) using a hall sensor measuring system, reel-to-reel winding system, geometrical measurement system and four terminal method I_c measurement system altogether. The resolution of the DDMS is 1 mm or higher.

By definition, magnetization is defined by

$$\mu_0 M = B - \mu_0 H_a,$$

where H_a is the applied external field, μ_0 is the permeability of a vacuum and B is the magnetic flux density that the hall sensor measured. If the defective areas are scanned with a hall probe array under the constant magnetic field $\mu_0 H_a$ and velocity, the demagnetizing field $\mu_0 M$ of the defective portion of the wire is smaller than the normal portion because the shielding current is reduced at the defective portion. Using this character of the superconductor, Bi, Pb(2223) long length wire can be inspected and defects can be detected with a higher resolution compared with the sequential critical current measurement system. Therefore, the uniformity of the Bi, Pb(2223) was observed using this system. Here, the critical current I_c of the hall sensor method is defined as[5]

$$I_c = c_f \sqrt{\sum_{i=1}^{N-1} \frac{(B_{i+1} - B_i)^2}{N-1}},$$

where c_f is the calibration factor, B_i is the magnetic flux density at hall sensor i and N is the number of hall sensor. In the superconductor, the currents are carried out by the field gradient, which is supported by pinning force. The field gradient given by $\boldsymbol{rot}\ B = \mu_0 J_c$. In one dimension, this is reduced to $dB/dx = \mu_0 J_c$. Therefore, the critical current I_c is given by

$$I_c = \frac{s}{\mu_0}\frac{dB}{dx},$$

where s is the current pass area of the sample. Suppose the distance of each hall sensor is dx, I_c is given by $I_c = c_f \cdot dB$ where $c_f = \mathrm{s}/\mu_0 dx$ and this is a constant value as the calibration factor. Therefore, it is defined that the current deduced from the field gradient might be corrected with the calibration value with the current from the transport critical current.

Using DDMS, uniformity of the critical current of the Cu alloy laminated Bi, Pb(2223) wires are evaluated. The unit length of about 100 m wires are measured. Figure 4 shows the results of DDMS and SCMS critical current. Resolution of the DDMS is 4 mm and that of SCMS is 4 m. c_f is calculated as the total average of the critical current becomes the same as the average critical current of the SCMS.

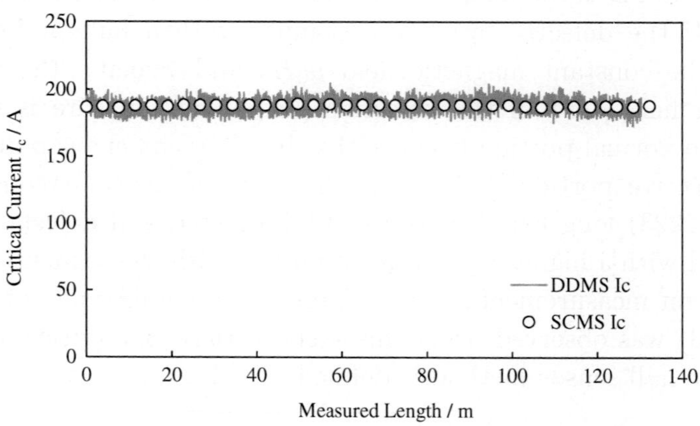

Fig. 4. Critical current distribution of Cu laminated Bi, Pb(2223) with the length of 130 m at 77K. Data compared DDMS I_c and SCMS I_c.

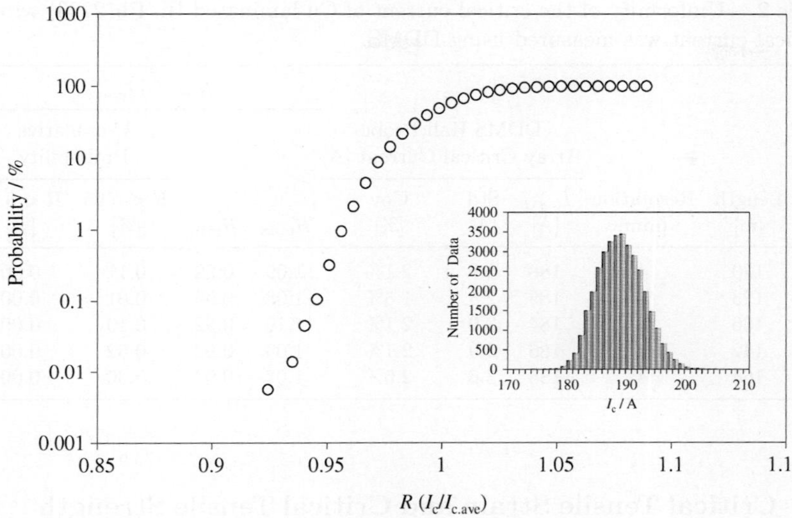

Fig. 5. Cumulative distribution of $R = I_c/I_{c.ave}$ and histogram of I_c.

The average critical current of SCMS of this wire is 188A and the standard deviation is 0.7A (COV 0.36%). For the DDMS measurement, the critical current of this wire is 188A and the standard deviation is 3.9A (COV 2.1%). For evaluating the uniformity of the critical current, calculate the cumulative distribution of $R = I_c/I_{c.ave}$ as shown in Fig. 5. The fluctuation range of R from average $I_c(R = 1.0)$ is between 0.92 to 1.09. The distribution of the critical current is within 10% from the average of the critical current. Furthermore, performing the distribution analysis with inverse function plot of the cumulative distribution, the plot shows linearity. This result indicates that the distribution of the critical current is Gaussian distribution.

Five samples of the uniformity of the critical current of the Cu alloy laminated Bi, Pb(2223) wires are evaluated. The unit length of those samples are 120 m to 160 m. The results are listed on Table 2. The average of those wires is 186A. The average of the standard deviation of those samples is 3.6A (COV 2.0%). It is proven that uniformity of the Bi, Pb(2223) wire is high enough for superconductor application.

Table 2. Uniformity of the critical current of Cu laminated Bi, Pb(2223) wires. Critical current was measured using DDMS.

			DDMS Hall Probe Array Critical Current [A]			$R = I_c/I_{c.ave}$			
								Cumulative Probability	
Lot #	Length [m]	Resolution [mm]	$I_{c.ave}$ [A]	Std [A]	Cov [%]	R_{max}	R_{min}	$R < 0.95$ [%]	$R < 0.90$ [%]
1	130	4	188	3.9	2.1%	1.09	0.92	0.11	0.00
2	123	4	185	2.8	1.5%	1.08	0.95	0.01	0.00
3	166	4	184	3.9	2.1%	1.10	0.92	0.19	0.00
4	142	4	186	3.9	2.1%	1.09	0.92	0.52	0.00
5	164	4	185	3.6	2.0%	1.07	0.93	0.30	0.00

5. Critical Tensile Strain and Critical Tensile Strength

The mechanical properties of the superconducting wire such as critical tensile strain and critical tensile strength at 77.3 K are listed on the specifications. Such mechanical properties are difficult to measure over the entire length. Therefore, off-line sampling inspections are carried out as typical data for shipping wire.

The tensile strain and strength tolerance tests were carried out at liquid nitrogen temperature by using the tensile machine with load. For detecting the strain of the sample, a Nyilas type double extensometers with the length of 25 mm was attached at the center of the sample. For measuring the critical current, the voltage taps were soldered on tapes outside of the Nyilas type double extensometers. Before loading the strength or stress, the critical current I_c of the unloading sample was measured as I_{c0}. After a certain load is applied to the sample, the critical current of the loaded sample was measured as I_c. Tensile strain and tensile strength dependence of the normalized critical current I_c/I_{c0} are evaluated. Figure 6 shows the stress dependence of the normalized critical current for Bi, Pb(2223) without lamination and several kinds of lamination tapes. The critical current decreases gradually in the reversible region and drops sharply at a certain level.[6] When the strain or stress are reached to a certain level, the filaments of Bi, Pb(2223) wire begin to become

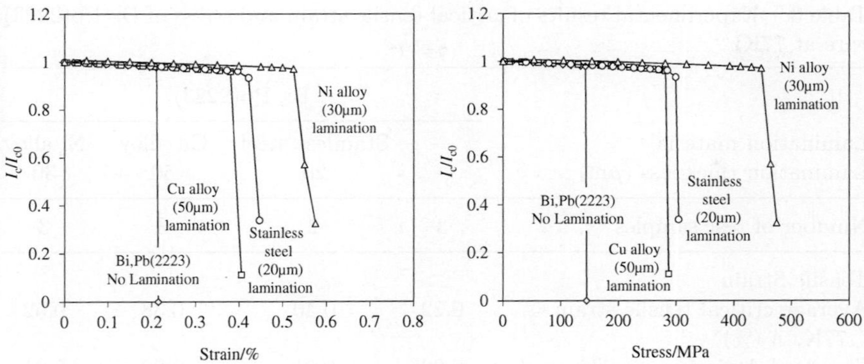

Fig. 6. Strain and stress dependence of the normalized critical current for various tapes at 77K.

broken and the critical current is decreased abruptly due to the weakness of the grain connectivity. The critical tensile strength and strain are determined as the strain or stress to retain 95% of I_{c0}. As seen in the figure, it is clear that the critical tensile strength increases with lamination.

Table 3 shows the experimental results of tensile strain and stress measurements of Bi, Pb(2223) wire. Enhancement of the mechanical properties is due to the lamination effect. Especially, average critical tensile strain and stress of the Ni alloy lamination wire is 0.52% and 444 MPa, respectively. For high field magnet application, a very large tensile load is added to the wire because of the hoop stress. Hoop stress is equal to the magnetic flux density B times the current density times the coil radius as BJR. Supposing B and R are constant for magnet design, wire having a high tensile stress means that the critical current density can be higher. It means that the magnet can reduce the size. This is a highly important requirement for nuclear magnetic resonance (NMR) application. In this view-point, Ni alloy lamination wires are thought to be applicable for NMR applications because of the maximum permissible tensile stress with low shielding current properties of Bi, Pb(2223) wire.

After a 400 MPa load is applied to the 100 samples of Ni alloy laminated wire, I_c retention of the critical current of the loaded sample are evaluated at the 77.3K. As shown in Fig.7, the distribution

Table 3. Experimental results of critical tensile strain and stress of Bi, Pb(2223) wire at 77KL.

	Bi, Pb(2223)			
Lamination material	—	Stainless steel	Cu alloy	Ni alloy
Lamination thickness (μm)	—	20	50	30
Number of test samples	3	2	3	3
Tensile Strain				
Average critical tensile strain @ 77K, A (%)*	0.22	0.40	0.38	0.52
Standard deviation σ_A (%)	0.02	0.01	0.00	0.01
$A - 3\sigma_{\mathrm{A}}$ (%)	0.17	0.36	0.37	0.50
Tensile Stress				
Average critical tensile stress @ 77K, R (MPa)*	146	290	282	444
Standard deviation σ_R(MPa)	2.9	1.4	0.8	4.2
$\mathrm{R} - 3\sigma_{\mathrm{R}}$ (MPa)	138	286	279	431

* 95% I_c retention, typical value

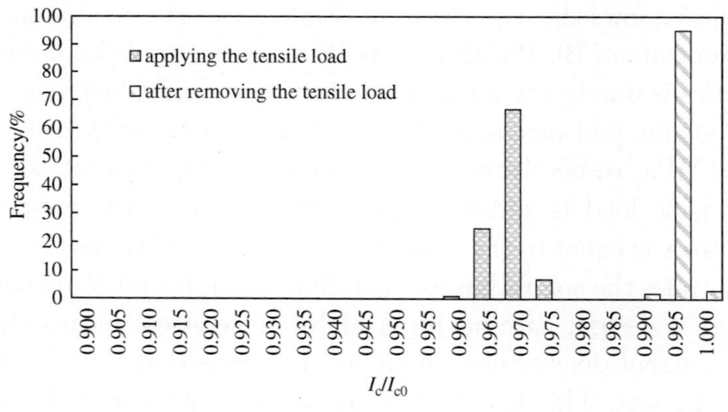

Fig. 7. I_c retention histogram of 100 Ni alloy lamination wires after loading the tensile stress of 400 MPa at 77K.

of the I_c retention when applying the tensile load and after removing the tensile load are 0.97–0.98 and 0.99–1.00, respectively. These results indicate that I_c retention behaves as reversible and there is no fracture of the filaments with loading 400 MPa at 77.3K.

6. Critical Double Bending Diameter

Specifications of the critical double bending diameters of Bi, Pb(2223) are listed in Table 1. When a wire is bent, tensile strain and compressive strain are respectively applied to the outside and inside of the neutral surface of the wire. Hence, the bending diameter becomes smaller, corresponding strain or stress becomes higher, and causes the fracture of the filaments of the superconductor. Therefore, considering the critical bending diameter is important to design the superconducting equipment. The relationship between I_c/I_{c0} and the double bending diameter of several kinds of the Bi, Pb(2223) wire are shown in Fig. 8. The critical double bending diameter is defined as that when I_c becomes 95% of its initial value.

Table 4 shows the experimental results of the double bending diameter measurements of Bi, Pb(2223) wire. Every wire used the same lot of Bi, Pb(2223) bare wire and three samples of each wires were tested for the double bending test. Enhancement of the bending properties is due to the lamination effect. As shown in Table 4, it is proven that laminated wire is flexible. Especially, the critical bending diameter of the Ni alloy wire is 32 mm which is lower than a half of

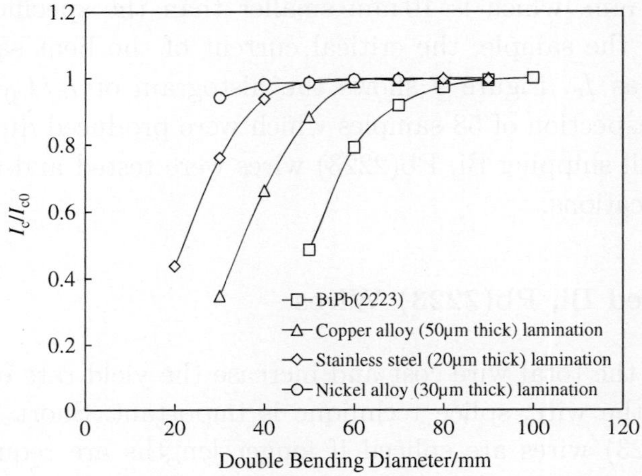

Fig. 8. Experimental data of the double bending diameter dependence of the I_c/I_{c0}.

Table 4. Experimental results of double bending diameter of Bi, Pb(2223) wire. Bending was performed at RT. I_c measurement was performed at 77K.

	Bi, Pb(2223)			
Lamination material	—	Stainless steel	Cu alloy	Ni alloy
Lamination thickness (μm)	—	20	50	30
Average width (mm)	4.2	4.6	4.5	4.5
Average thickness(mm)	0.23	0.31	0.33	0.31
Average critical double bending diameter, D_c (mm)	76	42	56	32
Standard deviation σ (mm)	1.1	1.4	0.2	1.3
$D_c + 3\sigma$ (mm)	79	47	57	36
Correnspoing critical bending strain (%)	0.29%	0.67%	0.57%	0.86%

*95% I_c retention, typical value

Bi, Pb(2223) wire. Accordingly, lamination wires are thought to be sufficiently applicable for compact race track coil magnet.

Off-line sampling inspections of the double bending test are carried out for the short length samples which are cut out from every shipment of Bi, Pb(2223) wires. Before bending, the critical current I_c of the sample was measured as I_{c0}. After a bending diameter of 70 mm, which is 10 mm smaller than the specifications, is applied to the sample, the critical current of the bent sample was measured as I_c. Figure 9 shows the histogram of I_c/I_{c0} after the bending inspection of 58 samples which were produced during a few months. All shipping Bi, Pb(2223) wires were tested and assured of the specifications.

7. Spliced Bi, Pb(2223) Wires

To reduce the total wire cost and increase the yield rate of the unit length of the wire, splice technique is important. Short lengths of Bi, Pb(2223) wires are spliced if longer lengths are required from the customer. The spliced wires are already installed in many of the superconducting cable PJ, transformer application as well as SMES application.

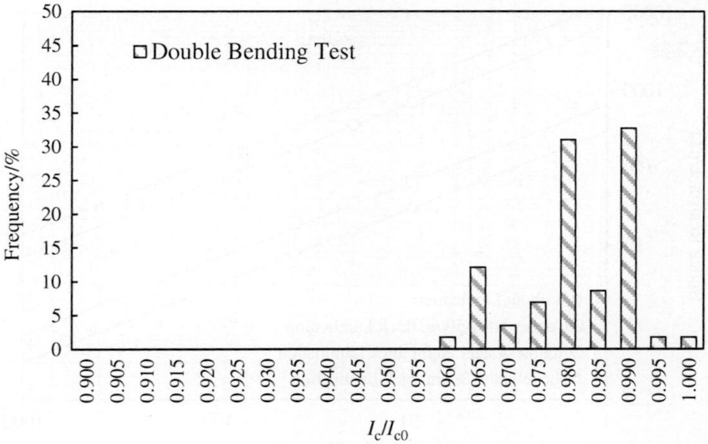

Fig. 9. Results of the off-line double bending test for shipping Bi, Pb(2223) wires ($N = 58$).

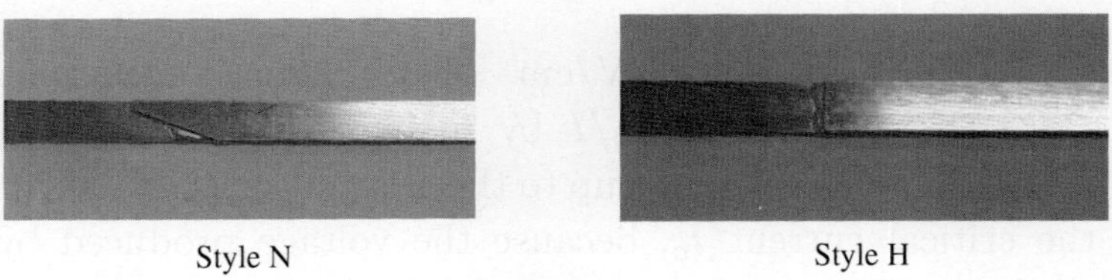

Fig. 10. Typical structure of the age of the spliced wire.

Bi, Pb(2223) splice wire basically uses the overlap method which is the so called shake hands lap splice.[7] A longer overlap length yields lower splice resistance. However, a shorter overlap length will give better flexibility at the splice point. The shape of the tip of the wires also changes the flexibility around the overlapped area. Figure 10 shows the typical structure of the tip of the splicing wires. Style N cuts the tip of the wires as a diagonal shape and usually is used for cable application for better flexibility. Style H cuts the tip of the wire vertically and is usually used for magnet application.

Splice resistance, R_s, can be calculated from gradient from $V - I$ curve. Voltage of the spliced wire, V_s, might be expressed as

$$V_s = R_s I + V_c \left(\frac{I}{I_c}\right)^n,$$

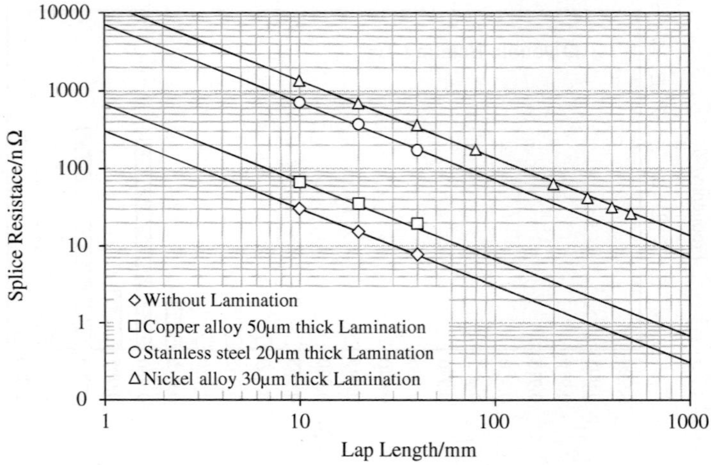

Fig. 11. The splice resistance of Style H spliced wire of the Bi, Pb(2223) wire and laminated wire with several laminated material.

where V_c is the voltage at $1\mu\text{V/cm}$. TheR_s value is usually calculated from the gradient as V_s/I by fitting with the least square method with the current range up to the current which is much lower than the critical current I_c, because the voltage produced by the superconducting current is much smaller than the voltage produced at the normal current passing through the solder between spliced wires.

Figure 11 shows the splice resistance of the Style H spliced Bi, Pb(2223) wires and laminated wires. The relationship between splice resistance and overlap length is linear in double logarithmic chart. It is clear that splice resistance is inversely proportional to the lap length. Splice resistance of the jointed parts can be calculated by the summation rule of the contact resistance in between the bare superconductor which means the resistivity and thickness of the reinforced material and solder decide the splice resistance.

Table 5 shows the typical specifications of Bi, Pb(2223) splice wire. Before shipping several hundred meters of Bi, Pb(2223) wire with spliced point, the splicing point measures the thickness, width and splice resistance at 77K, and assure the properties of those specifications.[3]

Table 5. Typical specifications of Bi, Pb(2223) splice wire.

	Bi, Pb(2223)		
Lamination	None	Stainless Steel	Cu alloy
Lamination Thickness (μm)	N/A	20	50
Splice Style	Style N	Style N	Style N
Overlap Length (mm)	20mm	20mm	20mm
Thickness of Splice (mm)	≦0.53	≦0.69	≦0.81
Width of Splice (mm)	≦4.6	≦4.8	≦4.8
Splice Resistance (nΩ)	≦40	≦600	≦90
Critical Wire Tension@RT (N)	≧80	≧230	≧280
Critical Double Bending Diameter @ RT (mm)*	180	140	140

*95% I_c retention typical value

8. Summary

Experimental and statistical data related to the specification of several kinds of Bi, Pb(2223) wires and the development of the spliced technique of wires are introduced. The hall probe measurement method has been developed for detecting the defect. The method of connecting between the current deduced from the field gradient and the critical current deduced from the transport critical current is corrected with the calibration value. Uniformity of the critical current of Bi, Pb(2223) is discussed. The standard deviation of the critical current of Bi, Pb(2223) laminated wire is 3.6A (COV 2.0%). Average critical tensile strain and stress of the Ni alloy lamination wire is 0.52% and 444 MPa, respectively. Ni alloy lamination wires are thought to be applicable for NMR applications because of the maximum permissible tensile stress with low shielding current properties of Bi, Pb(2223) wire.

References

1. K. Sato, T. Hikata, H. Mukai, M. Ueyama, N. Shibuta, T. Kato, T. Masuda, M. Nagata, K. Iwata, and K. Mitui, High-J_c silver-sheathed bi-based superconducting wires. *IEEE Trans. Mag.* **27**(2), 1231–1238 (1991).

2. A. Otto, L.G. Masur, J. Gannon, E. Podsburg, D. Daly, G.J. Yurek, and A.P. Malozemoff, Multifilamentary Bi-2223 composite tapes made by a metallic precursor route. *IEEE Trans. Appl. Supercond.* **3**, 915 (1993).
3. http://global-sei.com/super/hts_e/index.html.
4. http://www.theva.com/prod/tapestar.
5. Document of THEVA Tapestar, 3.6.2, 3–12.
6. K. Osamura, S. Machiya, H. Suzuki, S. Ochiai, H. Adachi, N. Ayai, K. Hayashi, and K. Sato, Improvement of reversible strain limit for critical current of DI-BSCCO due to lamination technique, *IEEE Trans. Appl. Supercond.* **19**, 3026 (2009).
7. Y. Iwasa, *Case Studies in Superconducting Magnets*, Chapter 7, p. 266. New York: Springer Science & Business Media.

Chapter 2.5

Silver Contribution to Bi-2223 Wires in Terms of Cost

Kan Kinoshita
Sumitomo Electric Industries, Ltd.
1-3-13, Motoakasaka, Minatoku, Tokyo

Bi-2223 High Temperature Superconductivity (HTS) wires use certain amount of silver. Recently there is a new wire which has less silver than before. People say that Bi-2223 wire is expensive, because it uses silver. However, silver contributes a lot in terms of stable quality of Bi-2223 HTS wires and silver is necessary for its performance. This paper summarized the actual value of silver in the wire and compares the copper value in the copper cable system and the silver value in the HTS cable system using an example of cable project in Germany.

1. Introduction

High Temperature Superconductivity (HTS) is one of the basic, innovative and great discoveries of the 20th century. After the discovery of the superconductivity of mercury in 1911, superconductor specialists have been trying to adopt a variety of metals, alloys, and compounds to produce high critical current HTS wire which can be produced under commercial production process. Since the beginning years of 2000, Bi-2223 HTS wires became commercially available and the wires were applied to many developmental projects such as power cables, motors, and magnets. After experiencing for more than a decade such worldwide developmental projects, it still remains an open question whether Bi-2223 HTS wires could be economically competitive electrical wires in the long run. Such question is brought easily because Bi-2223 HTS wires use certain amount of silver (Ag) in the wire.

In this chapter, I describe the contribution of Ag in Bi-2223 HTS wires in terms of quality and the cost.

2. Ag Content in Bi-2223 HTS Wires

Bi-2223 HTS wires are produced from a mixture of BiSrCaCuO compound (black color portion in Fig. 1) packed in Ag pipes (grey color portion of Fig. 1) using PIT method. The final shape of this wire is flat. Several different kinds of Bi-2223 HTS wires have been sold into the market. For example, those named as Type H, Type G, Type AC, and Type HT.

Silver content of Bi-2223 HTS wire Type H is approximately 6 g/m. In case of Type G, which is mainly used for the current lead application, the silver content is 6 g/m with 0.34 g/m of gold (Au). Type AC wire which is mainly used for AC power cable application has approx. 3.8 g/m of silver. As Type HT is the metallic enforced wire of Type H, silver amount in the wire is same as that of Type H.

In 2014, a new design of Bi-2223 HTS wire, which is called New Type H, was developed and prepared to be released in the market. This wire was designed and produced to have less thickness (0.20 mm) to create higher critical current density (J_c). The New Type H has 5.2 g/m of silver which is 13% less silver than the Type H.

3. Silver Contribution in Terms of Quality and Stability in HTS Wire

Why do Bi-2223 HTS wires have to use silver? There are mainly two reasons to use silver, one is for mechanical reason of the wire

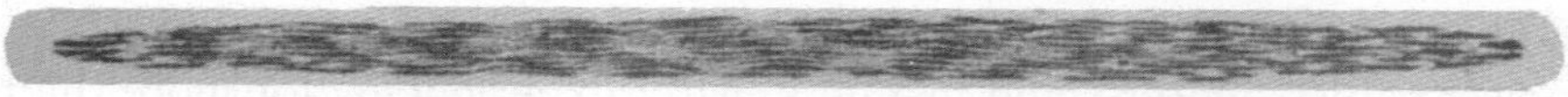

Fig. 1. Photograph of Bi-2223 HTS wire (Type H). Width is approximately 4.3 mm and the thickness is approximately 0.23 mm.

strength and the other is for helping the chemical reaction of bismuth compounds. During the PIT method, bismuth compounds are packed into a silver tube or silver alloy tube and sealed to form a billet. The billet is then drawn and re-bundled to a multifilament composite and transformed into a wire (flat) shape. Silver is suitable metallic material which has enough mechanical strength to form the wire shape. This silver strength will provide the certain protection needed against the delamination of the wire when it is used in extreme high magnetic field coils.

Also during the production of the wire, several heat treatment processes are applied to the wire, so that the Bi-2223 grains and filaments could grow by acquiring or releasing the oxygen through the silver layer easily. As of today, no other metallic material than silver can perform such beneficial function to the bismuth compound.

As the material of silver has the best thermal conductivity in metals, it will be beneficial to prevent quenching of the wire when there is non-uniform heating or hot spot. For power cable applications, Bi-2223 HTS Type HT-CA wires (copper alloy tapes are laminated with the Type H) are commonly used. Combination of silver material in the Type H and the laminated copper alloy tape, both of which have superior thermal conductivity, can spread the heat localization quickly and avoid dangerous phenomenon, such as melt down of HTS wires.

Over current loading testing (rectangular current testing) of Type HT-CA (sample name was CA1−10 in Fig. 2) was carried out in 2011 at JSC "The All-Russian Scientific Institute of Development and Technology in Cable Industry" (VNIIKP), Moscow, Russia (Fig. 2). The HT-CA sample had the critical current of 193.6 A at 77 K, self-field condition. Through this testing, the stable and unstable current regions were determined and thermal runaway current (thermal quench current) I_q was also determined. You can see the current I_q of this sample was 487 A (the I_c when stable to unstable regime happens). Due to the superior thermal characteristics of silver and copper alloy tape, Type HT-CA could survive up to almost 2.5 times of its critical current. Rectangular current pulse determines the I_q that shows the maximum current at which superconductor (or

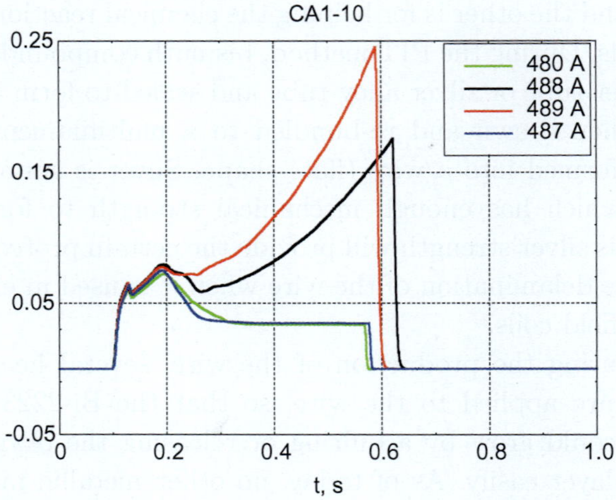

Fig. 2. Example of rectangular current testing of Type HT-CA (I_c = 193.6 A at 77 K, self-field). Tested at VNIIKP in 2011. Due to the superior thermal characteristics of silver and copper alloy material, Type HT-CA could survive up to almost 2.5 times of its critical current.

Table 1. Ag, Au contents in a HTS wire.

Ag, Au Contents (gram per wire meter)		
• Wire Type	• Silver	• Gold
• Type H	• 6.0g	
• Type G	• 6.0g	• 0.34g
• Type HT	• 6.0g	
• Type AC	• 3.8g	

superconducting device) can work stably. It is a real parameter of a superconductor and cooling conditions. Below this I_q current, superconductor remains superconducting infinitely.

4. Silver Contribution in Terms of Cost in HTS Wire

As described in Tables 1 and 2, Type H wire has 6 g/m or new Type H wire has 5.2 g/m of silver in it. Furthermore, Type G has 0.34 g/m

Table 2. Ag content in New Type H.

Ag Content (gram per wire meter)		
Wire Type	Silver	Gold
New Type H	5.2g	

Table 3. Silver, gold, and copper price history (Yen per gram).

Year	Silver (Ag) (Yen/g)	Copper (Cu) (Yen/g)	Gold (Au) (Yen/g)	Ag Value in Type H & G (Yen/m)	Ag Value in New Type H (Yen/m)	Au Value in Type G (Yen/m)
1999	21	0.22	1,069	125	108	363
2000	19	0.23	1,014	113	98	345
2001	19	0.23	1,105	113	98	376
2002	20	0.24	1,296	122	105	441
2003	20	0.25	1,399	119	103	476
2004	25	0.36	1,472	151	130	500
2005	28	0.46	1,619	168	145	550
2006	46	0.83	2,287	277	240	778
2007	54	0.89	2,659	323	280	904
2008	54	0.78	2,937	321	278	999
2009	47	0.53	2,951	280	243	1,003
2010	59	0.70	3,477	354	307	1,182
2011	94	0.75	4,060	561	487	1,380
2012	83	0.68	4,335	498	432	1,474
2013	78	0.76	4,461	467	405	1,517
2014	68	0.77	4,305	405	351	1,464
(Average)	(47)	(0.56)	(2,528)	(275)	(238)	(859)

Note: Also, you can see the silver and gold total value in each meter of Type H and Type G.
Source: Data are taken from the web site of Tanaka Kikinzoku Kogyo KK (silver, gold) and JX Nippon Mining & Metals (copper).

of gold in it, in addition to the silver. Silver and gold prices fluctuate based on the market price every day as same as the copper price. Table 3 shows the yearly price history of silver, gold, and copper since 1999 in Japanese Yen. These data are taken from the web site of Tanaka Kikinzoku Kogyo KK (silver, gold) and JX Nippon Mining & Metals (copper). So on an average, in the last 16 years, Type H has 275

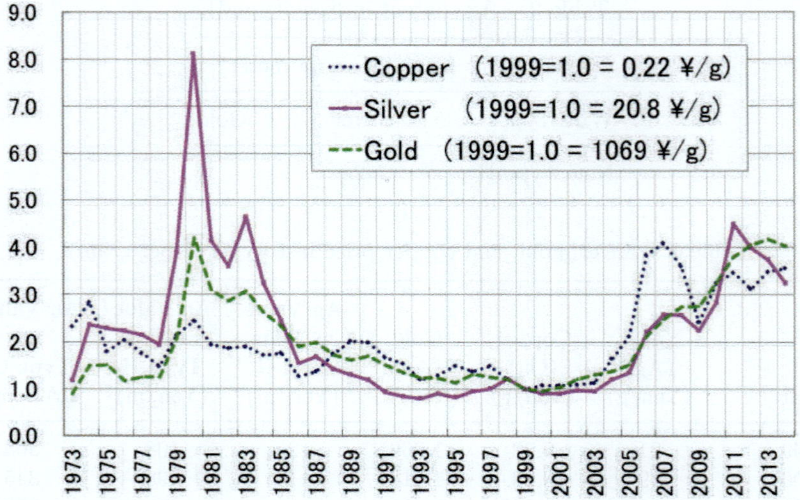

Fig. 3. Historical price trend of silver, gold, and copper since 1973, normalized to each material's prices in 1999. Data are taken from the web site of Tanaka Kikinzoku Kogyo KK (silver, gold) and JX Nippon Mining & Metals (copper).

Yen/m and New Type H has 238 Yen/m of silver. Due to the design improvement, new Type H has approx. 13% less silver (cost) in the wire than Type H. In case of Type G, total metallic value (silver and gold) would be 1,134 Yen/g. This is one of the reasons that the market price of Type G wire is more than the Type H. Based on the 2014 price, when the New Type H wire was sold at 2,380 Yen/m; the silver contribution to the price in 2014 was only 15% (68 × 5.2/2380).

Figure 3 shows the long-term price fluctuation of silver, gold, and copper when those prices normalized in 1999 as the value 1.0. As you can see, in recent decade, silver, gold, and copper prices are basically rising up with the same trend. Compared to the prices in 1999, silver and copper prices are getting almost 3.5 times and gold is close to 4.0 times. Around the year of 1980, you can see the speculative rise of silver price.

Nowadays, silver has various industrial uses, including as a component of cell phones and its demand is steady; however, since the development of digital camera, silver consumption for photographic demand declined steadily. Photography's share of total silver demand

Table 4. Supply and demand of silver in weight (ton).

(Ton)

	2009	2010	2011	2012
Supply	28,898	33,474	32,330	32,604
Mining	77%	70%	73%	75%
Recycle	22%	21%	25%	24%
Other	1%	9%	2%	1%
Demand	28,898	33,474	32,330	32,604
Industrial	43%	47%	47%	44%
Photo	9%	7%	6%	6%
Jewelry	34%	32%	34%	31%
Other	14%	14%	13%	19%

Source: These data are taken from World Silver Survey 2013.

fell to 6% in 2012 compared to 26% in 1999.[1] Most of the decline came from consumer color film. On the supply side, recycled silver had certain share of 24% of supply in 2012. Table 4 shows the supply and demand of silver in the world.

5. Recycle of Scrapped HTS Wire, Coil, and Cables for Silver

Bi-2223 HTS wire manufactures regularly sell the scrapped HTS wires, HTS coils, and HTS cables to the recycling companies. They normally evaluate and examine the silver contents of scrapped materials and they pay for it based on the silver's market price. One of the Bi-2223 HTS wire manufacturers highly recommends the HTS wire users to recycle the scrapped materials locally. For example, Fig. 4 is the logo made by Sumitomo Electric to encourage the users for recycling and to save their cost of production.

At the recycling company, scrapped HTS materials are [crushed Fig. 5(a), dissolved Fig. 5(b), recovered Fig. 5(c) and refined Fig. 5(d)].[2]

[1]World Silver Survey 2013, GFMS, Thomson Reuters.

[2]Courtesy of Tanaka Kikinzoku Kogyo K. K.

Fig. 4. Recycle logo made by Sumitomo Electric to encourage HTS wire users for recycling.

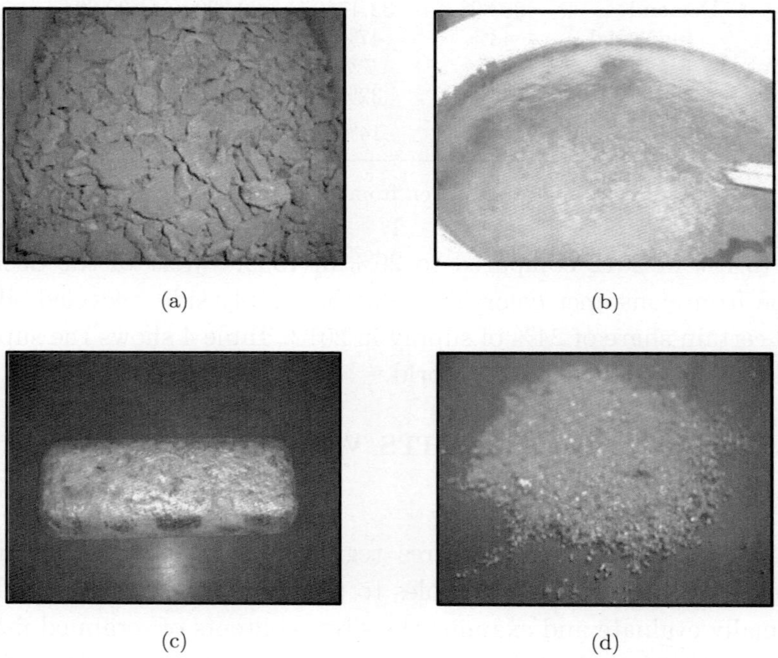

Fig. 5. (a) Crush (b) Dissolved in acid dissolution (c) Recovery (d) Refining.

6. Accounting Process of HTS Equipment which Includes the Silver

As same as existing equipment made of copper materials, once HTS equipment, such as HTS cables, are bought and installed in a company, such as utility company, it will be processed and accounted for fixed assets including the silver amount contained in the HTS wire. In Japan, depreciation period of underground power cables is

25 years. In future, when utility company scraps the old HTS cables which reaches the end of its life and recycles the silver in HTS wires in the old HTS cables, the profit on sales of the silver amount will be processed as non-operating profit. So, in the long run, as long as we do surely recycle the silver material in the HTS wires, coils, and cables, you can buy a new HTS wire with the same silver price which you sell as scrap.

6.1. *Case study: Comparison between the copper value in copper cables and the silver value in HTS cables which may replace the copper cables*

In April 2014, German utility RWE and Nexans announced the success integration of the world's longest HTS cable, one kilometer long, at Essen's power grid[3] which was financially supported partially by the German Federal Economics Ministry for the Environment and Energy (BMWi). Before they started this AmpaCity pilot project, which connects two sub-stations in Essen's city center, they did some feasibility study among 110 kV, 10 kV copper cables and 10 kV HTS cables[4] based on a peak load of 40 MVA. Table 5 shows the various cables which were compared in this project. For the purpose of comparison, we assumed that approx. 80 kilometers of HTS wires (equivalent to New Type HT) are used in 10 kV HTS cables. The silver content in New Type HT wires is 5.2 g/m as described in Table 2, and the copper density of 8,960 kg/m^3 is used in following calculation.

Each cable has following amount of copper and silver in the total system. 110 kV copper cable system has 8,064 kg, 10 kV copper cable system has 84,672 kg of copper and 416 kg of silver is present in 10 kV HTS cable systems.

Using the copper and silver price of 2014 in Table 3, i.e. 0.77 Yen per gram of copper and 68 Yen per gram of silver, the total copper

[3] http://www.nexans.com/eservice/Corporate-en/navigatepub_142506_33667_297_40_2579/World_premiere_in_Essen_RWE_integrates_superconduc.html.

[4] Cigre 2012 B1-301, Superconducting medium-voltage cables for urban power supply as an alternative scenario to 110 kV installations.

Table 5. Performance parameters of three kinds of cables for comparison.[4]

	110 kV copper cable	10 kV copper cable	10 kV HTS cable
Cable Type	1×300 mm^2	1×630 mm^2	Nexans HTS 10/40
Number of systems	1	5	1
Nominal voltage (kV)	110	10	10
Continuous current (A)	591	2310	2310
Continuous power (MVA)	113	40	40

Note:

110 kV copper cable: $1 \times 300\,\text{mm}^2 \times 10^{-6} \times 8,960\ \text{kg/m}^3 \times 3$ phases $\times 1$ set $\times 1$ km $= 8,064$ kg
10 kV copper cable: $1 \times 630\,\text{mm}^2 \times 10^{-6} \times 8,960\ \text{kg/m}^3 \times 3$ phases $\times 5$ set $\times 1$ km $= 84,672$ kg
10 kV HTS cable: 80,000 m $\times$ 5.2 g/m $=$ 416,000 g $=$ 416 kg

Table 6. Total amount of copper and silver in the cable system.

	110 kV copper cable	10 kV copper cable	10 kV HTS cable
Amount of copper	8,064 kg	84,672 kg	
Amount of silver			416 kg
Copper value (0.77 Yen/g)	¥6,209,280	¥65,197,440	
Silver value (68 Yen/g)			¥28,288,000

and silver amount of each system are as described in Table 6. 110 kV system has 6.2 million Yen; 10 kV copper system has 65.2 million Yen of copper, while 10 kV HTS cable has 28.3 million Yen of silver. From this table, we know that although the silver price is almost 88 times more than the copper price, as a total cable system, the price difference between the silver and copper amounts to only 4.6 times.

So, recycling the silver material in HTS equipment at the end of its useful life or the silver material in scraps during the production is important in terms of total expense.

7. Conclusion

In typical commercial Bi-2223 HTS wires, for example, in the case of Sumitomo Electric, Type H has 6 g/m and New Type H has 5.2 g/m of silver in it. Silver contributes a lot in terms of HTS wire's stability against strength and hot spot. For costing, Type H has 238 Yen per meter of silver. Due to the design improvement, new Type H has approx. 13% less silver (cost) in the wire than Type H. In case of Type G, total metallic value (silver and gold) would be 1,134 Yen per gram. These prices are based on the averaged prices of each material during the past 16 years. Based on the 2014 price, when the New Type H wire was sold at 2,380 Yen per meter, the silver contribution to the price in 2014 was only 15%. Although silver is 88 times expensive than copper per weight (as of 2014), one example of power cable systems comparison between 110 kV copper cable and 10 kV HTS cable shows that the silver amount is only 4.6 times expensive than the copper amount. But still, recycling activity of silver is important to save the cost of the systems.

of silver in it. Silver contributes a lot in terms of HTS wire's stability against strength and hot spot. For costing, Type H has 225 Yen per meter of silver. Due to the design improvement, now Type H has approx. 13% less silver cost) in 50 Yen more than Type H. (In case of Type G, total metal value (silver and gold) would be 1,341 Yen per meter. These prices are based on the averaged price of each material during the past 16 years. Based on the 2014 price, when the New Type H wire was sold at 2,850 Yen per meter, the silver contribution to the price in 2014 was only 7.5%. Although silver is 80 times expensive than copper per weight (as of 2014), one example of power cable systems comparison between 11 kV copper cable and 10 kV HTS cable shows that the silver amount is only 4.6 times expensive than the copper amount. But still recycling activity of silver is important to save the cost of the system.

PART 3
Applications

Chapter 3.1

Current Leads for ITER

Kaizhong Ding and HTS team of ASIPP
Tokamak Design Division, ASIPP,
No. 350, Shushanhu Road, Hefei (230032), China

According to the procurement agreement between International Thermonuclear Experimental Reactor International Organization (ITER IO) and China Domestic Agency (CNDA), Institute of Plasma Physics, Chinese Academy of Sciences (ASIPP) is responsible for the design, fabrication and testing of a series of high temperature superconducting (HTS) current leads (CLs) for ITER since 2007. Three types of HTS CLs have been developed, manufactured and tested, which contain a cold section based on the use of Bi-2223/Silver−Gold alloy tapes and helium forced cooling resistive part optimized for the rated current operations at 68 kA, 55 kA, and 10 kA. In this chapter, an outline of the design of the leads is presented, together with the manufacture and testing at ASIPP.

1. Overview

Since the discovery of high temperature superconductivity (HTS) in 1986, there has been a tremendous progress in research and development (R&D) of HTS applications.[1,2,3] Especially for current lead (CL) application, HTS provides considerable economic benefits because the external field is usually well below 500 mT, allowing an operating temperature not much lower than the critical temperature. Superconducting magnets cooled by 4.5 K super-critical helium require conductor connection to the power supply located at room temperature. Using HTS material the cooling power consumption needed for a HTS CL is notably reduced.

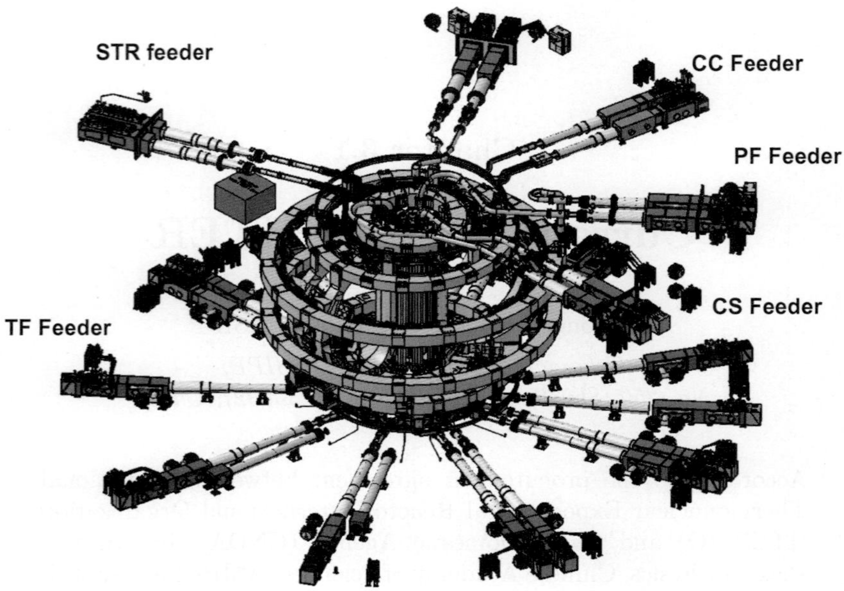

Fig. 1. Overview of ITER feeder system.

The International Thermonuclear Experimental Reactor International Organization (ITER)[4] magnet system (see Fig. 1) requires 31 sets of feeder system containing a total of 60 single CLs: 18 for the Toroidal Field (TF) feeders, 12 for the Central Solenoid (CS) feeders, 12 for the Poloidal Field (PF) feeders and 18 for the Correction Coil (CC) feeders. The TF leads conduct large current of 68 kA, but are operated in steady state. The PF and CS leads also carry large currents of 55 kA, but are operated in pulsed mode. The CC leads are smaller, but are also operated in pulsed mode. Total feeding current to ITER magnet is 2.7 MA which is maximum capacity for Tokamak device in the world.

Before 2005, ITER CL original design was the gas-cooled copper leads without incorporating HTS material. Because of the successful development and testing achievement of 13 kA LHC HTS leads in CERN,[5] 15 kA EAST HTS leads in ASIPP,[6] 68 kA HTS leads in KIT and CRPP,[7,8] 60 kA HTS leads in JAERI,[9] these HTS lead demonstrator has brought ITER to the decision to use HTS Lead for the ITER magnet system in 2005. This is an important milestone for HTS CL application.

According to the feeder procurement agreement of International Organization (IO) and ITER China, ASIPP is responsible for design, manufacturing, and cold testing of all HTS CLs. During 2008, the 68 kA and 52 kA HTS CLs trials for ITER TF and PF feeders were developed and tested,[10] and the maximum operating current of 68 kA trial leads reaches 90 kA (world records of HTS CL at present) in five minutes. The 10 kA HTS trial leads were developed for ITER CC feeder in 2010.[10,11] The mockup qualification and documentation preparation of prototypes were performed from 2012 to 2013.[12] Since 2014, prototype work was being completely carried out.

This chapter mainly presents general description of ITER leads design, manufacturing, and testing.

2. ITER Leads Design

The design of ITER CLs is implemented through the parameters listed in Table 1.

2.1. *General design description*

The design parameters of HTS leads for ITER are listed in Table 1. All CLs include five main parts: (1) A conventional gas-cooled copper section carrying the current between this intercept and the room temperature terminal, (2) A HTS part between the cold end and an intermediate intercept temperature, (3) A low temperature superconducting (LTS) part at the cold end, (4) Insulation flange and insulation layers fully covering vacuum section. (5) Instrumentation and control system. A sketch of the HTS leads is given in Fig. 2. This paper is mainly focused on part 1 and part 2. The details of design are discussed in Taylor *et al.* (2012).[14]

2.2. *HEX design*

ASIPP has manufactured and successfully tested two types of heat exchangers (fin type and copper foil stacked type) in the range of

Table 1. Design parameters of ITER CLs.

Parameters	TF	PF/CS	CC
Max heat load per lead at 4.5 K end (W)	15	12	3
Max flow rate in HEX section (g/s)	4.8	3.85	0.7
Max pressure drop in 50 K GHe circuit (MPa)		0.1	
Max allowable working pressure 50 K GHe circuit (MPa)		0.6	
Max pressure drop in 4.5 K SHe circuit (MPa)	<0.01	0.004	
Max test pressure 4.5 K SHe circuit MPa	3		
RT terminal to busbar joint maximum resistance ($n\Omega$)	~100	~150	
HEX to HTS joint max resistance ($n\Omega$)	10	25	
HTS to LTS joint max resistance ($n\Omega$)	1	2.5	
LTS to busbar joint max resistance ($n\Omega$)	2	5	
Max. length end-to-end including joint box (m)	<3		
Maxdiameter of the CTB hole for CL insertion (mm)	300	150	
HTS quench detection delay (s)	2		
Min HTS overheating time(s) (containing 2×2 s delay)	15	18	18
Heater power per lead at top end (kW)	3		0.5
Water cooling inside room temperature termination	No		
Vacuum insulation CL neck-section outside CTB	Yes		
Design of joint CL to busbar	Twin-Box-Type		
Max axial travel with respect to CTB surface (mm)	<5		

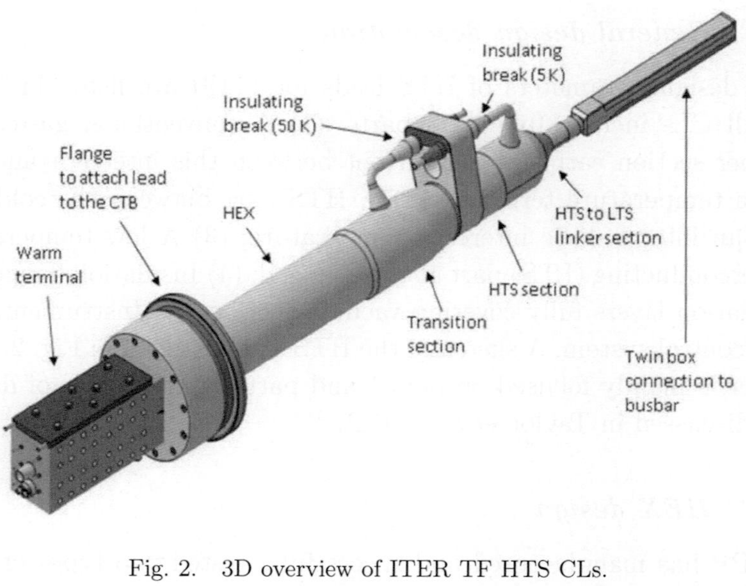

Fig. 2. 3D overview of ITER TF HTS CLs.

Table 2. The main parameters of heat exchanger.

Dimension	CC 10 kA	PF/CS 55 kA	TF 68 kA
Actual Length (mm)	900	950	950
Outer Diameter (mm)	110	174	188
Core Diameter (mm)	39.3	82.6	92.4
Cut (mm)	10	13	13

65 K to room temperature for ITER 68 kA, 52 kA, and 10 kA trial CLs.[10–12] The copper foil stacked type HEX has advantage of larger wet perimeter and heat efficiency of ∼98.4%[15]; and the fin type HEX has advantage of large thermal mass during the loss of flow accident (LOFA) and repeatable manufacture technology. Considering the balance of manufacture and heat efficiency, finally the fin type HEX is the choice.

The key parameters of HEX are listed in Table 2. The modeling of fin HEX for CL application has been developed in Rizzo *et al.* (2013)[16] and Heller *et al.* (2010).[17] 1D simulation model and results of prototype are shown in Rizzo *et al.* (2013).[18] The main flow consumption of HEX is ∼0.066 g/s/kA @ 50 K inlet temperature. In order to get an excellent heat exchanger with the gas, the tolerance between outer tube and fin heat exchanger is required at couple of micro meters. The challenge of manufacturing HEX is the stainless steel assembling with fin type copper rod as shown in Fig. 3.

3. HTS Module of ITER CLs

A few HTS materials have been discussed in the HTS application for fusion device.[19–21] For use in CLs, it is found to be practical to use the fragile Bi-2223 HTS tape using Ag–Au alloy in the form of stacks. The technology of vacuum soldering of the tapes had been successfully developed at ASIPP for the 68/55/10 kA trial leads and as prototype of the ITER project,[22,23] the 6/12 kA series CLs for accelerator application of JINR.[24] And hence, there was no need to change this component substantially.

Fig. 3. Picture of stainless steel tube assembly with heat exchanger.

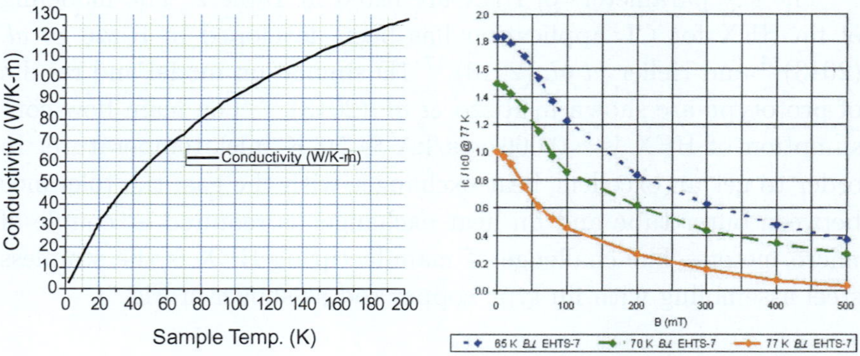

Fig. 4. Classic thermal conductivity of HTS tape (left), current density is a function of temperature and magnetic field (right).

In the present global industrial market of hundreds meters class Bi-2223/Ag–Au alloy tape, SEI of Japan[25] and InnoST of China[26] are the only two available vendors.

The classic thermal conductivity of Bi-2223/Ag–3.8% Au alloy tape is shown in Fig. 4, and the thermal conductivity difference between tape and stacks is discussed in Schwarz *et al.* (2009)[27] and Ballarino *et al.* (2007).[28] The perpendicular field on the width surface

Table 3. Parameters of HTS model for ITER CL.

Dimension	CC 10 kA	PF/CS 55 kA	TF 68 kA
Type	Bi-2223/Ag ∼4% Au alloy tape		
Tape's dimension	∼4 mm/0.25 mm		
Stacks no.	36	90	90
Layers of stack	4	7	8
I_c (self, 77 K) (A)	110	180	180
Effective length (mm)	310	310	310
OD of shunt(mm)	76	148	148
Groove of shunt(mm)	18	18	18
Overlap length warm/cold (mm)	70/50	70/50	55/45

Fig. 5. HTS module of 55 kA prototype (left), 10 kA prototype (right).

Table 4. Main interlock protection parameters.

Parameter	Value
Threshold Voltage Across HTS Module, V_{HTS} (mV)	3
Threshold Voltage Across HEX, V_{HEX} (mV)	120–150
Max. Temp. at Top of HTS, $T_{\text{HTS,max}}$ (K)	75
Max. Temp. at Bottom of LTS, $T_{\text{LTS,max}}$ (K)	8

causes a considerable degradation in Fig. 4 (right), the operation current sharing temperature and maximum operating current, as well as overheating time, etc. need to consider the applied field. Therefore, designer needs to consider the balance of heat load and overheating time for safety. According to the simulation result,[29,30] the parameters of ITER HTS module are mainly listed in Table 3. The ITER HTS prototypes are illustrated in Fig. 4.

3.1. *Testing of ITER CLs*

ASIPP have established a new complex for testing of large capacity HTS CLs and coils application, which contains a set of Linde 900W@4.5 K helium refrigerator system, a 80 kA power supply, a 20 kA power supply for CC type CLs (see Fig. 6), and a control system based on the ITER CODAC standard. The test setup of CC prototype is shown in Fig. 7. This test setup had achieved the CC prototype cold test in February 2015. For CC prototype lead, the temperature and voltage evolution of steady state operation of CC 10 kA prototypes under different boundary conditions and loss of flow accident (LOFA) is shown in Fig. 8. The LOFA time is about 13 minutes after stopping the 50 K flow. The overheating time is far

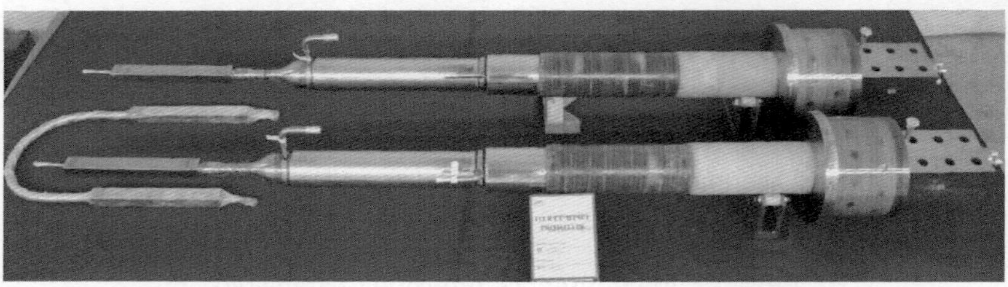

Fig. 6. Picture of CC 10 kA HTS CL prototypes for ITER (before assembling insulation layer).

Fig. 7. Test setup of 10 kA HTS CL (left); RT flexible busbar connection (right).

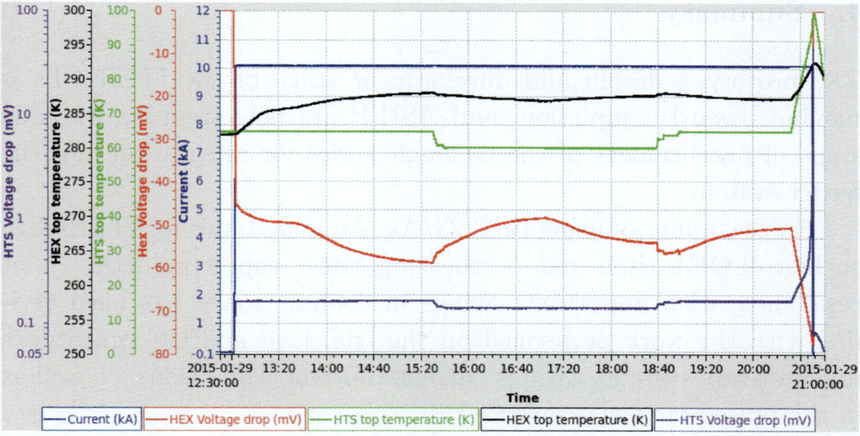

Fig. 8. Steady state operation of 10 kA prototypes under different boundary conditions and LOFA.

longer than the expected 18 seconds. The test results of CC prototype are still being under-analyzed and will be reported soon.

3.2. *Plan of manufacture and testing*

The HTS leads are being in line with the integration schedule foreseen for the installation of the feeder project of ITER. Thermal performance cold test of the CC 10 kA prototype was successfully performed at ASIPP in January 2015. According to the present plan, the thermodynamic test of the prototype of TF 68 kA and PF/CS 55 kA leads will be run in 2015. After thermodynamic test of prototype leads, the insulation layer manufacture will be constructed to verify the Paschen tight test. Prototype work, including the test of specific features of the leads, is being carried out in parallel with the final design work of the series. Hefei Juneng Inc. will be responsible for the manufacture of TF/CS/PF series, and Hefei Keye Inc. will be responsible for the manufacture of CC series. According to the integration schedule of ITER feeder procurement agreement, all the series will be manufactured and the cold test will be performed during the coming five years.

4. Summary

The prototype design and development stage of the ITER CLs is moving toward completion, and ASIPP are now stepping into the stage of procurement of the 60 single series for integration into the feeder system.

The feasibility of using Bi-2223/Ag–Au alloy tapes for ITER CLs, such as LOFA time and overheating time, superconducting joint resistance, with important savings in cooling power has been verified with the work performed on the trial lead and CC prototypes under steady state at various currents, in fault conditions, as well as in the pulsed current scenarios test.

Author believes that ITER HTS CLs will be a successful demonstration of commercial Bi-2223/Ag–Au alloy tapes in a large scale application following EAST, LHC, W7-X, and JT-60SA CLs.[31]

References

1. B. Seeber, ed., *Handbook of Applied Superconductivity* (1998).
2. J. G. Weisend II, *Handbook of Cryogenic Engineering* (1999).
3. M. N. Wilson, *Superconducting Magnets*. Oxford, Oxford University Press (1983).
4. ITER international organization, www.iter.org.
5. A. Ballarino, HTS current leads for the LHC magnet powering system, *Physica C* **373–376**, 1413–1418 (2002).
6. S. Wu and the EAST Team, An overview of the EAST project, *Fusion Eng. Des.* **82**, 463–471 (2007).
7. R. Heller, G. Friesinger, W. Goldacker, M. Quilitz, and M. Tasca, Status of the development program of a 60 kA HTSC current lead for the ITER toroidal field coils, *IEEE Trans. Appl. Supercond.* **9**(2), (1999).
8. R. Heller *et al.*, Experimental results of a 70 kA high temperature superconductor current lead demonstrator for the ITER magnet system, *IEEE Trans. Appl. Supercond.* **15**(2), 1496 (2005).
9. T. Isono *et al.*, Test result of 60 kA HTS current lead for fusion application, *Physica C: Supercond.* **392–396**, 1219–1224 (2003).
10. Y. Bi *et al.*, R&D towards HTS current leads for ITER, *IEEE Trans. Appl. Supercond.* **19**(3), 1500 (2009).
11. Y. Bi *et al.*, Test results of 52/68 kA trial current leads for ITER, *IEEE Trans. Appl. Supercond.* **20**(3), 1718 (2010).
12. P. Bauer *et al.*, Test of a 10 kA HTS current lead for ITER, *IEEE Trans. Appl. Supercond.* **21**(3), 1074 (2011).

13. T. Zhou, K. Lu, Q. Ran, K. Ding, H. Feng, H. Wu, C. Liu, Y. Song, E. Niu, and P. Bauer, Design and technology preparation for the ITER HTSCL, *IEEE Trans. Appl. Supercond.* **24**(3), (2014).
14. T. Taylor *et al.*, Design of the HTS current leads for ITER, *IEEE Trans. Appl. Supercond.* **22**(3), (2012).
15. Y. Bi, R&D on resistive heat exchangers for HTS high rated current leads [J], *PST* **13**(6), 757–764 (2011).
16. E. Rizzo, R. Heller, L. S. Richard, and R. Zanino, CtFD-based correlations for the thermal–hydraulics of an HTS current lead meander-flow heat exchanger in turbulent flow, *Cryogenics* **53,** 51–60 (2013).
17. R. Heller, A. Class, A. Batta, R. Lietzow, H. Neumann, and M. Tischmacher, Modelling of the fin type heat exchanger for the HTS current leads of W7-X and JT-60SA, *Cryogenics* **50**, 222–230 (2010).
18. E. Rizzo, P. Bauer, R. Heller, L. S. Richard, and R. Zanino, 1D thermal-hydraulic analysis of the high temperature superconducting current leads for the ITER magnet system from 5 K to 300 K, *Fusion Eng. Des.* **88**, 3125–3131 (2013).
19. R. Heller, W. H. Fietz, A. Kienzler, and R. Lietzow, High temperature superconductor current leads for fusion machines, *Fusion Eng. Des.* **86**, 1422–1426 (2011).
20. J.-L. Duchateau, High temperature superconductors (HTS) in fusion technologies, *Woodhead Publishing Series in Energy*, 345–368 (2012).
21. T. Isono, K. Hamada, and K. Okuno, Design optimization of a HTS current lead with large current capacity for fusion application, *Cryogenics* **46**, 683–687 (2006).
22. K. Ding *et al.*, Development of ITER HTS current lead at ASIPP [J], *Phys. Procedia* **36**, 931–936 (2012).
23. C. Liu, Y. Bi, *et al.*, Investigating the 10 kA HTS module of the ITER current leads [J], *IEEE Trans. Appl. Supercond.* **22**(3), 4800204 (2012).
24. K. Ding, Y. Bi, H. Feng, C. Liu, H. Wu, Y. Song, and T. Zhou, Development of 12 kA HTS current lead for accelerator magnet test application, *IEEE Trans. Appl. Supercond.* (2012). doi:10.1109/TASC.2012.2230675.
25. InnoST BSCCO 2223 wires, http://www.innost.com/index.php?m=index&a=actionClass&actype=channel&fid=35.
26. SEI BSCCO 2223 type G wires, http://global-sei.com/super/hts_e/index.html.
27. M. Schwarz, K.-P. Weiss, R. Heller, and W. H. Fietz, Thermal conductivity measurement of HTS tapes and stacks for current lead applications, *Fusion Eng. Des.* **84**(7–11), 1748–1750 (2009).
28. A. Ballarino, L. Martini, S. Mathot, T. Taylor, and R. Brambilla, Large scale assembly and characterization of. Bi-2223 HTS conductors, *IEEE Trans. Appl. Supercond.* **17**(2), 3121 (2007).
29. R. Heller, Safety analysis of the 70 kA ITER HTS current lead demonstrator, *IEEE Trans. Appl. Supercond.* **19**(3), (2009).

30. K. Ding, Y. Bi, *et al.*, Safety research of the 68 kA trial-HTS current lead for ITER [J]. *IEEE Trans. Appl. Supercond.* **20**(3), 1729–1732 (2010).
31. W. H. Fietz, R. Heller, A. Kienzler, and R. Lietzow, Status of HTS current leads for WENDELSTEIN 7-X and JT-60SA, Fusion Eng. Des. **84**(2–6), 776–779 (2009).

Chapter 3.2

Current Leads for Conduction-Cooled Magnets

Takehisa Tsurudome

Sumitomo Heavy Industries, Ltd, Technology Research Center, 19 Natsushima-cho, Yokosuka-shi, Kanagawa-Ken, Japan

A current lead (CL) for conduction-cooled magnets or cryocooled superconducting magnets (CSM) is required to have small heat leakage from high temperature stage to low temperature stage and sufficient strength against electromagnetic force (Lorentz force) and thermal stress. A high temperature superconductor (HTS) has high critical temperature, and the thermal conductivity is lower than the normal conductor. This chapter will show the configuration and design of the CL for CSM using Bi-2223 HTS wire.

1. Introduction

High temperature superconductor (HTS) has been used as a current lead (CL) for various superconducting magnets because of the small heat leakage and large transport current without Joule heating.[1–4] For example, a Bi-2223 bulk superconductor CLs have ever been developed for conduction-cooled magnets (CSM).[1–3,5,6] In the present study, various kinds of HTS wires and HTS CLs have been developed and applied for the CSM.[7–11] They can easily facilitate modification of the transport current capacity and the length of CSM.

However, the CL is required to have sufficient strength against the Lorentz force and thermal stress caused by difference of thermal expansion coefficient between the HTS-wire and other materials. In addition, it is important to suppress the heat leakage from higher temperature stage to lower temperature stage and Joule heating because of the restriction of cooling capacity.

This chapter will show a configuration of the CSM and the CL using Bi-2223 HTS wire. In addition, it will also show the mechanical and thermal design and example of the HTS wire CL for the CSM.

2. CL Design for CSM

2.1. *Basic configuration of CSM*

The CSM consists of a cryocooler, a superconducting coil, a thermal shield, a vacuum chamber, a power supply terminal block, normal conductor current leads (NCLs), superconducting bus bars (SC bus bars), and the HTS CLs. Figure 1 shows a schematic view of the CSM.[1]

For example, the cryocooler is typically using a 2 stage GM cryocooler. Table 1 shows the typical specification of th GM cryocoolers. The 2nd stage in which the superconducting coil and

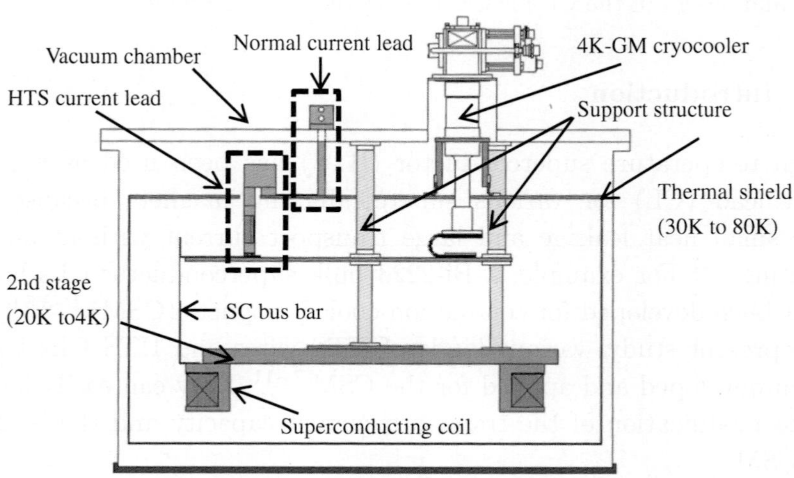

Fig. 1. A schematic view of the CSM, which consists of a cryocooler, support structures, a thermal shield, three types of CLs 2nd stage, and superconducting coil.

Table 1. A specification example of the GM cryocooler.[14]

Type	4 K-GM (RDK-408)	10 K-GM (RDK-408 S2)
2nd stage Capacity	1.0 W at 4.2 K	5.4 W at 10 K (50 Hz) 6.3 W at 10 K (60 Hz)
1st stage Capacity	40 W at 43 K (50 Hz) 50 W at 43 K (60 Hz)	35 W at 45 K (50 Hz) 40 W at 45 K (60 Hz)

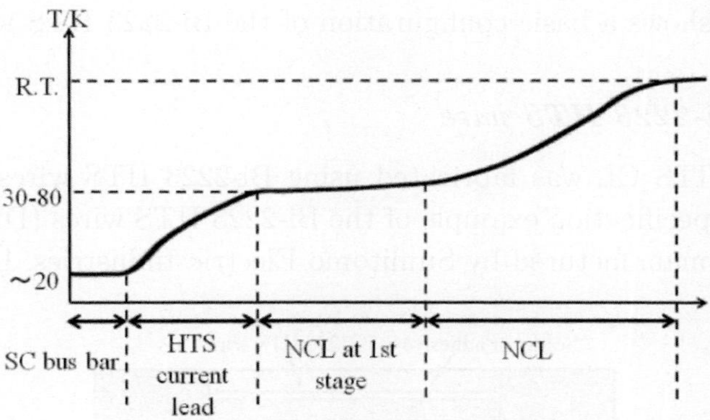

Fig. 2. A typical temperature distribution for the CL in the CSM from a room temperature to the SC bus bar connected at the SC coil (R. T. = Room temperature).

SC bus bars are thermally connected is about 4 K to 20 K. The 1st stage in which the thermal shield is thermally connected is 30 K to 80 K.

In the CSM, the CL consists of four parts and are connected between the power supply terminal block and the superconducting coil in series. The NCLs, typically made of copper are connected to the power supply connector at room temperature and to the other NCLs at the 1st stage. The HTS CLs are set between 1st stage and 2nd stage and are connected to the NCLs at the 1st stage and SC bus bars at 2nd stage. Figure 2 shows the temperature distribution of the CL in the CSM.

2.2. *Configuration of Bi-2223 HTS wire CL for CSM*

The Bi-2223 HTS wire CL consists of the Bi-2223 HTS wires, two copper terminal blocks and support structure. The quantity of the wire is designed from the required current capacity. The Bi-2223 HTS wires and copper terminal blocks are soldered together.

The support structure reinforces mechanical strength and holds the distance of two copper terminal blocks. Terminal blocks and support structure are soldered or bonded together. At both ends of the unit, NCLs and SC bus bars are connected with low resistivity. Figure 3 shows a basic configuration of the Bi-2223 HTS wire CL.

2.3. *Bi-2223 HTS wire*

Bi-2223 HTS CL was fabricated using Bi-2223 HTS wires. Table 2 shows a specification example of the Bi-2223 HTS wires (DI-BSCCO Type G) manufactured by Sumitomo Electric Industries, Ltd.[13]

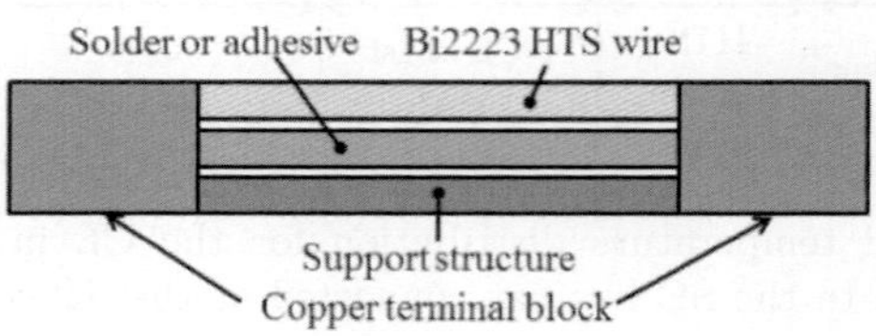

Fig. 3. A basic configuration of the Bi-2223 HTS wire CL.

Table 2. A specification of the Bi-2223 HTS wire (DI-BSCCO Type G).[13]

Specification	
Width	4.3 mm ± 0.3 mm
Thickness	0.23 mm ± 0.03 mm
Sheath material	Ag–5.4 wt.% Au
Maximum critical current at 77 K	200 A
Tensile force at R. T.(*)	≦50 N
Tensile stress at 77 K (*)	≦90 MPa
Bending radius at R. T. (*)	≦110 mm

Note: (*) = Degrade of 95 % critical current.

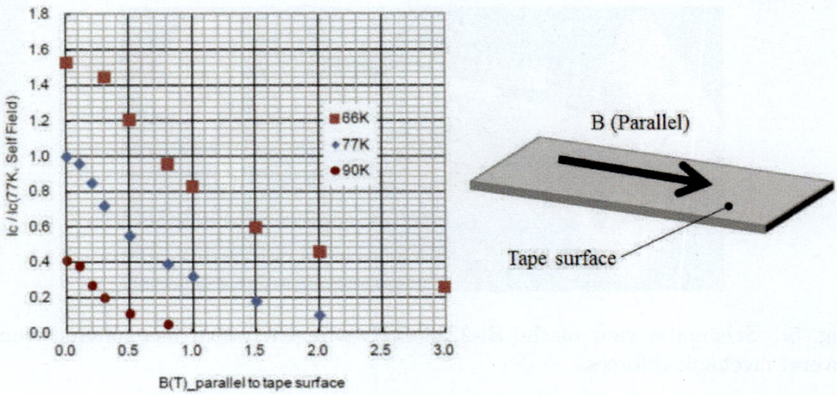

Fig. 4. The characteristics of critical current in several temperatures *versus* magnetic field in which the direction is parallel to the tape surface, and schematic view of the Bi-2223 HTS wire and magnetic field direction.[13]

A current capacity of the Bi-2223 HTS wire is 200 A at 77 K, self-field. The sheath metal is Ag-5.4 wt% Au because of a low thermal conductivity.[13] The tensile stress for the wire at 77 K is below 90 MPa, and it is the mechanical criterion for the HTS CL design.

A quantity of wire for the CL is designed from a required current capacity, operating temperature, and stray field, especially higher temperature. Figure 4 shows the characteristics of critical current (I_c) *versus* temperature and magnetic field direction parallel to the tape surface.

2.4. *Mechanical force*

The Bi-2223 HTS wire CL suffers from a mechanical force caused by Lorentz force, thermal stress caused by thermal contraction difference between Bi-2223 HTS wires and other components, and other external forces (bending, tensile, or compression). Until the excitation of the CSM, these forces are applied to the CL. Figure 5 shows the schematic view of the CL which has suffered from mechanical forces.

2.4.1. *Thermal contraction difference*

After cooling down the CSM, all components' length, support structure, Bi-2223 HTS wire CL, and the superconducting coil, etc., will

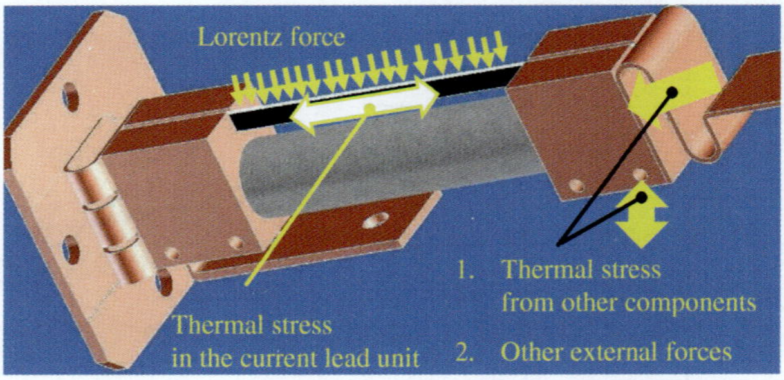

Fig. 5. Schematic view of the Bi-2223 HTS wire CL which has suffered from several mechanical forces.

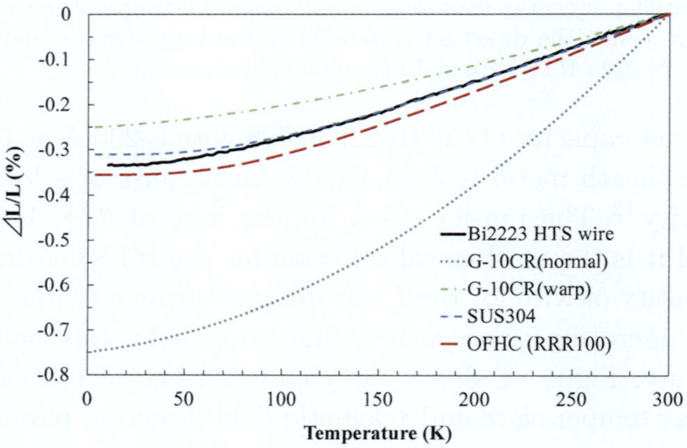

Fig. 6. Linear expansion *versus* temperature for Bi-2223 HTS wire, SUS304, OFHC, and G-10 CR (normal and warp directions).[13,14]

be changed. The thermal contraction difference ΔL is

$$\Delta L = L \int_{T_L}^{T_H} \alpha \cdot dT, \tag{1}$$

where T is the temperature, L is the length at room temperature, T_H is the room temperature, T_L is the temperature after cooling, and α is a thermal expansion coefficient. Figure 6 shows a linear expansion for a typical material in the CSM. If the CL is connected rigid in the CSM, it suffers from the thermal stress, after cooling caused by the thermal contraction difference. Figure 7 shows a schematic view

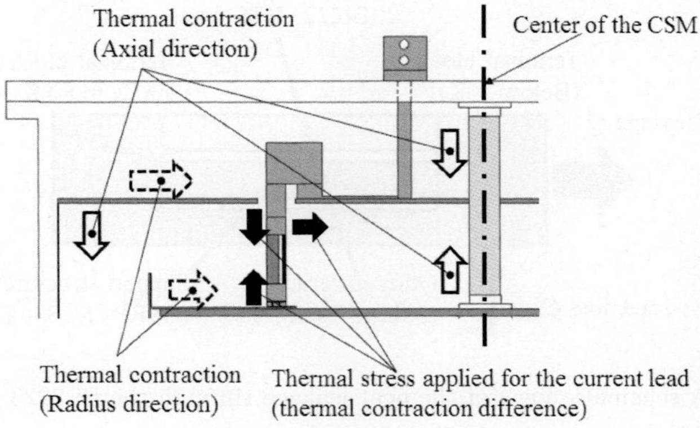

Fig. 7. A schematic view of the thermal stress due to the thermal contraction difference in the CSM.

of the thermal stress applied to the CL in the CSM. In addition, the Bi-2223 HTS wire suffers from the support structure in the CL (see Fig. 5). To relieve the thermal stress, it is effective to set a flexible plate between the terminal blocks and the CSM and to design the support structure in the CL in which the thermal expansion is approximate to the Bi-2223 HTS wire.

2.4.2. *Lorentz force*

During the excitation of the CSM, the Bi-2223 HTS wire suffers from bending stress caused by the Lorentz force. It is generated by stray field and operating current (see Fig. 5). The Lorentz force, F is

$$F = \int_{L_2}^{L_1} I \cdot B(x) dx, \tag{2}$$

where L_1 is a position of the terminal block 1, L_2 is the position of the terminal block 2, I is transport current, $B(x)$ is stray field perpendicular to the current.

Against the Lorentz force, the Bi-2223 HTS wire CL is required to be of sufficient strength. However, if the CL could not have sufficient strength; it should be set in the small stray field or designed in small length between L_1 and L_2.

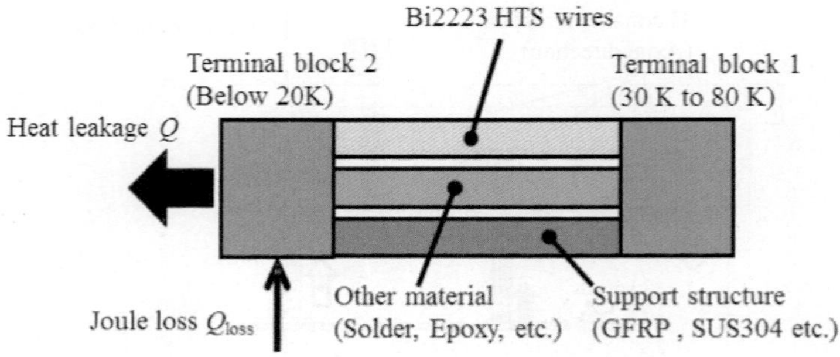

Fig. 8. A schematic view of the heat leakage through the Bi-2223 HTS CL.

2.5. *Heat leakage*

The CSM is required to have a small heat leakage from higher temperature stage to the superconducting coil because of a restriction of the cooling capacity as regards the cryocooler (see Table 1).

Figure 8 shows a schematic view of a heat leakage through the Bi-2223 HTS CL. The heat leakage Q is

$$Q = \frac{S_1}{L}\int_{T_L}^{T_H} \lambda_1(T)dT + \frac{S_2}{L}\int_{T_L}^{T_H} \lambda_2(T)dT + \frac{S_3}{L}\int_{T_L}^{T_H} \lambda_3(T)dT + Q_{Loss}, \tag{3}$$

where S is cross-sectional area of each parts, L is the length between terminal blocks, $\lambda(T)$ is the thermal conductivity, T_H is the terminal block 1 temperature which is typically 30K to 80K. T_L is the terminal block 2 temperature, connected to the superconducting coil which is below 20 K using HTS wire (Bi-2223, Y123, etc.) and about 4 K using low temperature superconducting wire (NbTi, Nb_3Sn, etc.). Q_{Loss} is the joule loss at terminal block 2 (Subscript "1" indicates the Bi-2223 HTS wire, "2" indicates the support structure, and "3" indicates solder, epoxy resin, etc.). Figure 9 shows the temperature dependence of the thermal conductivity, $\lambda(T)$,for various materials typically using the Bi-2223 HTS wire CL. Figure 9 and Eq. (3) indicate that the less quantity of Bi-2223 HTS wires and soft solder or the larger length between terminal blocks, the less heat leakage.

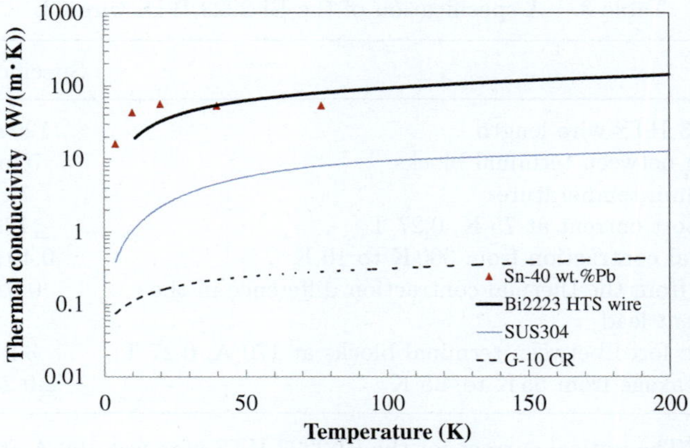

Fig. 9. Temperature dependence of the thermal conductivity, $\lambda(T)$, DI-BCCCO Type G, SUS304, G-10 glass fiber epoxy (normal direction), and soft solder (Sn-40 wt% Pb).[13–15]

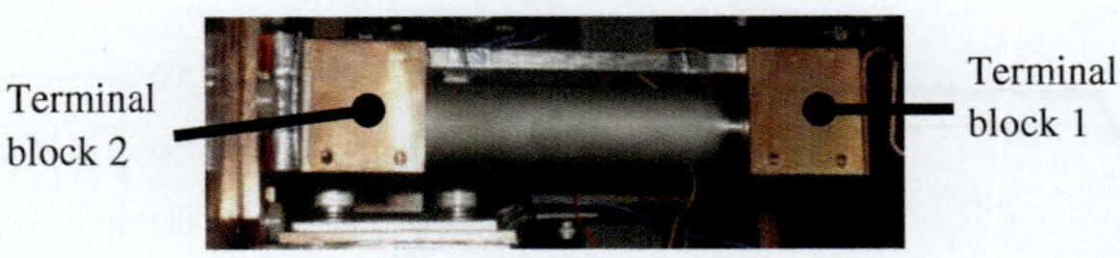

Fig. 10. A photograph of the Bi-2223-HTS wire CL for the CSM (2010).[11]

3. Example of the CL for the CSM

Figure 10[11] shows the photograph of the Bi-2223 HTS wire CL for the CSM which was constructed in 2010 and the specifications of the HTS CL are shown in Table 3. The CL consists of five Bi-2223 HTS wires, a stainless steel support tube, and two copper terminal blocks.

In 2010, the critical current of the Bi-2223 HTS wire was 100 A at 77 K, self field.[16] The Bi-2223 HTS wires and copper terminal blocks were soldered together. At both ends of the unit, flexible copper plates were placed for current transport and to relieve stress. The support tube reinforced the mechanical strength.

Figure 11 shows the result of the 930 cycles of Lorentz force test at 170 A, 0.27 T. The result shows that measured voltage was below 0.5 mV, and indicated that the CL was not degraded after 930 Lorentz force cycles.

Table 3. A specification of the Bi-2223 HTS wire CL.

Item	Specifications
Bi-2223 HTS wire length	120 mm
Length between terminal blocks	70 mm
Maximum temperature	75 K
Transport current at 75 K, 0.27 T	≧170 A
Thermal contraction from 300 K to 10 K	0.40 mm
Strain from the thermal contraction difference in the current lead	0.02%
Lorentz force between terminal blocks at 170 A, 0.27 T	3.2 N
Heat leakage from 55 K to 4.5 K	≧0.23 W

Note: *The critical current for the Bi-2223 HTS wire was 100 A at 77 K, self-field.[16]

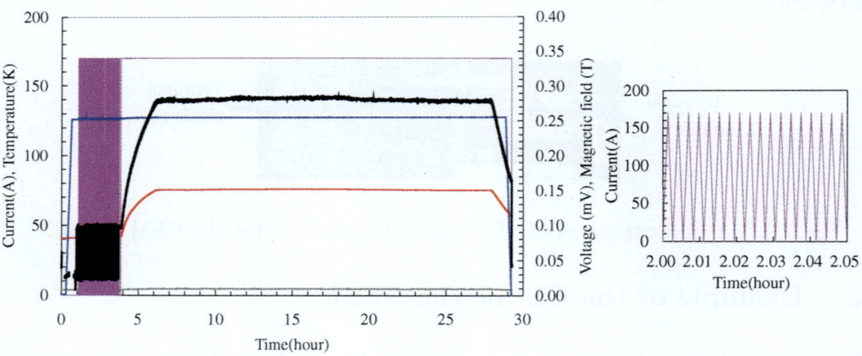

Fig. 11. Results of the Lorentz force cycle test of the HTS CL and transport current pattern during the Lorentz force cycles.[11]

Figure 12 shows the estimated and measured values of heat leakage. The estimated value from Eq. (3) omitted the joule loss. The measured value was close to the estimated value. For example, the estimated value was 0.20 W, and the measured one was 0.21 W from 55 K to 4.5 K. This result indicated that the heat leakage was below 0.23 W.

These results indicate that it is possible to use the Bi-2223 HTS wire CL for the CSM.

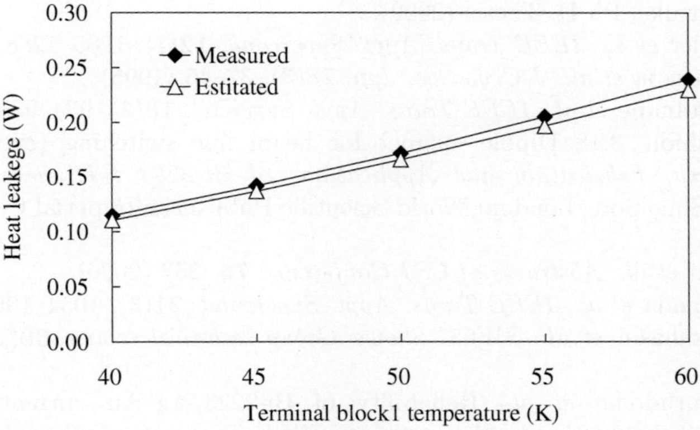

Fig. 12. Results of the heat leakage when the terminal block 2 temperature was 4.5 K.[11]

4. Summary

This article has focused on the design and example of the CL for CSM. The importance of CL for the CSM is the current capacity, less heat leakage, and sufficient mechanical strength against the Lorentz force, thermal stress, and the external force. These components are trade off.

For example, if the current capacity is increased, or the distance between terminal blocks is decreased, the mechanical strength will be improved, but the CSM temperature will rise. However, on decreasing the current capacity or increasing the distance between the terminal blocks, the heat leakage will be decreasing, but the mechanical strength will be insufficient. In addition, if the CL size is larger, the position of the CL is limited in the CSM.

It is important to adjust the CL design (current, thermal, mechanical strength, size, and position, etc.) for the purpose of the CSM.

References

1. Y. Yamada *et al.*, Superconducting current leads of Bi-based oxide, *IEEE Trans. Appl. Supercond.* **3**(3), 923–926 (1993).
2. M. Ishizuka *et al.*, Critical current density (J_c) and mechanical characteristics of a Bi-2223 bulk current lead for cryocooler-cooled superconducting magnets, *Physica C* **433**, 173–181 (2006).

3. M. Ishizuka, Ph D. Thesis (2009).
4. R. Heller *et al.*, *IEEE Trans. Appl. Supercond.* **12**(1), 1285–1288 (2002).
5. J. Sakuraba *et al.*, *J Cryo Soc. Jpn.* **28**(3), 37–45 (1995).
6. K. Watanabe *et al.*, *IEEE Trans. Appl. Supercod.* **16**(2), 934–939 (2006).
7. J. Yoshida, 3.18 Dipole magnet for beam line switching (cyclotron), In *Research, Fabrication and Applications of Bi-2233 HTS wires*, K. Sato (ed.). Singapore/London: World Scientific Publishers/Imperial College Press (2015).
8. Y. Ishii *et al.*, *Abstracts of CSJ Conference* **75**, 257 (2006).
9. Y. Yamada *et al.*, *IEEE Trans. Appl. Supercond.* **21**(3), 1054–1057 (2011).
10. K. Takahashi *et al.*, *SWCC Showa Group technical review* **60**(116), 25–28 (2014).
11. T. Tsurudome *et al.*, Reliability of Bi-2223/Ag–Au current leads for cryocooled superconducting magnet, *IEEE Trans. Appl. Supercond.* **23**(3), 4802204 (2013).
12. http://www.shicryogenics.com/products/.
13. http://global-sei.com/super/.
14. http://cryogenics.nist.gov/index.html.
15. J. W. Ekin, Appendix data tables from experimental techniques for low-temperature measurements. In *Cryostat Design, Material Properties, and Superconductor Critical-Current Testing, Fourth Printing*, 141 (2011).
16. Sumitomo Electric Industries, Ltd., Specification data for DI-BSCCO Type G (2010).

Chapter 3.3

Saturated Iron Core Type Fault Current Limiter

Ying Xin

School of Electrical Engineering and Automation, Tianjin University, No. 92 Weijin Road, Nankai District, Tianjin, China 300072

Superconducting Fault Current Limiter (SFCL) is one of the most promising applications of HTS for electric power industry. Through years of innovative improvement, saturated iron-core type SFCL is one of the most prevailing technologies in SFCL R&D around world.

In this chapter, the working principle, typical structure and key components of a saturated iron-core type SFCL is introduced. Functional characteristic, grid operation performance and test results are also presented. This kind of device is applicable for both transmission and distribution networks and can satisfy a wide range of fault current limiting impedance requirements.

1. Introduction

An ideal fault current limiter should meet the following criteria: (a) sufficiently low impedance throughout normal operation, (b) sufficiently large impedance under fault conditions, (c) rapid detection and initiation of limiting action, (d) fast recovery to normal operation after the clearing of a fault. For commercial applications, ideal devices would also be compact, lightweight, inexpensive, fully automatic, reliable, and have long lifespan.

SFCL's are the short-circuit current limiting devices capable of associating a low impedance at normal operation to a large impedance at short-circuit conditions, although the technology of these devices still requires further development and techno-economic studies.

After about 20 years of R&D and practice evaluations, there have been two primary classes of SFCL's chosen for grid deployment by utilities worldwide so far, namely the resistive type and the saturated iron-core type.

The concept to use superconducting DC bias for a current limiting device was first suggested by B. P. Raju *et al.* in early 1980's.[1] T. Verhaege and Y. Laumond had quite a detailed discussion on the concept in 1998.[2] M. Noe and M. Steurer **made some summarized comments on this topic in their review paper.**[3] The schematic configuration of the circuits and the iron-cores for a single phase saturated iron-core type SFCL is given in Fig. 1. The superconducting DC coil is used here for its high capability to supply permanent ampere-turns to magnetize the iron-cores almost without losses. Two iron-cores are used here for the cancellation of the electromotive forces and high voltage hazard induced by the AC windings through the transformer effect on the superconducting DC coil.

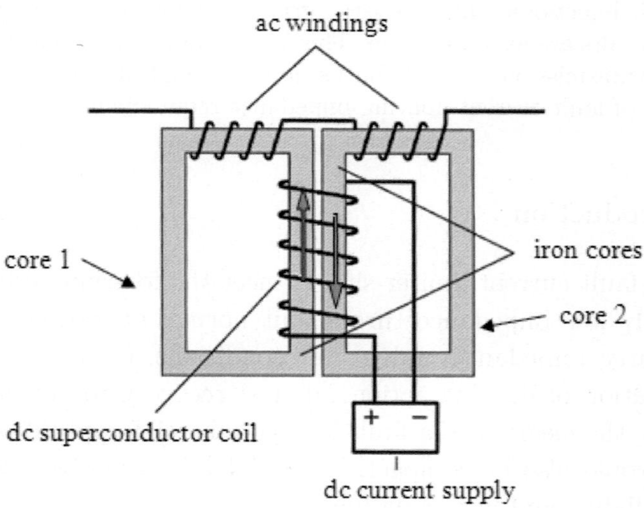

Fig. 1. The schematic circuit configuration of a typical single phase saturated iron core type SFCL.

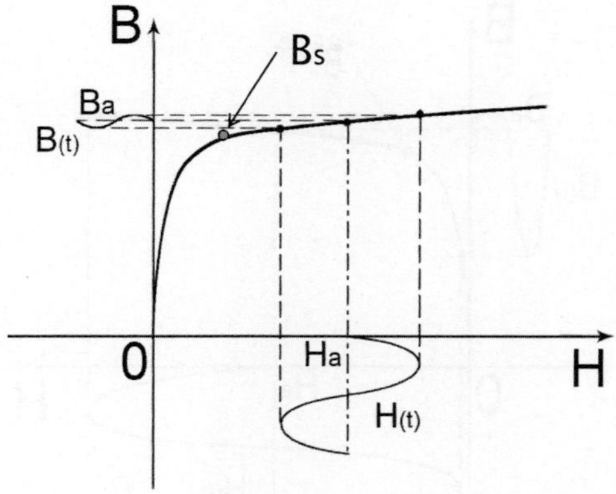

Fig. 2. The *B–H* behavior for a saturated iron-core type SFCL during normal power transmission.

The AC windings are configured the way that can make the above mentioned cancellation and are serially connected to the power transmission line.

Figure 2 shows the *B–H* curve of an iron-core in a saturated iron-core type SFCL. The curve is composed by a non-saturated zone (before the saturation induction, B_s), where B changes quickly with the change of H, and a saturated zone (after B_s), where B changes only slightly with the change of H.

During the normal operation of power transmission, the DC current is adjusted to make the iron-cores in deeply saturated state (at about the point H_a in Fig. 2), so that the AC amplitude of the line normal current is insufficient to take the iron-cores out of their saturated state. The saturated iron-core type SFCL behaves as if there were no iron-core and flux variations are low, i.e.

$$\left|\frac{dB_1}{dt}\right| \to 0 \quad \& \quad \left|\frac{dB_2}{dt}\right| \to 0 \tag{1}$$

(where B_1 and B_2 are magnetic flux density in iron-core 1 and iron-core 2 respectively, refer to Fig. 1), so that the voltage drops on the AC windings are low.

As demonstrated in Fig. 3, when a short-circuit fault takes place, the AC amplitude of the fault current is great enough to de-saturate one of the iron- cores in a part of the cycle. The magnetization states

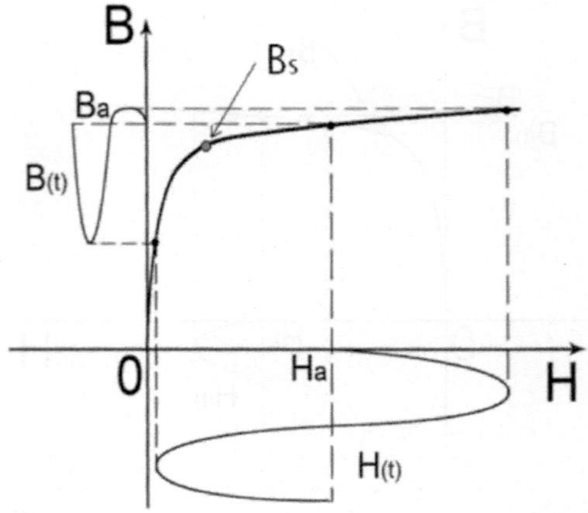

Fig. 3. The B–H curve demonstrating the function of current limitation of a saturated iron core type SFCL.

of the iron-cores can be described as

$$\left|\frac{dB_1}{dt}\right| \ll \left|\frac{dB_2}{dt}\right| \quad \text{or} \quad \left|\frac{dB_1}{dt}\right| \gg \left|\frac{dB_2}{dt}\right|. \tag{2}$$

Under such a condition, significant inductive voltage appears on the AC coils. This voltage is able to slow the evolution of the current and then to reduce the level of fault current. The limitation can be compared to that of a reactance made up of the primary transformer and its magnetic core, provided that the secondary current is kept constant.

2. Structure and Key Components of a Saturated Iron-Core Type SFCL

2.1. *The core structure*

Figure 4 is a drawing of the new core structure developed in recent years for a three phase saturated iron-core type SFCL.[4,5] The major components of the core structure include iron-core, AC coils, and a DC superconducting coil which is contained in a cryostat.

Six iron-core frames are conjugated in a hexagonal structure. Each frame comprises a top yoke, a bottom yoke, and two vertical limbs.

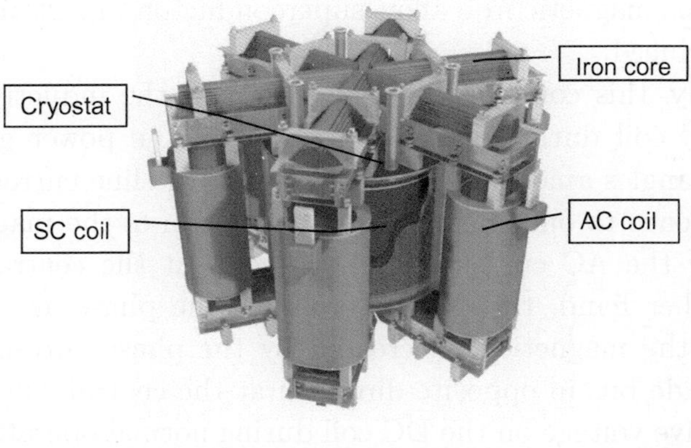

Fig. 4. The new core structure for a three phase saturated iron core type SFCL.

Six vertical limbs, one of each frame, are closely arranged together to form the central column at the center of the hexagon. The superconducting DC coil is installed inside the cryostat and is entwined with the central column. The edge vertical limb of each iron-core frame is surrounded by an AC coil. The cross-sectional areas at different sections of an iron frame differ, to ensure a sufficient magnetization potential on the edge vertical limb. It should be emphasized that using the cross-section variation approach to achieve a deep magnetic saturation in the edge vertical limb in such a core structure is not trivial. A pair of AC coils is connected in series as one phase element of the three phase device.

This core configuration has a number of advantages. First, it maximizes efficiency in the use of space and greatly reduces the material costs. Having a central column that consists of six limbs from each of all six iron frames making use of only one DC superconducting bias coil and one cryostat possible. The sectional geometry of the six limbs forming the central column is fan-shaped with an injection angle of 60°, which allows these six limbs to be put together in such a way to minimize the diameter of the column. Decreasing the diameter of the central column makes it possible to decrease the diameters of the superconducting coil and the annular-cylinder-shaped cryostat. As a result, the length of the yokes of the frames can be abridged. Therefore, the physical size of core and the materials used for building the core, such as magnetic iron alloy, superconductor, and stainless steel, etc. are reduced.

Secondly, this configuration can minimize the inductive voltage on the DC coil during normal operation of the power grid. Since the phase angles among the three phases of the line currents have a 120° difference to one another, the vector sum of the magnetic flux created by the AC currents is always zero at the central column. On the other hand, the two AC coils in one phase are connected such that the magnetic flux created by the phase current is equal in magnitude but in opposite direction at the central column. Thus the inductive voltage on the DC coil during normal operation can be greatly minimized if not eliminated.

Finally, this type of configuration makes electric insulation (can be dry type or oil, depending on the voltage level) of the AC coils simple and convenient for manufacturing and maintenance.

2.2. *The superconducting DC coil*

The superconducting DC coil is used to supply ampere turns to magnetize the iron core into a deeply saturated state with almost no losses. No conventional coil is capable of taking the place of the superconducting coil for this kind of application. For a conventional coil to have the same magnetizing capacity, the size and weight would be formidable and the energy loss in the form of joule heat would be intolerable. Thus, it is the high temperature superconductor that makes the saturated iron-core type SFCL practically feasible.

Double pancake rings made of high temperature superconducting tapes are commonly used to build the superconductor coil. So far, the Bi-2223 tape has been mostly employed for making the double pancake rings. Figure 5 is the picture of superconducting coil prepared for a 220 kV saturated iron core type SFCL.

2.3. *The magnetization control circuit*

The magnetization control circuit for a kind of saturated iron-core type SFCL is schematically depicted in Fig. 6. The circuit consists of

Fig. 5. Photo picture of a Bi-2223 coil prepared for a saturated iron core type SFCL.

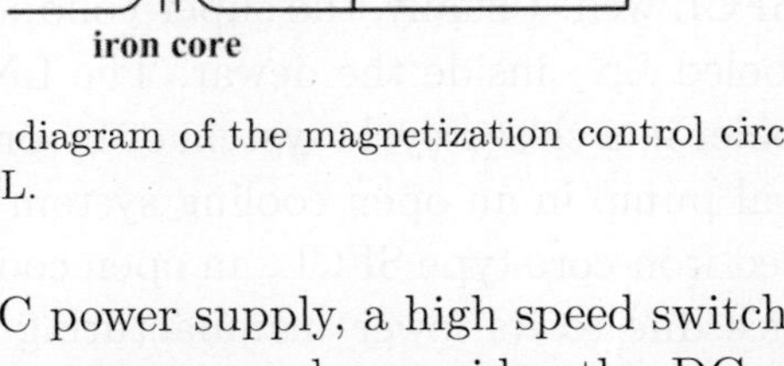

Fig. 6. Schematic diagram of the magnetization control circuit for a saturated iron-core type SFCL.

main parts: a DC power supply, a high speed switch, and an energy release unit. The power supply provides the DC coil with a constant current for magnetizing the iron-core. When a short-circuit fault is detected, the circuit control will send a signal to the high-speed switch and the DC circuit is switched to open. High Voltage Insulated Gate Bipolar Transistors (HVIGBT) are usually used for the high speed switch. The magnetic energy stored in the iron-core is then released through the energy release unit. Piezoresistors in the energy release unit also suppress the voltage surge caused by the quick opening of the magnetizing circuit, protecting the elements of the circuit. The total time from the fault detection to the completion of the required magnetic energy release is within several milliseconds. Taking the measure of breaking the DC circuit is very beneficial, since it not only keeps the elements in the DC circuit from damage by the sudden increase of voltage and current due to the fault current, but also enhances the fault current limitation function.

2.4. *The AC coils*

Copper wires or bars are usually used for winding the AC coils of a saturated iron-core type SFCL. Generally, the number of turns of the AC coils is much smaller than that of the super-conducting DC coil. Since the AC coils will be connected to the power transmission lines, reliable electrical insulation is required. Mechanical strength of the conductors and structure also needs to be assured for withstanding the huge electromagnetic force acted on the coils when a short-circuit fault takes place.

2.5. *The cryogenic system*

Either a closed cycle or an open cooling system can serve a saturated iron- core type SFCL well. Usually, the super-conducting coil is bath cooled by sub-cooled LN_2 inside the dewar. The LN_2 sub-cooling is done by a cryocooler in a closed cycle system or by pressure reduction with a mechanical pump in an open cooling system respectively.

For a saturated iron-core type SFCL, an open cooling system may be a better choice due to its lower manufacturing and installation cost, higher reliability, and easier maintenance.[5]

3. Grid Deployment, Test, and Operation

Saturated iron-core type SFCLs have been developed and deployed in power grids around the world.[5–7]

In 2007, a 35 kV/90 MVA saturated iron-core type SFCL developed by Innopower was installed in Puji substation of Kunming, China. After the installation of the SFCL in the substation, a series of field tests were carried out. This device demonstrated having very low impedance in normal power transmission and no adverse effect on the power network. To verify its capacity in fault current limiting, compliance with utility requirement, and fail-safe, current limiting tests were conducted in live-grid under artificially imposed short-circuit conditions in 2009.

The test results indicate that the SFCL has a reliable current limiting capability and its performance is in good agreement with the design expectations.[8] Figure 7 gives the comparison between the design expectation and the test results in fault current limitation of the device. Test results also show that the device satisfies the requirements of grid routine protective procedures. The result of the fail-safe test demonstrates the reliability and robustness of this device.

In the October of 2012, a 220 kV/300 MVA saturated iron-core type SFCL with the same developer was installed at the Shigezhuang substation of Tianjin, China. After passing a series of filed acceptance

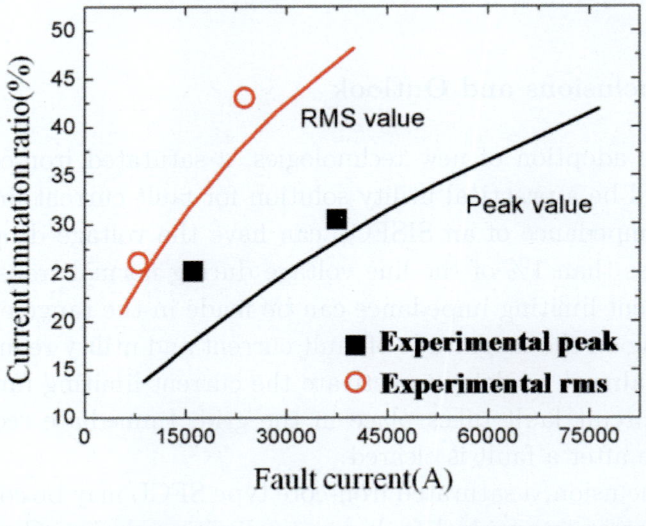

Fig. 7. Comparisons between the test results and computer simulations in fault current limitation for the 35 kV SFCL.

Fig. 8. Photographs of the saturated iron-core type SFCLs developed by Innopower and installed in the power grids. (a) The 35 kV/90 MVA device at Puji substation, Kunming, China; (b) The 220 kV/300 MVA device at Shigezhuang substation, Tianjin, China.

tests, it has been in operation for more than two years. Test and operation results show that the performance of this device satisfies design expectations. Functional performance, usability, reliability, and operation economics of the device will be examined through grid operation.

Figure 8 is the photo pictures of these two devices at their installation sites.

4. Conclusions and Outlook

With the adoption of new technologies, a saturated iron-core type SFCL will be a practical utility solution for fault current limiting.

The impedance of an SISFCL can have the voltage drop on the device less than 1% of the line voltage during normal transmission. The current limiting impedance can be made in the range of 1–20 Ω depending on the magnitude of fault current and utility requirement. There is almost no delay to activate the current limiting function as a short-circuit fault takes place in the grid. Immediate recovery is realizable after a fault is cleared.

In conclusion, a saturated iron-core type SFCL may be considered as the closest to an ideal fault-current limiting device. Commercial application of this kind of device can be expected in the near future.

References

1. B. P. Raju, K. C. Parton and T. C. Bartram, *IEEE Trans. Power Appar. Syst.* 3173 (1982).
2. T. Verhaege and Y. Laumond, "Fault Current Limiters", In *Handbook of Applied Superconductivity.* Bristol: IOP Publishing, 1691 (1998).
3. M. Noe and M. Steurer, *Supercond. Sci. Technol.* R15 (2007), R15.
4. Y. Xin *et al.*, *IEEE Trans. Appl. Supercon.* 1760 (2007).
5. Y. Xin *et al.*, *IEEE Trans. Appl. Supercon.* 1934 (2009).
6. M. Franco *et al.*, *IEEE Trans. Appl. Supercon.* 1288 (2011).
7. Y. Xin *et al.*, *Supercond. Sci. Technol.* 105011 (2012).
8. Y. Xin *et al.*, *IEEE Trans. Appl. Supercon.* 1294 (2011).

Chapter 3.4

Resistive-Type Fault Current Limiter

L. Martini, M. Bocchi and G. Angeli
T&D Technologies Department,
Ricerca sul Sistema Energetico — RSE S.p.A.
Via Rubattino 54, 20134 Milano, Italy

Among the wide range of High-Temperature Superconducting (HTS) materials presently known Bismuth Strontium Calcium Copper Oxide (BSCCO) is a very suitable candidate for power applications either at low temperature (e.g. <30 K) at any field or at high temperature (e.g. 77 K) in self-field conditions. This is due to several advantages of BSCCO from an electrical, thermal, mechanical and economic point of view. In particular, BSCCO has been proven to be particularly suitable for hybrid current leads and HTS cables. However, BSCCO-based Superconducting Fault Current Limiter (SFCL) applications have been an important issue within the Ricerca sul Sistema Energetico (RSE) S.p.A. R&D portfolio in the last decade. The SFCL project, funded in the framework of a R&D national project, started focusing on a preliminary single-phase device, which was submitted to dielectric and short-circuit current testing. The first success paved the way for the finalization of the remaining two phases and the final result was a three-phase resistive-type 9 kV/3.4 MVA SFCL device, based on first generation (1G) BSCCO tapes that was installed in the S. Dionigi substation, belonging to the Italian utility A2A Reti Elettriche S.p.A. (A2A), in the Milan MV distribution grid. The in-field activity lasted for more than two years, demonstrating the SFCL capability to cope with the grid in every-day operating conditions. Moreover, at the end of the experimentation, the SFCL device was able to perform a true limitation during a three-phase fault, thereby becoming one of the first SFCL devices in the world (the first in Italy) installed in a real grid and to have limited a real short-circuit current.

1. Introduction

Among Bi–Sr–Ca–Cu–O (BSCCO) compounds, Bi-2223 or BSCCO-2223 has the highest critical temperature with Tc equal to 110 K and can be suitably cooled by means of liquid nitrogen.[1] Thus, there is no need for costly liquid helium, which is used for conventional intermetallic superconductors such as the niobium-titanium alloy and niobium-tin compound. Also, the BSCCO compound does not contain expensive rare-earth and toxic elements. This compound has attracted a great deal of attention and efforts to fabricate long wires using the BSCCO compound and were made under worldwide industrial competition. Among the 1G wire's major benefits, it is worthwhile pointing-out that the silver matrix guarantees a good thermal stability and a significant mechanical straightness, that makes the soldering procedure easier and the winding manufacturing definitely feasible. The unit price is surely affordable, particularly if compared with 2G materials, making the purchase of large amount of wire viable. Among the large number of BSCCO-based applications, it seems Superconducting Fault Current Limiter SFCL devices are able to take several benefits from the usage of this material. The SFCL is an innovative device that, when properly designed and placed in appropriate positions in an electrical grid, is able to limit the short-circuit current I_{SC} to values compatible with safety and reliability of the installed network components. A SFCL device combines unique features such as transparency in the network under normal conditions, quick response and recovery without external intervention. By exploiting their intrinsic properties, BSCCO tapes can meet several typical requirements of resistive-type SFCL devices: for example, besides the unique property of resistivity increase upon the current crossing the critical current (I_c) (that is common to all the superconducting materials), BSCCO tapes guarantee also an easier manufacturing procedure and a major electro-mechanical strength thanks to their particular layout (superconducting filaments inserted in a silver matrix).

2. SFCL Design: The Italian Case Study

Several distribution utilities already experienced high fault current levels at different position within their MV grid. Similarly, the Italian utility, A2A Reti Elettriche S.p.A. (A2A), identified a few very suitable locations in its 23 kV and 9 kV networks where a SFCL device could be efficiently installed. According to the considered grid location and associated protection scheme, for three-phase faults the maximum steady-state prospective short-circuit current I_{SC} can be as high as $12\,kA_{rms}$ corresponding to peak I_{SCP} exceeding $30\,kA_p$ with a fault duration of 400 ms. In this case, an SFCL device which enables a limiting factor $LF = 1.7 \div 2$, being this LF value the ratio between the prospective I_{SC} in the absence of SFCL and the limited current I_{Lim} in the presence of the SFCL device ($LF = I_{SC}/I_{Lim}$), is highly desirable. The network specifications corresponding to this installation site and relevant for the SFCL device design, are as follows:

- Rated voltage $V_{nom} = 9.3\,kV$ and current $I_{nom} = 220\,A$;
- Power factor at nominal conditions $\cos\phi = 0.92$;
- Steady-state short-circuit current $I_{SCP} = 12.3\,kA_{rms}$ without SFCL (maximum prospective I_{SC} peak $= 32\,kA_p$);
- Power factor of short-circuit current I_{SC} without SFCL $\cos\phi = 0.1$;
- Ungrounded three-phase fault duration $t_{fault} = 400$ ms;
- Limiting factor $LF = 1.7 \div 2$.

In the past, RSE developed MVA-class single and three-phase SFCL prototypes with solenoid winding design, constituted by HTS layers anti-inductively wound on fiberglass cylinders.[2] In our design the resistive-type SFCL device includes the following components: three series connected HTS windings per phase, a cryostat containing the HTS windings immersed in a liquid nitrogen pool at 65–77 K, the cryostat top flange with the MV bushings, the feed through for instrumentation wiring, a safety valve and a burst disk, a closed-loop refrigeration system and a data acquisition, control and communication system, suitable for remote long-term operation.

The selection of the type of the HTS conductor for a SFCL device must be seeking the following main features[2]:

- High critical current I_c value at operating temperature,
- Mechanical and thermal properties such as to withstand thermal and electrodynamic stresses that occur during fault transients,
- Commercial availability in useful quantities at affordable cost.

Among the available HTS conductors, both first and second generation have been considered in the SFCL design.

In first generation (1G) multifilamentary BSCCO tapes with silver matrix, the 1G tapes taken into consideration are 4.6 mm wide and 0.35 mm thick, that has a self-field critical current exceeding 180 A at 77 K, stainless steel reinforcement and are insulated by Kapton overlapped ribbons. In second generation (2G) conductors with a 1.1μm thick HTS film deposited on a metallic substrate, the considered 2G YBCO-based conductors are typically 12 mm wide and have a 100μm thick Hastelloy C276 substrate, that has a self-field critical current of about 380A at 77 K and a 1–4 μm thick silver coating layer.[3] Figure 1 shows the comparison between the experimental critical current measurements for 1G and 2G HTS conductors at 65 K and 77.3 K.

3. SFCL Design: General Criteria and Methodologies

The main parameters to be defined for the optimized design of a SFCL device are:

- The HTS operating temperature under nominal condition.
- The ratio between the peak value of the rated current and the HTS I_c value.
- The maximum allowable temperature of the HTS conductor at the end of the most severe fault event.

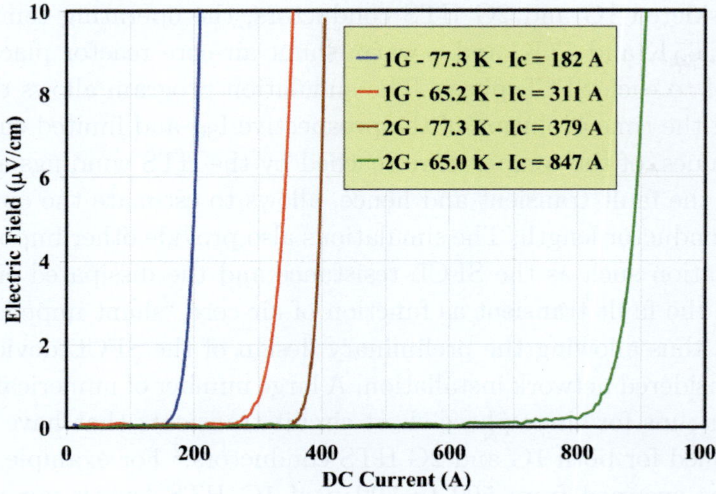

Fig. 1. Experimental measurements of electric field as a function of applied current for 1G and 2G HTS conductors (Ec = 1μV/cm).

Operating temperature normally is appropriate, given the significant increase of the critical current value of HTS conductors as the temperature decreases, to work at the lowest possible temperature. As regards, the increase in the allowable maximum temperature for 1G HTS has been estimated a conservative upper limit value of $T = 280\,\mathrm{K}$ to preserve its integrity. As it will be shown in the following, the two most important design parameters, whose variation leads to significant changes in the limiting factor and in the HTS conductor temperature at the end of fault, are the length of HTS conductor for each phase-winding of the SFCL device and the impedance value of the air-core reactor placed in parallel to each SFCL winding phase. Hence, a sensitivity analysis to identify the optimal SFCL design has been performed on these parameters. The single-line equivalent electrical circuit used for short-circuit numerical simulations comprises the presence of an air core reactor placed in parallel to the HTS windings to achieve the desired limiting factor and to significantly reduce the needed HTS winding length and its final temperature. Three-phase short-circuit numerical simulations have been performed by taking into account the network specifications, the actual property of considered 1G and 2G HTS conductors, the operating temperature of 65 K and 77 K, and a warm shunt air-core reactor placed in parallel to each SFCL phase. The simulation program allows determining the time evolution of the prospective I_{SC} and limited current I_{Lim} values, of the temperature reached by the HTS windings at the end of the fault transient and hence, allows to estimate the optimal HTS conductor length. The simulations also provide other important information such as the SFCL resistance and the dissipated energy during the fault transient as function of air-core "shunt impedance" values, thus allowing the preliminary design of the SFCL device for the considered network installation. A large number of numerical simulation runs for three-phase short-circuit transients that have been performed for both 1G and 2G HTS conductors.[2] For example, simulations spanned from 500 to 700 m of 1G HTS length per phase and from 0.3 to 0.5 Ω for the shunt impedance value in parallel to the SFCL phases. Figure 2 summarizes the results of many simulation runs as function of shunt impedance and HTS tape length for

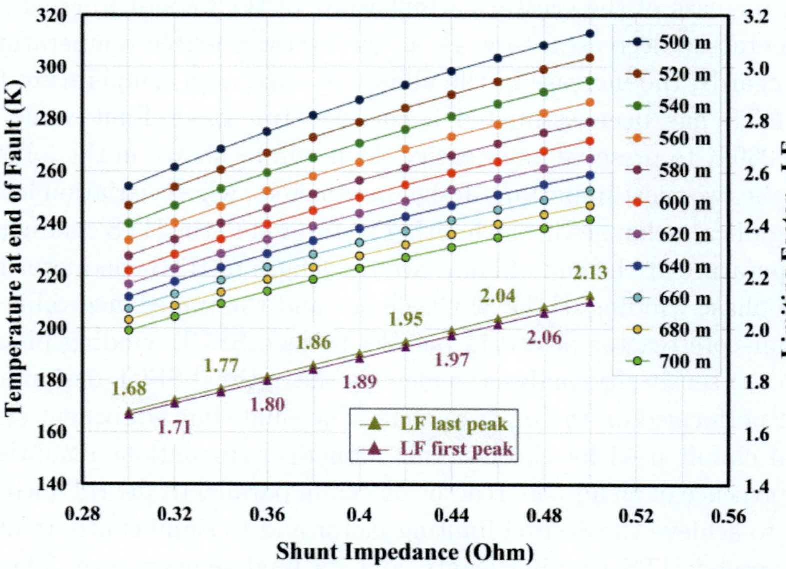

Fig. 2. Temperatures at the end of fault and limitation factors as function of air coil impedance and 1G HTS length.

phase.[4] As it can be seen, the simulation results for $Z_{shunt} = 0.5\Omega$ show the highest Limitation Factor (LF = 2.13). However, at this shunt impedance value corresponds to the highest HTS conductor final temperatures, being the $T = 280\,K$ threshold satisfy only for HTS lengths larger than 580 m per phase. A more appropriate shunt-impedance value Z_{shunt} of 0.4Ω and a HTS conductor conservative length per phase equal to 620 m, corresponding to a limiting factor of 1.9, has been selected for the SFCL design.

4. Construction and Characterization of a MV Single-Phase SFCL Device

The single-phase SFCL components are constituted by three series connected coaxially-arranged 1G HTS windings. The three windings were made by two anti-inductively wound 1G HTS layers with a total HTS conductor length of 620 m immersed in liquid nitrogen at 65 K. Critical current I_c measurement results at 77.3 K for the three HTS coils range from 195 A to 220 A, being $I_c = 201$ A for the series connected coil assembly. This assembly exhibited an I_c value as high as 340 A when measured at 65 K. AC loss measurements on the three individual 1G HTS coils and on the final assembly have been performed at both 77.3 and 65 K for different 50 Hz AC currents up to $I_{nom} = 250\ A_{rms}$. AC losses[5] of 0.3 W/m have been measured at 65 K at the nominal current value $I_{nom} = 220\ A_{rms}$ (peak value 311 A_p), being this value in line with the estimated values and compatible with the refrigeration system cooling capacity. The single-phase SFCL device was moved to the CESI Medium Power Laboratory for carrying out short-circuit tests for qualification according to the utility technical specifications. The short-circuit tests were carried out with $I_{nom} = 225\ A_{rms}$ at two different voltage values, and with different fault durations and parallel shunt impedances in order to qualify the behavior of the SFCL device. It is recalled that since the test circuit is single-phase an applied test voltage of 5.4 kV corresponds, for a 3-phase device to 9.3 kV, being this value the nominal voltage of the network where the SFCL will be finally installed. The single-phase SFCL, having a warm air-core reactor with $Z_{shunt} = 0.5\Omega$ in parallel, was tested against short-circuit during a validation test lasting for 200 ms. As a result, comparing the magnitude of unlimited short-circuit current I_{SC} (i.e. without SFCL) and the limited current I_{Lim}, the first current peak was reduced from 26.6 kA to 13.7 kA (LF = 1.94). According to numerical estimations, during the fault transient the HTS windings temperature steadily increases, being however lower than T = 210 K after 200 ms at the end of fault. The HTS winding temperature has been calculated under adiabatic assumptions, i.e. neglecting the heat exchange with the surrounding liquid nitrogen bath, thus leading to a conservative HTS temperature estimation. It is interesting comparing the results of AC and DC tests carried-out on the three SFCL phases in order to check the AC losses level and the critical current value at 77 K respectively. Figure 3 shows that AC losses values for phases S and T are lower than those recorded for phase R, accordingly the same two phases feature higher critical current values.[5] Figure 4 shows the behavior of the HTS resistance during a short-circuit tests performed at the reduced voltage of 1.6 kV and at the temperature of 65 K. It is interesting to note the typical superconductive behavior, in fact

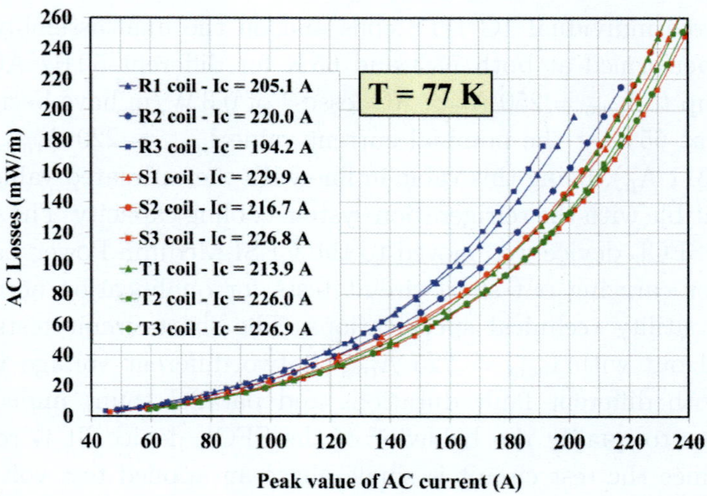

Fig. 3. Comparison among AC losses and phases critical currents.

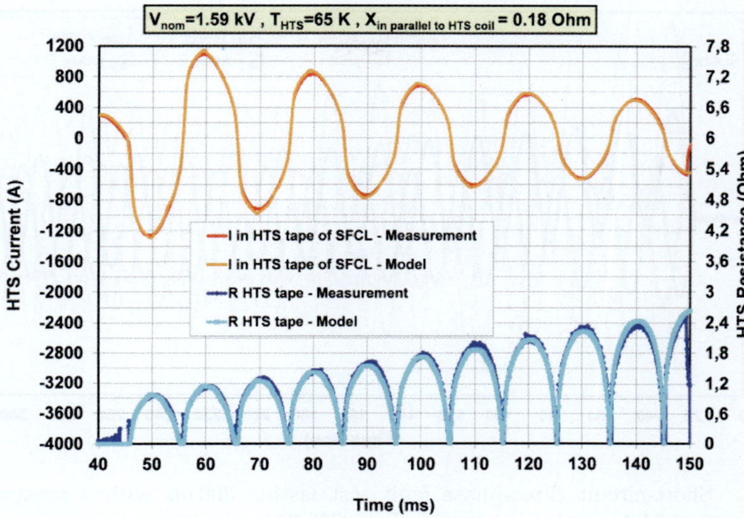

Fig. 4. Resistance time-behavior during fault.

the resistance features a pulsating time-behavior due to its dependence on the actual current value; moreover the amplitude increases because of the temperature effect, that makes the limitation even more effective while time proceeds.

5. Experimental Activity in the Milano MV Distribution Grid

The characterizations performed on BSCCO tapes discussed in the previous paragraphs, along with the lab-testing activity carried-out on the HTS windings, paved the way for the installation of the final three-phase 9kV/3.4 MVA SFCL device in the MV substation S. Dionigi belonging to the Milano distribution grid of A2A, the second largest utility in Italy. This is a very important achievement that RSE has accomplished in applied superconductivity, in fact this is officially the first SFCL device successfully installed in a real grid in Italy and one of the first worldwide. The in-field experimental activity consisted in a period of more than two years (March 2012–June 2014) of SFCL operation facing the MV grid nominal

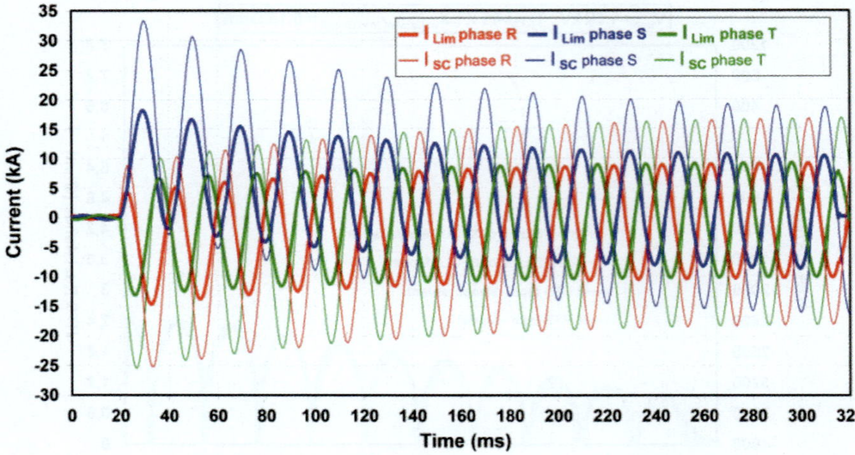

Fig. 5. Short-circuit three-phase fault test lasting 300 ms with a prospective current of 12.3 kA_p under a voltage value of 10.2 kV.

conditions, was finally rewarded in May 2014 by the limiting action during a real three-phase short-circuit. This is the first time in Italy that a SFCL device installed in a real grid is able to demonstrate its fault limiting capability during a real short-circuit, in agreement with the hosting utility requirements.[6]

Briefly recalling the last steps towards the in-field installation, it is worth starting from the validation tests performed on the three-phase SFCL device in the CESI laboratory in Milano on January 2011. Figure 5 shows the main findings stemming from this preliminary short-circuit tests: the SFCL device was able to limit the prospective first peak of about 33.3 kA_p down to 18.2 kA_p, i.e. featuring a LF of about 1.83.

After the aforesaid power tests, it was mandatory to subject the SFCL device to the dielectric tests foreseen by the relevant international standards,[7–10] (partial discharge test, short-duration power-frequency withstand voltage test, lightning impulse withstand voltage test), at the same time taking into consideration also the unique features of the device and adapting the standard procedures to it. The successful outcomes of the validation tests allowed to install the SFCL device in the S. Dionigi substation as a 9 kV single feeder protection in December 2011, fixing the temporary connections needed to preliminarily energize the system. The first commissioning was organized in March 2012, thereby launching a more than two years in-field activity with the SFCL device acting as a single feeder protection. During the two-year long field testing activity in the S. Dionigi substation, no short-circuit events characterized by high fault currents thus implying a significant limiting action by the SFCL device occurred. This is the reason why RSE and A2A teamed-up to organize a real short-circuit event in the grid in order to fully ascertain the limiting performances of the SFCL. For the sake of a reliable and safety, A2A chose a dedicated time slot, moreover configured the grid and tuned the protection relays to avoid any inconveniences to customers and grid devices. Figure 6 shows the main findings of the test during the in-field three phase short-circuit test, that lasted about 70 ms under a voltage level of 9 kV, the SFCL device was able to limit the first prospective peak of 20.84 kA down to 11.87 kA, i.e. performing a limiting capability with LF = 1.76.

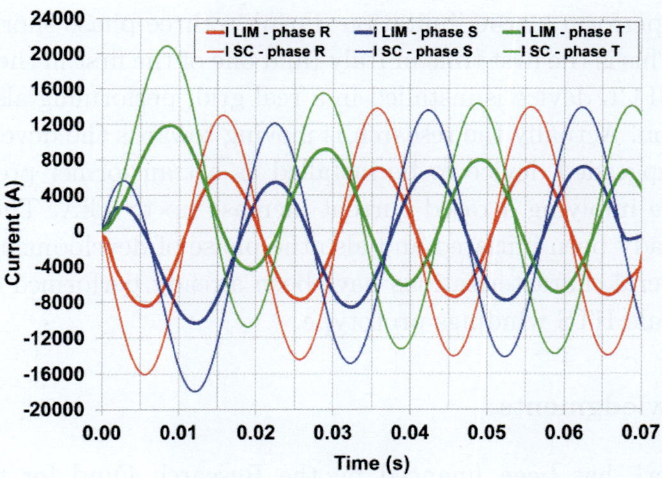

Fig. 6. Prospective I_{SC} and limited I_{LIM} fault current time-behavior.

6. Conclusions

Ricerca sul Sistema Energetico S.p.A. (RSE) developed a more than five years long project, funded in the framework of the Italian R&D fund Ricerca sul Sistema Energetico (RdS), for the design, development, characterization and field-testing of a resistive-type 9kV/3.4 MVA Superconducting Fault Current Limiter (SFCL) device. This project has greatly benefited from the 1G BSCCO tapes mechanical, electrical and thermal performance. Starting from the development of the single-phase device, composed by three coaxially-arranged BSCCO-based series-connected windings, each composed of two layers anti-inductively wound, it was realized that this type of architecture could have been the optimal choice for the final device. The complete three-phase SFCL device is made of a total amount of 1860 m of BSCCO tape and, after having successfully undergone the validation tests, it was finally installed in the S. Dionigi substation inside the 9 kV distribution grid belonging to the Italian utility A2A Reti Elettriche S.p.A. (A2A) in the Milano area. The in-field activity was launched in March 2012 and lasted for more than two years (until June 2014), during which the SFCL device faced the everyday grid conditions acting as a single feeder protection and it was also able to perform a true limitation during a three phase short circuit event. This is the first time in Italy (and one of the first in the world), that a SFCL device is installed in a real grid, performing also a real limitation. Actually the research is moving towards the development of the upgraded device to be installed as a transformer protection, therefore implying a rated current increase up to 1 kA. The design has already been initiated and also the phase of development, moreover several characterizations have been already performed on some small-scale HTS windings prototype.

Acknowledgments

This work has been financed by the Research Fund for the Italian Electrical System under the Contract Agreement between RSE

S.p.A. and the Ministry of Economic Development in compliance with the Decree of March 8, 2006.

References

1. K. Sato, S. Kobayashi and T. Nakashima, Present status and future perspective of Bismuth-based high-temperature superconducting wires realizing application systems, *Jpn. J. Appl. Phys.* **51** (2012).
2. L. Martini, C. Ravetta and M. Bocchi, Development and testing of innovative fault current limiters for distribution system applications, In *Proc. CIRED2011, 21th Int. Conf. Electricity Distribution*, Frankfurt, 6–9 June (2011), Paper 0339.
3. A. Hobl, W. Goldacker, B. Dutoit, L. Martini, A. Petermann, and P. Tixador, Design and production of the ECCOFLOW resistive fault current limiter, *IEEE Trans. Appl. Supercond.* **23** (3), (2013).
4. L. Martini, M. Bocchi, M. Ascade, A. Valzasina, V. Rossi, G. Angeli and C. Ravetta, Development, testing and installation of a superconducting fault current limiter for medium voltage distribution networks, *Phys. Procedia* **36**, 914–920 (2012).
5. W. T. Norris, Calculation of hysteresis losses in hard superconductors carrying ac: isolated conductors and edges of thin sheets, *J. Phys. D Appl. Phys.* **3**, 489–507 (1970).
6. L. Martini, M. Bocchi, G. Angeli, M. Ascade, V. Rossi, A. Valzasina, C. Ravetta, S. Fratti and E. Martino, Live grid field-testing final results of the first italian superconducting fault current limiter and severe 3-phase fault experience, *IEEE Trans. Appl. Supercond.*, **25** (3), (2015).
7. IEC 60270 High-voltage test techniques — Partial discharge measurements.
8. IEC 60071-1 Insulation co-ordination. Part 1: Definitions, principles and rules.
9. IEC 60076-3:2000 Power transformers — Part 3: Insulation levels, dielectric tests and external clearances in air.
10. IEC 60076-6 Power transformers — Part 6: Reactors.

Chapter 3.5

AmpaCity Project — World's First Superconducting Cable and Fault Current Limiter Installation in a German City Center

Mark Stemmle[*,§], Frank Merschel[†] and Mathias Noe[‡]

[*] *Nexans Deutschland GmbH, Hannover, Germany*
[†] *RWE Deutschland AG, Essen, Germany*
[‡] *Karlsruhe Institute of Technology, Karlsruhe, Germany*
[§] *mark.stemmle@nexans.com*

This chapter will give an overview on the German AmpaCity project, which started in September 2011. The objective of the project is developing, manufacturing and installing a 10 kV, 40 MVA HTS system consisting of a fault current limiter and of a 1 km cable in the city of Essen. It is the first time that a one kilometer HTS cable system is installed together with an HTS fault current limiter in a real grid application. In addition, it is the longest installed HTS cable system worldwide. Within the project the development phase was finished in March 2013 with successfully completing the type test of the cable system. Subsequently, all system components were manufactured for the installation on site in Essen. The installation took less than three months finishing at the end of November 2013. Afterwards, the commissioning test of the system was performed in December and the system was finally commissioned beginning of 2014.

1. Introduction

In most European countries the power supply within cities is predominantly ensured through high, medium and low voltage power cables. A large fraction of these cables as well as the associated substations are approaching the end of their lifetime and therefore need to be refurbished in the upcoming years. Usually, old power devices will be simply replaced by new ones and if there are major load changes, substations will be adapted accordingly by up- or downgrading.

The application of medium voltage HTS systems consisting of a concentric three phase HTS cable connected in series with an HTS fault current limiter (FCL) offers attractive alternatives to conventional systems. Replacing conventional high voltage cables by medium voltage HTS systems with the same power rating, enables a considerable reduction of the number of inner city substations. Since HTS cables are in general more compact than conventional cables, the required right of way is much smaller and the installation is easier. Moreover, there are many other advantages of HTS cables. Besides the increased power density there is no thermal impact on the environment. In addition, HTS cables do not exhibit outer magnetic fields during normal operation and in combination with HTS fault current limiters the operating safety in the grid is increased as a result of reduced fault current levels.

Over the last few years, high temperature superconductors have matured, especially due to the technical progress achieved in the manufacturing of these materials, and are now on the verge of industrial scale production. Furthermore, several superconducting cables,[1–5] as well as superconducting fault current limiters,[6–8] for power systems have been tested during the last years in real grid applications worldwide and are now on the threshold of commercialization. The experience gathered in these tests shows that all technical requirements are fulfilled so far and a high reliability of these new power devices can be achieved. At present, the main reason preventing wider use of the HTS technology is the higher capital cost compared to conventional devices. However, today an economic advantage of the HTS technology can be realized wherever positive secondary effects are present. Regarding the design of electricity distribution networks, this would mainly include higher power density and space savings especially in congested urban areas.

Between December 2009 and December 2010 a feasibility study had been conducted under guidance of the Karlsruhe Institute of Technology (KIT) on behalf of the German utility RWE.[9,10] Together with superconductor cable and fault current limiter specialists from Nexans as well as other partners, it was investigated whether the electric power supply with medium voltage superconductor systems in city centers offers technical and economical advantages compared to conventional high voltage technology. The German city Essen was chosen for the study and the key result is, that four out of ten 110/10 kV transformer substations become dispensable in the downtown area by using 10 kV cable systems. Conventional 10 kV cables do not constitute a viable alternative because of large routing requirements and high losses involved. In comparison to 110 kV cables, 10 kV HTS systems allow a much simpler grid structure, which requires less space for cable routing and smaller areas for equipment installations. In addition, the grid with 10 kV HTS systems exhibits lower overall costs than the grid with conventional 110 kV systems.

Subsequent to the positive results of the feasibility study RWE, Nexans and KIT started a pilot project, also referred to as "AmpaCity", with the objective to install an HTS system in the downtown area of a German city to demonstrate the technology under technical and economical aspects.[11]

2. AmpaCity Project

Due to the innovative character of the AmpaCity project an application for funding was submitted to the German Federal Ministry of Economics and Technology by the consortium consisting of RWE, Nexans and KIT. After being approved by the ministry, the demonstration project started in September 2011.

As for the feasibility study the downtown area of the city Essen was chosen for the installation of the HTS system consisting of a concentric three phase cable and a fault current limiter. The system will

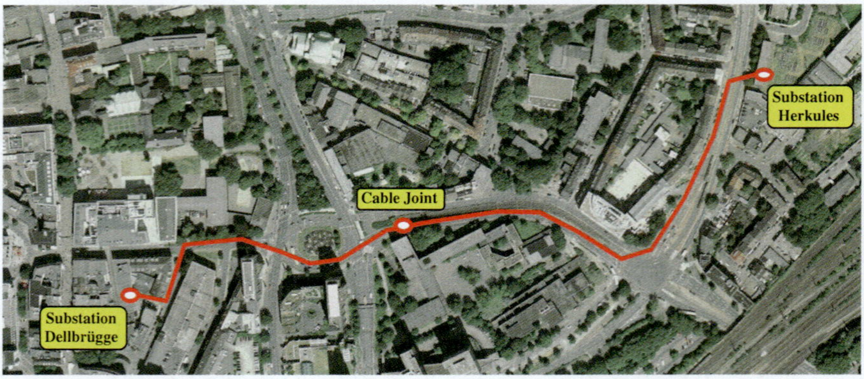

Fig. 1. Route for the installation of the superconductor cable in the city of Essen.[12]

be introduced as a connection between the medium voltage busses of two substations replacing a conventional high voltage system. The HTS system is rated for a transmission power of 40 MVA (2310 A) at 10 kV and the route for the installation with a length of approximately 1 km is shown in Fig. 1. Substation Dellbrügge is located in the downtown area of Essen in close proximity to the main pedestrian shopping street and the main railway station, whereas substation Herkules with an outdoor switchyard is further away from the city center. The superconducting fault current limiter as well as the cooling system will be installed in substation Herkules in order to maximize the space reduction in substation Dellbrügge.

Within the AmpaCity project RWE was responsible for the system specification, the location for the field test, and the implementation of the system in the grid. Nexans was in charge of the development, testing and manufacturing of the HTS system consisting of cable and fault current limiter. The HTS system development was supported by Karlsruhe Institute of Technology with characterization and tests of HTS materials as well as the establishment of a simulation model for AC losses of three phase concentric cable systems and a test setup for precise measurements of these losses on short sample cables.

3. HTS System Design

3.1. *Cable system*

The general specification which is basis for the design of the AmpaCity HTS cable system is shown in Table 1.

The superconductor cable design for the AmpaCity project is presented in Fig. 2. The cable core is based on a corrugated tube former which is used as a cooling channel for liquid nitrogen (LN_2). All three phases and a common screen are concentrically arranged around the former, each of them separated by a lapped dielectric of polypropylene laminated paper (PPLP). The three phase layers consist of stranded Bi-2223 wires, and the common screen is made of stranded copper wires. The cable core is placed into a cryostat, which is composed of two corrugated tubes in concentric arrangement with vacuum insulation in between. Another cooling channel for LN_2 is provided by the annulus between the cable core and the cryostat. Subcooled LN_2 is circulated through the cable system, keeping the cable core at its operating temperature of below 77 K.

For medium voltage applications, the concentric arrangement of all three phases and the connection in series with a superconducting fault current limiter allows a very compact cable design. Other major benefits of the compact design are the integrated return channel for the cooling medium as well as reduced fault current levels in the downstream grid due to the FCL application. During normal

Table 1. AmpaCity cable system specification.

Parameter	Value
Nominal voltage	10 kV
Design current	2.3 kA
Transmission capacity	40 MVA
Length	1 km
AC withstand voltage	30 kV
Lightning impulse voltage	75 kV

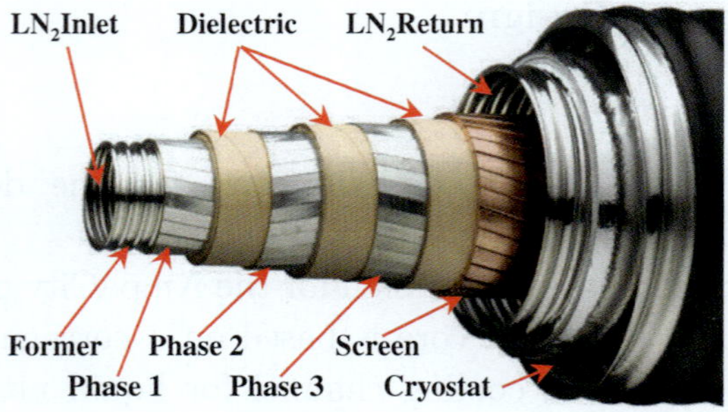

Fig. 2. Superconductor cable design with concentric arrangement of phases.

operation the currents in the three concentric phases are balanced exhibiting the same absolute value at a 120° phase shift and no current flows in the screen. Therefore, no magnetic stray field appears outside the HTS cable system. Additionally, due to active cooling with LN_2 inside the thermal insulating cable cryostat, the system is thermally independent from the environment. This unique thermal behavior and its very high electromagnetic compatibility lead to a simplified siting as well as an easier installation.

3.2. *Fault current limiter*

The general specification which is basis for the design of the AmpaCity HTS fault current limiter is shown in Table 2.

Superconducting materials exhibit a strongly nonlinear, voltage–current characteristic. Usual superconductor applications operate below the critical current, where virtually no voltage drop across the superconductor is measured for direct currents. Once the critical current is exceeded, an increasing voltage drop is observed up to the full transition to the normal conducting state. Resistive fault current limiters utilize this transition during normal operation the amplitude of the alternating current stays below the critical current without voltage drop and during faults the short circuit currents lead to a high resistance of the device with associated heating of the superconductor. Hence, a fault current limiter introduces a nonlinear

Table 2. AmpaCity fault current limiter specification.

Parameter	Value
Rated voltage	12 kV
Rated current	2.3 kA
Rated power	40 MVA
Limitation time	100 ms
AC withstand voltage	28 kV
Lightning impulse voltage	75 kV
Prospective peak short circuit current	50 kA
Limited peak short circuit current	<13 kA
Prospective short circuit current	20 kA
Limited short circuit current	<5 kA

resistor in the circuit with the transition being triggered by the current. The transition between the superconducting (low impedance) and the normal conducting (high impedance) state is fully reversible. After the limitation the material automatically cools back to its operating temperature. This time interval is typically in the range of a few seconds.

For resistive fault current limiters HTS tapes are typically assembled to limiter components. Several of these components are then connected in series and are assembled into a cryostat which provides the low temperature operating conditions. Three vacuum insulated LN_2 vessels, one for each phase, keep the limiter components stable at the operating temperature of 77 K. A picture of the fault current limiter for AmpaCity is shown in Fig. 3. The device is designed to limit a 38 kA peak short circuit current to about 10 kA.

3.3. *Cooling system*

The general specification which is basis for the design of the AmpaCity cooling system is shown in Table 3.

A cooling system for HTS power applications consists of different components. Most of the systems require forced flow of the coolant which is provided by a circulation pump. Due to electrical insulation requirements and height differences sub-cooled cryogens are

Fig. 3. Superconducting fault current limiter delivered in substation Herkules.

Table 3. AmpaCity cooling system specification.

Parameter	Value
Inlet temperature	67 K
Cooling power	4 kW
Mass flow	450 g/s

commonly used. Therefore, a pressure regulation system inside the closed cooling circuit is required. The heat input to the circulation system is re-cooled in a heat exchanger. The primary side of the heat exchanger is either cooled by cryocoolers,[13] (closed system) or from bulk liquid supply (open system). Vacuum pumps may be applied to the heat exchanger in order to reduce the operating temperature below the atmospheric pressure boiling point of the cooling liquid. An open system features lower complexity and potentially high reliability, but requires re-filling of a nitrogen storage tank in regular intervals. After initial filling, a closed system only consumes electrical power but therefore requires higher capital investment and specific methods to ensure availability and reliability. The cooling power of HTS systems is influenced by different factors like system size and total losses. Further, cooling systems not only have to enable normal operation of superconducting systems, but also need to be safe and efficient in all operating modes like cool down and warm up as well as during any failure mode.

The schematic of the cooling system for AmpaCity is shown in Fig. 4. Due to financial reasons it was decided to utilize an open cooling system with re-filling of a storage tank. The storage tank is installed in a separate section of substation Herkules which can be accessed by the LN_2 truck drivers without utility personnel being present.

While the fault current limiter is directly supplied from the storage tank, the nitrogen circulating in the cable is re-cooled in the heat exchanger. In the heat exchanger, vacuum pumps reduce the nitrogen gas pressure and therefore the nitrogen bath temperature, resulting in an inlet temperature of 67 K for the cable. The cooling system is designed for a total heat load of 4 kW at an inlet temperature of 67 K. Inside the cable Ln_2 circulates with a mass flow of 400 g/s and a minimum pressure of 6 bar. In order to achieve a high reliability and availability the cooling system is fully automated and also redundant with respect to the circulation and vacuum pumps. In total,

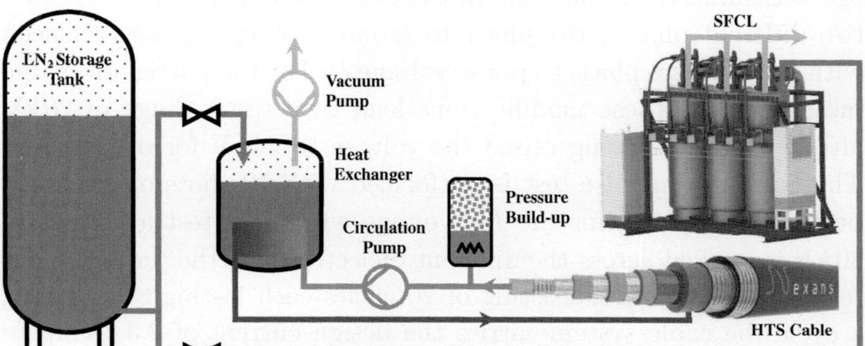

Fig. 4. Schematic of cooling system for AmpaCity HTS cable and fault current limiter.

the system includes two circulation pumps and three vacuum pumps each with separate valves.

4. Prototype Test

The beginning of the project was dedicated to the design phase and specification of the system requirements. In May 2012, a pre-prototype cable was manufactured in order to qualify the manufacturing process and the cable design mainly with respect to high voltage constraints. Since the superconducting material is still quite expensive, only a few HTS tapes were used for each phase layer of the cable and copper tapes exhibiting the same dimensions were used to replace the remaining HTS tapes. All mechanical tests as well as all high voltage tests were successfully completed on the pre-prototype cable. Therefore, the cable design as well as the manufacturing process was validated. Subsequently, in October 2012 a prototype cable was manufactured with each of the phase layers containing only HTS tapes exactly as the cable to be installed in Essen.

In the project consortium, a type test procedure was designed in agreement with the German standard DIN VDE 0276-620 with respect to the voltages applied during type testing. For the testing basically three different tests with high voltage are mandatory — the partial discharge test, the lightning impulse test, and the AC voltage withstand test. Since two of the dielectrics in the cable separate two different phases, the phase to ground voltage U_0 was replaced with the nominal phase to phase voltage U. For the partial discharge measurements, these modifications lead to a test voltage of 20 kV after previously having raised the voltage to 24 kV for one minute. The lightning impulse test is performed with 10 shots of 75 kV for both polarities and for the AC voltage withstand test a voltage of 30 kV is applied across the different dielectrics. Further, a load cycle test is also foreseen consisting of 20 cycles each lasting 8 h. During a cycle the cable system carries the design current of 2.3 kA up to 5 h and carries no current for the rest of the cycle. For type testing, the test sequence of partial discharge test, load cycle test, partial

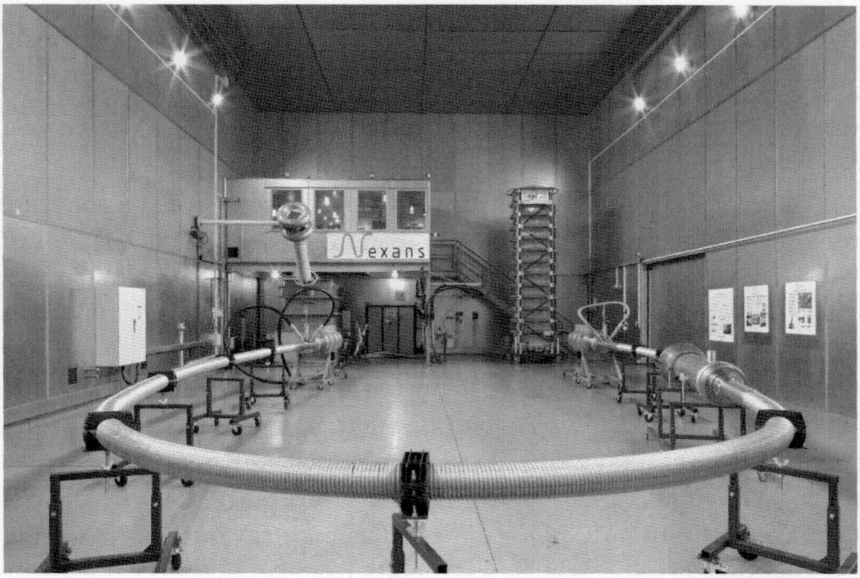

Fig. 5. Type test setup with prototype cable, terminations and joint.

discharge test, lightning impulse test and AC voltage withstand test was defined.

The type test of the AmpaCity cable system was performed on a 25 m loop including two prototype cable sections with cryostats, one joint and two terminations which were set up in Nexans' Hannover laboratory as illustrated in Fig. 5. All tests of the type test sequence were passed and the complete type test was finished successfully in March 2013. Afterwards, manufacturing of the HTS system components for the installation in Essen started promptly.

5. Installation

In parallel to manufacturing the system components, the civil works in Essen for the installation of the HTS cable started in April 2013. A polyethylene duct system with an outer diameter of 250 mm and a wall thickness of 12 mm was installed in the ground between each of the two substations and the joint bay. For manufacturing of the cable cryostat it was important to precisely determine the length of the duct system. The joint bay with a width of 2 m, a length of 6 m and a depth of 2.5 m was just accessible during the installation and was backfilled afterwards. The installation of the duct system was completed beginning of September.

The cable cryostat manufacturing was finished once the length of the duct system was determined and the vacuum was pumped for at least two weeks. Before shipping the two cable lengths to Essen, the cable cores were pulled into the two cryostat sections in the Nexans factory in Hannover. The complete cables were spooled on special transport drums with a flange diameter of 4.2 m and a width of 3.8 m for the bigger drum. At the end of September 2013 the two drums were loaded onto flat bed trucks and shipped to Essen.

The cable pulling was performed by a subcontractor with special equipment to properly handle the drums and the cable. On October 1st, the shorter cable length was pulled from a standard cable drum trailer (Fig. 6), which was positioned at the joint bay, to substation Dellbrügge. One day later, on October 2nd, the longer cable length was pulled from substation Herkules to the joint bay. The different pulling directions were determined by taking the forces into account

Fig. 6. Pulling of the first cable length from the joint bay to substation Dellbrügge.

on the cables during pulling and the space restrictions on site for the cable drum trailer. The cable cryostat was qualified to support pulling forces equivalent to at least 4 t (4.000 kg) and a sidewall pressure of more than 10 t (10.000 kg).

In the following weeks until the end of November 2013, the terminations and joint of the HTS cable, the fault current limiter, the cooling system as well as all the auxiliary equipment were installed on site in Essen.

6. Commissioning

After the installation on site was finished, the fault current limiter and the cable system were gradually cooled down to their operating temperature. For the HTS cable the cooling process was achieved within one week while for the FCL the process was performed in two days.

Subsequently, the commissioning test was conducted in December 2013 with a cable test van with very low frequency (VLF) testing equipment at 0.1 Hz. The commissioning tests comprised a partial discharge measurement of each dielectric with 20 kV (Fig. 7), a loss factor diagnoses with 10 kV, 15 kV and 20 kV as well as an AC voltage withstand test of each dielectric with 30 kV for one hour. All tests with the cable test van were passed successfully.

Beginning of 2014 the HTS system was finally connected to the grid. At first it was only connected to the bus bar in substation Herkules and afterwards to the bus bar in substation Dellbrügge as well.

7. Operation Experience

After the installation on site was finished, the fault current limiter and the cable system were gradually cooled down to their operating temperature. For the HTS cable the cooling process was achieved within one week while for the FCL the process was performed in two days.

Fig. 7. Partial discharge measurement of dielectric between phase 1 and phase 2.

In the beginning of the operation, an issue with the vacuum pumps of the cooling system was observed. Inside the vacuum pumps leaks were found which led to freezing of humidity in the exhaust system. Since in the cooling system design one vacuum pump is redundant, the pumps could be modified one by one during the operation of the system.

During a big thunder storm in June 2014, Whit Monday storm "Ela", automatic reclosings in the 110 kV network feeding substation Herklues led to a voltage sag in the power supply of the cooling system. As a result, the circulation pumps stopped briefly which was mistakenly interpreted by the control system as an invalid state of operation, therefore switching the HTS system off. Consequently, the control system was readjusted by reducing its sensitivity.

8. Conclusion

The application of concentric medium voltage HTS systems enables very attractive grid concepts for urban area power supply. From the current perspective expanding grids using HTS cables is the only technically and economically appropriate option for avoiding the expansion of inner city power grids using high voltage cables, and reducing the number of high voltage transformer substations in downtown areas. Further, concentric HTS cable systems for medium voltage applications are very compact, exhibit a very high electromagnetic compatibility and are thermally independent from the environment. For these systems the required right of way is much smaller and the installation is easier compared to conventional high voltage cable systems. In combination with superconducting fault current limiters the fault current levels in the downstream grid are reduced in addition, thus increasing the grids operating safety.

The authors are of the opinion that the AmpaCity project in Germany is world's first, in terms of both content and scope. Worldwide the project is the first implementation of an HTS cable in combination with a resistive HTS fault current limiter in a medium voltage grid replacing a conventional high voltage system. In addition, with a length of approximately 1 km, it is the longest superconducting cable system up to date. The type test including two sections of prototype cable, one joint and two terminations was completed successfully in March 2013. Afterwards, the manufacturing of the HTS system consisting of cable, fault current limiter and cooling system as well as the civil works in Essen progressed well. Further, the complete installation in Essen started in September and was finished in late November 2013. The commissioning test of the AmpaCity system was performed with a cable test van in December 2013. After complete commissioning of the HTS system in 2014 a field test period of at least two years is foreseen. The AmpaCity system has been in operation for more than one year now, supplying the city center of Essen and during this operation period no major issues occurred.

The HTS technology is still rather unknown, since today no large scale demonstration projects have been realized in particular in Europe. The AmpaCity project represents a worldwide unique reference, and therefore could act as a potential catalyst for initiating the expansion of production capacities in the HTS materials, cryogenic systems, and HTS cable technology sectors. When the target innovation objectives are reached, the electricity supply of major population centers requiring high power may be simplified in the mid to long term through partial substitution of high voltage to medium voltage transformer substations by introducing medium voltage HTS systems similar to the AmpaCity system.

References

1. Working Group D1.15 CIGRE, *Status of Development and Field Test Experience with High-Temperature Superconducting Power Equipment.* Technical Brochure, **418** (2010).
2. J.F. Maguire *et al.*, Development and demonstration of a HTS power cable to operate in the long island power authority transmission grid, *IEEE Trans. Appl. Supercond.* **17**(2), 2034–2037 (2007).
3. J.A. Demko *et al.*, Triaxial HTS cable for the AEP bixby project, *IEEE Trans. Appl. Supercond.* **17**(2), 2047–2050 (2007).
4. C.S. Weber *et al.*, Testing and demonstration results of the 350 m long HTS cable system installed in Albany, NY, *IEEE Trans. Appl. Supercond.* **17**(2), 2038–2042 (2007).
5. Y. Xin *et al.*, Introduction of China's first live grid installed HTS power cable system, *IEEE Trans. Appl. Supercond.* **15**(2), 1814–1817 (2005).
6. Working Group A3.23 CIGRE, *Application and Feasibility of Fault Current Limiters in Power Systems.* CIGRE Technical Brochure 497 (2012).
7. J. Bock *et al.*, Nexans' Superconducting Fault Current Limiters for Medium Voltage Applications — Status and Prospects. CIRED 2011, University of Applied Science, Frankfurt, Germany, Paper 0352 (2011).
8. M. Noe, M. Steurer, High temperature superconductor fault current limiters: Concepts, applications, and development status, *Supercond. Sci. Technol.* **20**, R15–R29 (2007).
9. M. Stemmle *et al.*, Novel grid concepts for urban area power supply, *Phys. Procedia* **36**, 884–889 (2012).
10. A. Breuer *et al.*, Superconducting Medium-Voltage Cables for Urban Area Power Supply as an Alternative Scenario to 110 kV Installations. CIGRE Session 2012, RWE Deutschland, Paris, France, Paper B1-301 (2012).
11. M. Stemmle *et al.*, AmpaCity Project — Worldwide First Superconducting Cable and Fault Current Limiter Installation in a German City Center. CIRED 2013, Stockholm, Sweden, Paper 0742 (2013).
12. Courtesy of Essen (Luftbild: Darstellung aus HK Luftbilder/Karten Lizenz Nr. 197/2012 mit Genehmigung vom Amt für Geoinformation, Vermessung und Kataster der Stadt Essen vom 13.02.2012).
13. A.T.A.M. De Waele, Basic operation of cryocoolers and related thermal machines, *J. Low Temp. Phys.* **164**, 179–236 (2011).

Chapter 3.6

AC Cable: Yokohama Project

Takato Masuda
Superconductivity Technology Division,
Sumitomo Electric Industries, Ltd.
1-1-3, Shimaya Konohana-ku, 554-0024, Osaka, Japan

High Temperature Superconducting (HTS) cables can transmit large amounts of electricity in a compact size with minimal losses. Therefore, they are expected to save the construction cost of underground lines in urban areas and decrease transmission losses. Several HTS cables have recently been demonstrated in networks around the world, and full-scale commercialization is expected in the near future. In Japan, the development of compact HTS cables suitable for urban deployment has been underway since the early 1990s. In 2007, a national project was started to verify their operational performance and long-term reliability in the grid. An HTS cable 240 m long was installed at the Asahi substation of the Tokyo Electric Power Company (TEPCO) in Yokohama; then a joint, terminations and cooling system was constructed in 2011. After successful performance tests, the cable was connected to the grid for the first time in Japan, and started to deliver electricity to 70,000 households in October 2012. This trouble-free in-grid service continued for over a year. We can conclude that the HTS cable system performs well and has the stability required for long-term in-grid operations.

1. Introduction

High Temperature Superconducting (HTS) cables can transmit large amounts of electricity in a compact size with minimal losses. They can thus reduce the construction cost of underground lines in urban areas and decrease transmission losses. Several HTS cables have recently been demonstrated in networks around the world, and full-scale commercialization is expected in the near future. In Japan, the development of compact HTS cables suitable for urban deployment has been underway since the early 1990s. In 2007, a national project was started to verify their operational performance and long-term reliability in the grid.[1] After construction of a 66 kV-200 MVA HTS cable system at a Tokyo Electric Power Company (TEPCO) substation, it was connected to the grid for the first time in Japan and has since then provided electricity to 70,000 households in the Tokyo area.

2. Outline of the Project

The project is supported by New Energy and Industrial Technology Development Organization (NEDO). TEPCO provides the real grid and studies the impact of connecting the HTS cable to the existing conventional facilities. Sumitomo Electric Industries, Ltd. and Mayekawa Mfg. Co., Ltd. are in charge of designing and manufacturing the HTS cable and the cooling system, respectively.

The two aims of this project are to develop a 66 kV, 200 MVA HTS cable with three cores for the cable system at TEPCO's Asahi substation, and to conduct in-grid operations for more than one year.[1]

The project was started in 2007 and during the first two years, technologies were developed for all elements of the HTS cable system, including the cable structure. The structure provides large current capacity with reduced transmission losses, and withstands short-circuit currents. At the same time, the demonstration site was determined and the grid conditions carefully scrutinized. In 2009, a 30 m cable system was manufactured to verify the design and several tests were completed successfully.[2] Then the cable for the Asahi substation was manufactured and shipped to the site in 2011. At the same time, a cooling system was developed and tested to check its operation, maintenance requirements, and efficiency. Both the HTS cable and the cooling system were installed at the site in 2011, followed by performance tests of the HTS cable in combination with the cooling system in 2012. In-grid operations began in October, 2012.[3]

Table 1. Specifications of the HTS cable.

Items	Specifications
Rated Capacity	200 MVA (66 kV, 1.75 kA)
Maximum current	2.63 kA
AC loss	<1 W/m/ph
Withstand voltage	AC 90 kV for 3 h Imp ±385 kV 3 repetition
Fault current	(i) No degradation against the Fault current of 31.5 kA for two sec. (ii) The rated capacity can be transmitted immediately after the fault current of 10 kA for two sec.

3. 66 kV-200 MVA HTS Cable System

(1) System specifications

The specifications of the HTS cable system are summarized in Table 1. From the rated 200 MVA capacity of the power line, the nominal current of the cable was determined to be 1.75 kA. The maximum overload current was determined to be 2.63 kA, corresponding to 1.75 kA × 1.5.

The HTS cable produces no Joule-heating losses because it has no electrical resistance at liquid nitrogen (LN_2) temperature. However, when an AC current is transmitted through the HTS cable, there is some hysteresis loss due to the changing magnetic field. This is called AC loss. Decreasing the AC loss is a technical subject and in this project, the target value is 1 W/m/ph or less. This level of AC loss is needed to reduce the HTS cable's transmission loss to half or less than that of conventional cables when the efficiency of the cooling system is taken into account.

The withstand voltage was determined in accordance with the Japanese standards for 66 kV class oil-filled (OF) cables (JEC-3401).[4]

The short-circuit withstand current requirements were considered for two cases. Case (i): According to TEPCO's regulations for equipment in a 66 kV power network, the short-circuit withstand current is defined as 31.5 kA for two seconds. On the other hand, the maximum

Fig. 1. HTS cable structure.

load at a short-circuit fault near the testing site was calculated to be 20 kA for two seconds. Therefore, to be on the conservative side, the current requirement selected was 31.5 kA for two seconds. In this case, facilities near the transformer would have failed severely, so the cable is not required to continue power transmission. Case (ii): According to the simulation, the maximum load at the through fault was calculated to be 10 kA for two seconds. In this case, the system should continue power transmission even when exposed to such a current.[5]

(2) Cable structure and some elementary tests

The structure of the Three-in-One HTS cable is shown in Fig. 1. The cable core consists of a former made from stranded copper wires, four layers of HTS conducting layers, an insulation layer, two layers of HTS shielding layers, and a copper shielding layer, all of which are coaxially wound around the former. A thermal insulation layer is provided by a high vacuum between two concentric stainless steel double-corrugated pipes that form a cryostat around the three cores.

To reduce the AC loss, low-loss Bi-2223 HTS wires were used in the outer two conducting layers. This wire contains twisted HTS filaments, reducing the AC loss to one-third of that of the standard Bi-2223 wires that are used in the inner two conducting layers and shielding layers. The load test showed that the AC loss of the cable was 0.8 W/m/phase at 2 kA as designed, thereby meeting the required specification of 1 W/m/phase or less.[5]

The electrical insulation layer is a composite construction of polypropylene laminated paper (PPLP) 7 mm thick, impregnated with LN_2.

When the three cores are cooled to LN_2 temperature, they shrink 0.3% lengthwise. To reduce cable size as much as possible, there is no slack along the cable. Thus three tons of tension is generated when it is cooled, but the cable is designed to be sufficiently robust to withstand this large tension.

(3) Demonstration site

The Asahi substation, located in Yokohama, was selected as the demonstration site for its available space, current capacity, and other factors. The Asahi substation is an outdoor type, with a triple set of 200 MVA, 154/66 kV step-down transformers that connect 154 kV bus lines with 66 kV bus lines. The layout of the HTS cable system is shown in Fig. 2. The system includes 240 m of HTS cable, a joint, two terminations, and a cooling system. Photographs of the system are shown in Fig. 3. One of the terminations is connected to the 66 kV end of a transformer and the other to a 66 kV bus line through several line switches and circuit breakers. For protection, the system relies on the conventional protective systems of the transformer. The

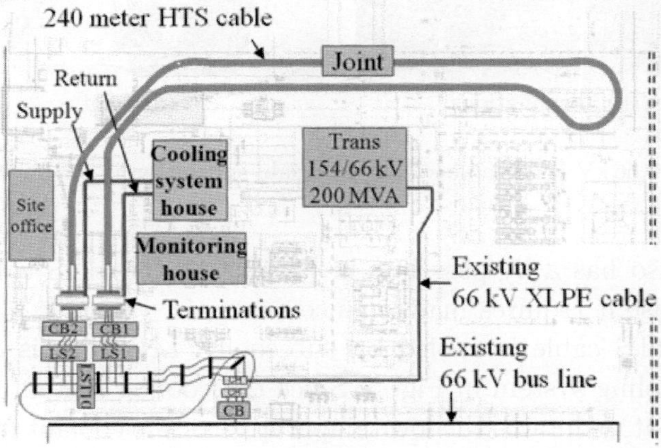

Fig. 2. HTS cable layout.

(a) HTS cable installed in a duct

(b) Terminations

(c) Joint

Fig. 3. A 66 kV HTS cable system at Asahi substation in TEPCO network.

system also has a by-pass line in parallel to the HTS cable, to be used in case it requires maintenance or repair.

The HTS cable is cooled by sub-cooled LN_2 that is circulated by the cooling system in Fig. 3. Since the cooling system is the key facility that maintains the low temperature of the coolant in the HTS cables, it is designed with some redundancy. The cooling system consists of six Stirling refrigerators, two circulation pumps, a reservoir tank, and other apparatus. Each refrigerator has a cooling capacity of 1 kW at 77 K or 0.8 kW at 67 K. As shown in Fig. 3,[6] these six refrigerators were arrayed in three rows with two units in each, while the circulation pumps were placed in parallel. The system is designed so that refrigerators and pumps may be removed and replaced for maintenance or repair, while the cable remains in operation.

4. Performance Tests before Connection to the Grid

After successful shipping tests in the factory, the HTS cable was shipped and installed at the test site as shown in Fig. 4. After installation at the site and a successful pressurized system test at 0.5 MPaG (design pressure), the whole cable was gradually cooled down to LN_2 temperature over three days and thereafter cooled with LN_2 circulated by the cooling system. During the cooling process, the cable tension was measured with load cells at the terminations. The tension generated by thermal contraction was measured as approximately 3.5 tons, which is within expectations. Then several cooling system tests were conducted, including temperature control, pressure control, and switching tests for backup equipment, relevant to

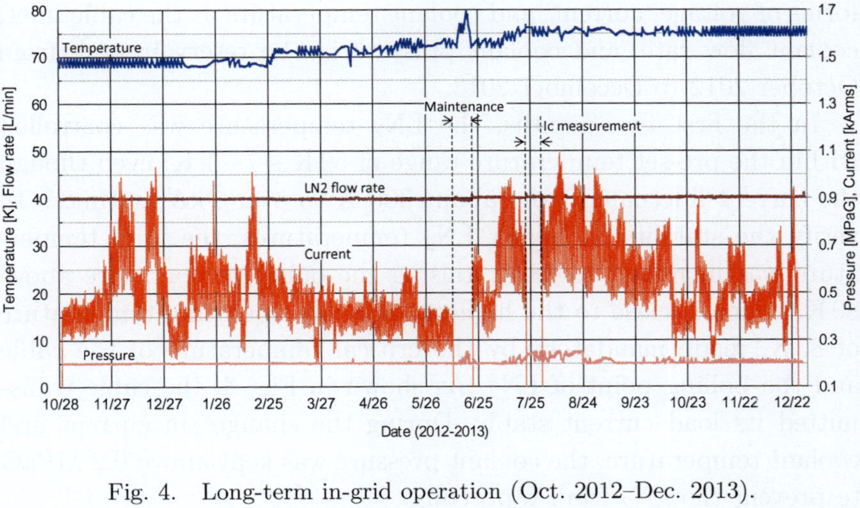

Fig. 4. Long-term in-grid operation (Oct. 2012–Dec. 2013).

its maintainability. After the cooling system tests, performance tests were conducted as follows.[7]

The critical currents of the full-length cable were measured to check the soundness of the HTS conductor layer. Measured critical current was 6.4 kA at 77 K, which is in good agreement with the design value in consideration of magnetic field effects. A successful DC withstand voltage test was carried out at 151.8 kV around $4\,\mu_0$ for 10 minutes, in accordance with the Japanese Interpretation of Technical Standards for Electrical Equipment, instead of applying AC voltage. Then one end of the cable was connected to the grid to check partial discharge (PD), and no sign of PD was observed at 50 pC sensitivity.

These successful test results showed that the cable performed well even after transportation, installation, and initial cooling. Finally, both ends of the HTS cable were connected to the grid and regular power transmission through the HTS cable began in October, 2012.

5. System Behavior after Grid Connection

Verifying the long-term stability of the cooling system is one of the themes of this in-grid demonstration. Figure 4 shows the waveforms of voltage, current, and coolant temperature at the cable inlet, coolant flow rate, and coolant pressure in the reservoir tank from October 2012 to December 2013.

In the first two months, the LN_2 temperature was controlled within the pre-set temperature range of 69 K +/− 1 K, even though the current fluctuated from about 300 A to over 1 kA at times. To verify the stability at higher LN_2 temperatures, the inlet temperature was increased to 78 K, raising the outlet temperature above 80 K, which is close to the highest allowable operating temperature of 83 K, made mandatory by the critical temperature of the cable and the boiling point of LN_2. As shown in Fig. 4, the cable transmitted its load current stably. During the changes in current and coolant temperature, the coolant pressure was kept above 0.2 MPaG to prevent the LN_2 from vaporizing.

In June and July 2013, transmission through the HTS cable was suspended twice for scheduled maintenance and critical current measurement to check its intermediate performance, and the critical current remained identical to its initial value. At the restart of power transmission, a PD measurement revealed no trace of PD. Thus, the system showed no sign of degradation.

6. Conclusion

A 66 kV-200 MVA HTS cable 240 m long was constructed at TEPCO's substation, together with a cooling system. This first-in-Japan in-grid operation served the electrical needs of 70,000 households without interruption for more than a year, with no indication of any degradation or problems. The temperature and pressure of LN_2 in the cable remained controlled within the target range. After nine months of operation, measurements showed no decline in the critical current of the cable. In addition, periodic checks showed the absence of PD.

Conclusion: The HTS cable system performs well, and with the stability required for long-term in-grid operation.

At a next step, a new Bryton cycle refrigerator developed by Mayekawa will be installed in Asahi Substation and connected to the grid again for verifying its high efficiency performance and stability in 2015 and 2016.

References

1. T. Masuda, H. Yumura, M. Ohya, T. Kikuta, M. Hirose, S. Honjo, T. Mimura, Y. Kito, K. Yamamoto, M. Ikeuchi, and R. Ohno, *IEEE Trans. Appl. Supercond.* **19**(3), 1735–1739 (2009).
2. H. Yumura, Y. Ashibe, M. Ohya, H. Itoh, M. Watanabe, K. Yatsuka, T. Masuda, S. Honjo, T. Mimura, Y. Kitoh, and Y. Noguchi, *Physica C* **470**, 1558–1562 (2010).
3. M. Watanabe, M. Ohya, H. Yumura, T. Masuda, N. Nakamura, R. Ono, H. Yaguchi, A. Machida, T. Nakano, O. Maruyama, and S. Honjo, *EUCAS 2013*, 1A-LS-07.
4. Standard of the Japanese Electrotechnical Comittee JEC: 3401-2006.

5. H. Yumura, Y. Ashibe, M. Ohya, H. Itoh, M. Watanabe, T. Masuda, H. Ichikawa, T. Mimura, S. Honjo, T. Hara, R. Ohno, M. Shimoda, N. Nakamura, T. Komagome, and H. Yaguchi, *IEEE Trans. Appl. Supercond.* **23**(3), (2013).
6. N. Nakamura, M. Shimoda, T. Komagome, R. Ohno, H. Yaguchi, H. Ichikawa, T. Mimura, T. Hara, M. Watanabe, and T. Masuda, *ICEC 24-ICMC 2012*, May, 693–697 (2012).
7. S. Honjo, In *World Energy Congress 2013*.

Chapter 3.7

Research and Development of Bi-2223-Based AC Power Cables in Russia

Vitaly Vysotsky*, Sergey Fetisov,* and Viktor Sytnikov†

*Russian Scientific Cable R&D Institute, Moscow, Russia

†R&D Center at Federal Grid Company United Energy System, Moscow, Russia

The Russian program for R&D of AC power cables has been started in 2005. During the program, the several 5 m prototypes and witness samples of Bi-2223 based cables have been tested, as along with the 30 m experimental three phase cable. Eventually 200 m full size AC power cable with ratings 20 kV 1.5/2 kA and 50/70 MVA has been developed and successfully tested. The review of cable designs and test results is presented.

1. Introduction

Superconducting power cables made of Bi-2223 superconductors were the very first successful devices operated at liquid nitrogen temperatures. Some reviews about history and present state of power cables development could be found in Refs. 1, 2. The success in power cables development could be attributed to the good physics of Bi-2223 tapes. In parallel magnetic field their critical currents decay much slower than in perpendicular field.[3] In round superconducting power cables, dominant field is parallel to the surface of Bi-2223 tapes, that is why even in early developments high operating parameters were achieved.[1]

In Russia in 2005 the special program for R&D of HTS AC power cables has been initiated by Russian company United Energy Systems.[4] R&D were based on theoretical developments for design of HTS power cables.[5–7] The Program began with 5 m prototype cable[8] and followed by 30 m experimental three phase power cable.[9,10] Eventually, 200 m, three phase HTS power cable with ratings 1.5/2 kA–20 kV–50/70 MVA has been developed and successfully tested[11,12]. In this chapter, we present the review of major results of R&D work performed on Bi-2223-based AC power cables in Russia.

2. 5 m Prototypes and Witness Samples Tests

In the framework of Russian program for R&D HTS power cables, several 5 m cables were developed and tested. Tests were performed at specially developed test facility, including 5 m flexible cryostat, set of DC (up to 6.5 kA) and AC (up to 3.6 kA_{rms}) power supplies and many measurement devices (Fig. 1).[6,11,12] The test program developed included (but not exhausted) DC critical current measurements, current distribution measurements among superconducting layers in DC and AC conditions, AC loss measurements, etc.

The first full scaled in the cross-section model tested in 2006, we used AMSC Bi-2223 based Hermetic^{TM} tapes. We put to the cable the two layers with 24 Hermetic^{TM} tapes in each, 48 tapes in total.

Fig. 1. The test facility to study 5 m cable models and witness samples.

The total length of the superconducting part of the cable model was 5 m. Copper terminations were attached to each end of the cable *via* soldering to connect the cable with current leads (CLs).

For scientific purposes, the model was heavily instrumented. Each layer has been provided by sets of Rogowsky coils (RCs). Numbers of voltage taps were attached to each layer along layer's length and to CLs. In some cases, sets of Hall sensors were placed to measure all magnetic field components (axial, radial, and transversal) in different locations. Thermocouples could be attached to measure a possible temperature rise. All of above instrumentation permitted us to collect detailed information about a cable model behavior during all tests and to get the proper feedback for further technology development. We used similar instrumentation for all 5 m cable models and witness samples, both 1G and 2G developed and tested after this first prototype.[8,9,14]

The results of DC test of the first 5 m prototype are shown in Fig. 2. They are the typical results of all tests of 5 m cable models and witness samples. In Fig. 2, left, an example of the current run is

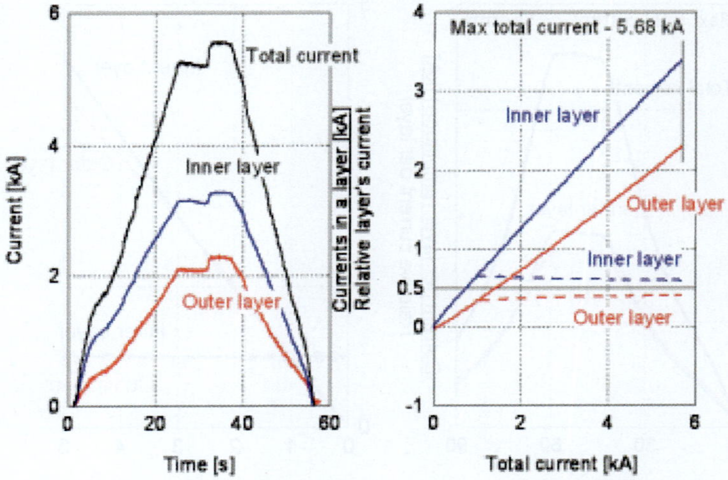

Fig. 2. The example of a current run up to ∼5.7 kA (left) and measured current distribution among layers in DC mode (right).[8] In the right graph: Solid lines- currents in layers, dashed lines- current in a layer divided by the total current.

shown and in Fig. 2, right, the current distribution among layers is shown.

Current distribution among layers is the important feature of power cables demonstrating if the full use of superconducting properties of basic tapes is implemented or not. In DC mode, it is determined by the difference in joint resistances, which lead to slightly more current in the inner layer for the model shown in Fig. 2. This is because the resistance between inner layer and termination is less than those for the second layer. The measuring of voltage–current characteristics in different points along the length of the model demonstrated that full use of current carrying capacity of tapes was achieved.[8]

In Fig. 3, the example of AC current ramp is shown (left) and measured current distribution in AC mode (right). One can see that in AC mode, current distribution among layers is much more uniform. The difference is less than ± 2.5%.

Unlike the DC mode, the AC mode current distribution among layers is determined by inductances of layers that in their turns are

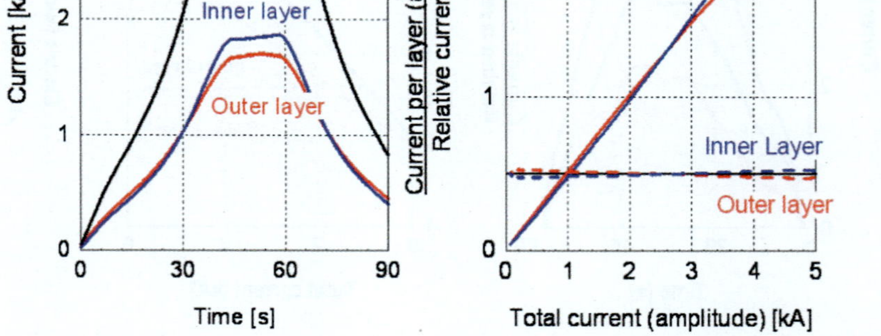

Fig. 3. An example of a current run up to ~3.5 kA_{rms} (left) and measured current distribution among layers in AC mode (right).[8] In the right graph- solid lines current in layers, dashed lines- current in a layer divided by the total current.

determined by twist pitches and directions. The twist pitches were calculated through the use of theories developed in VNIIKP.[5–7] During cabling, all measures were taken to strictly keep the prescribed twist pitches. The test results showed that all was done properly and AC current distribution was quite uniform.

The results shown in Figs. 2 and 3 demonstrate that designing and calculating methods and cabling technology are properly developed to provide the full use of superconducting properties of basic HTS tapes used for the first cable model in both AC and DC modes. These results are typical for all our developed and tested Bi-2223-based cables.[8,9,13,14]

AC losses are one of the most important operational parameter for all electro-technical HTS power devices, particularly for HTS power cables. AC losses are relatively small in superconducting devices, so their measurements are a difficult task. We used several methods to determine AC losses: Calorimetric by liquid nitrogen evaporation[8] and through the use of thermocouples for temperature measurements.[13] But these methods provided little accuracy. Thus, we used electrical method through digital measuring of current and voltage *versus* time and then through digital integration.[13,14] The accuracy of digital integration depends on the points measured per cycle. We usually used data acquisition from 400 to 1000 points per 50 Hz cycle. The accuracy of AC loss digital calculation from our measurements was ~10–20% with 1000 points per cycle. Example of AC loss measurements is shown in Fig. 4.

All tests of 5 m models and witness samples of Bi-2223 cables demonstrated good performance and full correspondence to the expected designed parameters.

3. 30 m, Three Phase Experimental HTS Cable

We considered the 30 m cable development mainly as a research project necessary to work through the technology and production of HTS power cables. That is why variations were used in basic Bi-2223 materials, cryostats and CLs for the cable design. The basic design of the cable is two layers of HTS tapes, cold insulation, copper, and

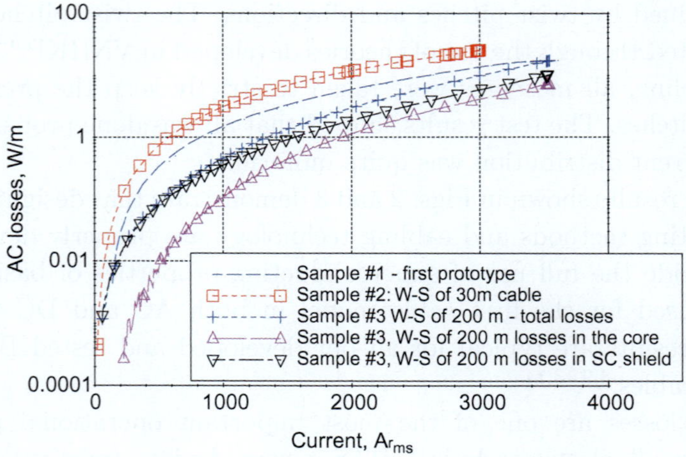

Fig. 4. AC losses *vs.* total r. m. s. current in 5 m cables made of 1G HTS wires. Sample #1 — first prototype[8], sample #2 — the witness sample from 30 m cable[9], sample #3 — witness sample from the 200 m cable.[13,14] Sample #3 was tested twice: With superconducting shield and without it and demonstrated different AC losses in two modes.

non-superconducting shield. The non-superconducting shield design has been accepted due to limited budget in this project. The sketch of the cable and its photo are shown in Fig. 5.[9,10]

For one phase, the CT-OP^{TM} tapes from Sumitomo Electric Industries, Ltd. were used. Two phases were made up of American Superconductor Corp.'s Hermetic^{TM} wires. Different tapes were used to verify the survival of superconducting properties of different HTS tapes during all technology routes for the power cable production with standard cabling machines.

The direction and twist pitches of HTS tapes were calculated by using computer codes for HTS power cables optimization developed in Cable Institute. The optimization methods are based on theories described in Refs. 5–7.

The 30 m flexible cryostats were produced by Nexans in Hannover (Germany). We used two of three cryostats with O.D. of 110 mm and one with O.D. of 92 mm to compare the heat losses in cryostats.[10]

Three pairs of CLs were developed and produced by two different Russian institutions. The CLs were developed for a 4 kA maximum

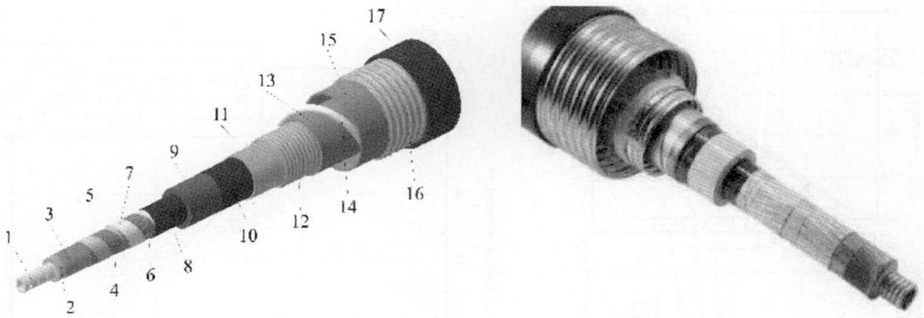

Fig. 5. The sketch of the cable phase design and the photo of the cable.[9,10] 1–3 — Former (SS spiral and copper wires); 4–7 — Two layers of Bi-2223 tapes, the HTS layers are separated by SS tapes; 8–10 — Insulation (standard cable paper and semiconductor paper); 11 — Copper shield (Cu tape); 12 — Cryostat–inner SS corrugated tube; 13–15 — Super-insulation; 16 — Outer SS corrugated tube; 17 — Protection PE cover.

current and 20 kV nominal voltages. All CLs were tested at cryogenic conditions at ~70 kV during 1 min.[10]

Three phases of the cable inserted into cryostats has been delivered and installed at the special test facility for tests of HTS power devices under full load in R&D Center at Federal Grid Company United Energy System.[9,10]

During 2008–2009, several test sessions were performed. In Fig. 6, the measuring methods and the dependence of critical currents of phases on temperature are presented. One can see that DC critical current of all phases at temperatures below 77 K is more than 4.2 kA.[9,10] It means that all Bi-2223 tapes used are intact after industrial manufacturing process.

The cable successfully passed high voltage DC test of 70 kV. The cable has been working for about 24 h with 30% overload current including ~1 h with 67% overload (2.5 kA). Several days tests at nominal load ~20 kV, 1500 A or ~50 MVA transporting power through three phases were successfully completed as well.[10]

The cable also successfully survived the overload current test with maximum overload current amplitude reaching 28 kA during ~0.4 sec, which is 10 times more than the nominal current amplitude.[10]

AC loss measurements demonstrated relatively high level of losses in a cable without superconducting shield: ~6 W/m at 1500 A and

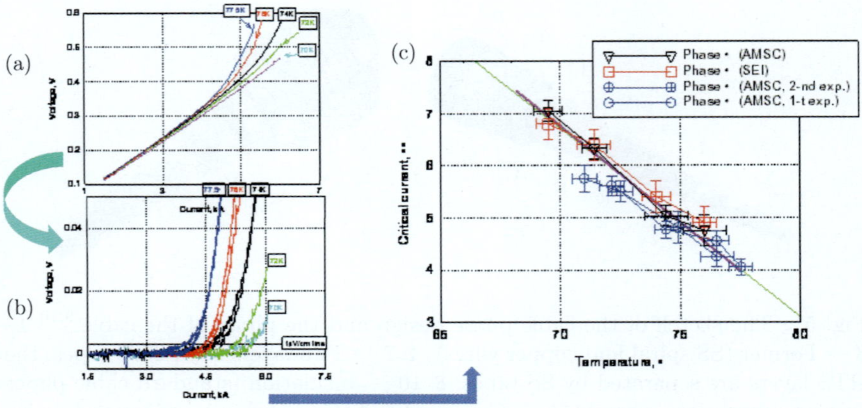

Fig. 6. (a) Measured voltage-current characteristics during DC critical current measurements; (b) Reduced voltage current characteristics during DC critical current measurements; (c) Critical current dependence on temperature for all three phases of 30 m Bi-2223-based power cable. Solid line- $I_c(T)$ dependence from producer's specifications.[17,18]

~12 W/m at 2000 A (see. Fig. 3).[9,14] To reduce AC losses, superconducting shield should be used in Bi-2223-based power cables, as was shown during our 200 m cable project.[11,12]

During all tests, the 30 m cable demonstrated good performance and all parameters corresponded to the designed one. The successful fulfillment of 30 m cable project paved the way for development of first Russian long Bi-2223-based power cable.

4. Full Size 200 m Bi-2223 AC Power Cable

After successful tests of 30 m experimental cable, the program to develop full size HTS power cable to be installed to a real grid has been started.[11,12,15] The cable nominal rating is the same as that of the 30 m cable: 20 kV–1500 A with possible 30% overload, i.e. 2000 A at 20 kV. This means 50–70 MVA of transmitted power. Many of the design features of this cable are similar to those of the 30 m cable described in Refs. 11, 12. The significant difference is that a superconducting shield is used in the 200 m cable. Other features are as follows: Basic wire: Sumitomo Electric DI-BSCCO^{TM} HT-CA

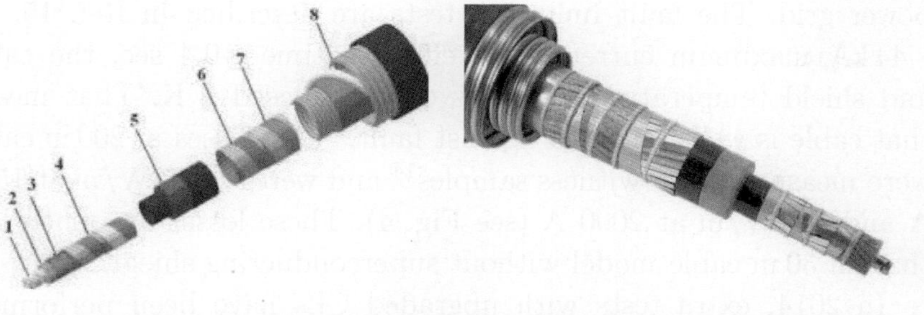

Fig. 7. 3D view of the cable (left) and the photo of the cable model (right). 1 — Former — a spiral of stainless steel, surrounded by a bundle of wires made up of copper and stainless steel, wrapped in copper tape — 2 and 3; 4 — Two layers of HTS tapes; 5 — high-voltage paper insulation; 6 — HTS shield; 7 — flexible copper strips to protect shield, wrapped by tapes of stainless steel; and 8 — flexible cryostat, 200 m long.

1G tapes with Cu-alloy lamination (Bi-2223-based); design: Central spiral, copper former, two layers of the cable core and one layer for the superconducting shield, copper protection for superconducting shield; high voltage insulation made from a standard cable paper; cryostats: Nexans flexible cryostats with 92 mm O.D., with PE protection cover extending up to the diameter of 102 mm.

3D view of the cable and the photo of the cable model are shown in Fig. 7. The cable has been made by same industrial technology as in previous 30 m experimental cable.[11,12] The test of the witness sample demonstrated full compliance with rated parameters; the critical current at 77.4 K was 4.2 kA, as designed. Current distribution between layers and AC losses were also measured in this sample.[14] In the AC mode, the difference between currents per layer was less than ± 0.5%.[14]

In December 2009, the acceptance test of the full length cable has been performed at the test facility at R&D Center at FGC United Energy System. During testing, we found that the cable has a cooling time of about 30 h. DC critical currents for all three phases are more than 5.2 kA at about 74 K. 50 kV DC voltage test has been passed successfully. Energy transfer of 50 MVA (1500 A and 20 kV for three phases) has been performed for a 72 h-period.

The 30% overload test, the fault current test, etc., are being performed to ensure the cable reliability before its installation in the power grid. The fault imitation tests are described in Ref. 15. At ~44 kA maximum current with effective time ~0.4 sec, the cable and shield temperature increases do not exceed 2 K. That means that cable is well protected against fault.[15] AC losses in 200 m cable were measured with witness samples[14] and were ~1.35 W/m at 1500 A and ~3 W/m at 2000 A (see Fig. 4). These losses are much less than in 30 m cable model without superconducting shield.

In 2014, extra tests with upgraded CLs have been performed. All parameters of the cable and CLs were confirmed the same as in 2009. Currently, the cable is stored and is expecting the decision to be installed at some substation in Moscow power grid.

5. Conclusions

In this chapter, we presented the results of Russian R&D program for Bi-2223-based superconducting power cables. Started with 5 m prototype cable, the Program followed with the 30 m experimental three phase power cable. Eventually, 200 m, three phase HTS power cable with ratings 1.5/2 kA–20 kV–50/70 MVA has been developed and successfully tested. To reduce AC losses, superconducting shield should be used in the cables. All cables survived high current overload and high voltage tests. The results of R&D works confirmed the good manufacturability and reliability of Bi-2223-based HTS wires and cables made of such wires. The wires could be used in industrial cabling machines without loss of superconducting properties.

Acknowledgments

This work is the review of big collective efforts and great teamwork and the authors have to acknowledge the great contribution of their co-authors and collaborators to this work: E. P. Volkov, A. V. Rychagov, K. A. Shutov, V. V. Zubko, A. A. Nosov, and many others from Russian Scientific R&D Cable Institute, R&D Center at Federal Grid Company United Energy System, Krzhizhanovsky Institute, Moscow Aviation Institute, and many others.

References

1. P. M. Grant, Superconducting lines for the transmission of large amounts of electrical power over great distances: Garwin–Matisoo revisited forty years later, *IEEE Trans. Appl. Supercond.* **17**(2), 1641–1647 (2007).
2. S. S. Fetisov, Superconducting cables for electrical energetic-state of works in the world and in Russia, *Kabeli I Provoda* **3**(346), 22–29 (2014) (in Russian).
3. V. E. Sytnikov, V. S. Vysotsky, *et al.*, 1-G *versus* 2-G-comparison from the practical standpoint for HTS power cables use, *J. Phys.: Conf. Ser.* **97**, 012058 (2008).
4. V. E. Sytnikov and V. S. Vysotsky, *Russian Development Program on HTS Power Cables*, available at: http://www.ewh.ieee.org/tc/csc/europe/news-forum/pdf/CR18. pdf (2011).
5. V. E. Sytnikov and V. S. Vysotsky, Basic principles for design of high temperature superconducting power cables, *Izvestiya Russian Acad. Sci.- Energ.* **1**, 89–107 (2008) (in Russian).
6. V. E. Sytnikov, P. I. Dolgosheev, G. G. Svalov, *et al.*, *Physica C* **310**, 357–362 (1998).
7. V. E. Sytnikov, N. V. Polyakova, V. S. Vysotsky, *Physica C* **401**, 47–56 (2004).
8. V. E. Sytnikov, V. S. Vysotsky, and A. V. Rychagov, *et al.*, The 5 m HTS power cable development and test, *IEEE Trans. Appl. Supercond.* **17**(2) 1684–1687 (2007).
9. V. E. Sytnikov, V. S. Vysotsky, and A. V. Rychagov, *et al.*, 30 m HTS power cable development and witness sample test, *IEEE Trans. Appl. Supercond.* **19**(3), 1702–1705 (2009).
10. V. E. Sytnikov, V. S. Vysotsky, and S. S. Fetisov, *et al.*, Cryogenic and electrical tests results of 30 m HTS power cable, *Adv. Cryog. Eng.: Trans. Cryog. Eng. Conf.- CEC 2009* **55**, 461–468 (2010).
11. V. E. Sytnikov, V. S. Vysotsky, and A. V. Rychagov *et al.*, Development and production of the power superconducting cable on the base of HTS technologies, *Kabeli I Provoda* **2**(321), 3–10 (2010) (in Russian).
12. E. P. Volkov, V. S. Vysotsky, and V. P. Firsov, First Russian long length HTS power cable, *Physica C* **482**, 87–91 (2012).
13. V. E. Sytnikov, K. A. Shutov, and N.V. Polyakova, *et al.*, The AC loss analysis in the 5 m HTS power cables, *IEEE Trans. Appl. Supercond.* **19**(3), 1706–1709 (2009).
14. V. S. Vysotsky, A. A. Nosov, S. S. Fetisov, and K. A. Shutov, AC loss and other researches with 5 m HTS model cables, *IEEE Trans. Appl. Supercond.* **21**(3), 1001–1004 (2011).
15. A. A. Nosov, S. S. Fetisov, and V. V. Zubko, *et al.*, Overload and high voltage tests of witness samples of 200 m HTS power cable, *Phys. Procedia* **36**, 1127–1130 (2012).

16. V. S. Vysotsky, V. E. Sytnikov, A. A. Nosov, *et al.*, AC loss of a model 5 m 2G HTS power cable using wires with NiW substrates, *J. Phys.: Conf. Ser.* **234**, 032061 (2010).
17. HTS brass hermetic wire. Available at: http://www.amsc.com/pdf/WFS_BHM_0509_A4_FINAL.pdf.
18. Sumitomo CT-OP BSCCO Wire (DI-BSCCO) Type HT. Available: http://global-sei.com/super/hts_e/type_ht.html.

Chapter 3.8

DC Cable for Data Center

Satarou Yamaguchi
Center of Applied Superconductivity and Sustainable Energy Research (CASER), Chubu University
Matsumoto-cho 1200, Kasugai, Aichi 487-8501, Japan
yamax@isc.chubu.ac.jp

The internet data center (iDC) is becoming one of the most important and widespread infrastructures in modern life, but it consumes electrical power plentifully. The reduction of its electric power consumption is an urgent subject. A Direct current (DC) power systems are now being applied. The DC high temperature superconducting (HTS) power cable is well-suited for such systems and is a suitable application in applied superconductivity. Subjects of the superconducting power cable system are discussed, especially about the cable design and the heat leakage of the current lead. Several ideas are proposed and discussed for the future iDC using HTS power cable and renewable energy (RE) system.

1. Introduction

The internet data center (iDC) is becoming one of the most important and common infrastructures in modern life. Almost all the information for modern life is exchanged, stored, and served for people, governments, companies and international organizations through iDCs. Centers with huge numbers of computer servers and data storage systems, connected by optical fiber systems are rapidly spreading all over the world. Information traffic through iDCs is increasing 20% annually and this trend will continue for the next decade. The "Internet of Things (IoT)" and "Big Data" are the new trends of technology integration. The iDC is the key locus of these technologies and will therefore be one of the most important infrastructures of the future. However, an iDC consumes a huge amount of electric power. In order to keep the data safety, iDCs are also built to duplicate data and other functions. Therefore two important subjects for the iDC, from the energy point of view, are how to supply the electric power to the iDC, and how to save power. These subjects have been discussed for the past 10 years. The direct current (DC) power system has been proposed to reduce the consumption of electric power.

The configurations of a DC powered iDC (DC-iDC) and an alternative current (AC) powered iDC (AC-iDC) are shown in Fig. 1. The loads are lighting loads such as LED, electronic loads such as Central Processing Units (CPUs) and motors for hard discs, and air conditioners. Since the present AC-iDC power system needs many power converters as shown in Fig. 1, the facility cost is high and the electric power loss of the power converter is much greater than that of the DC-iDC power system. A DC power system was proposed and was tested at the Sakura Internet Corp., Japan, from 2011 to 2014.

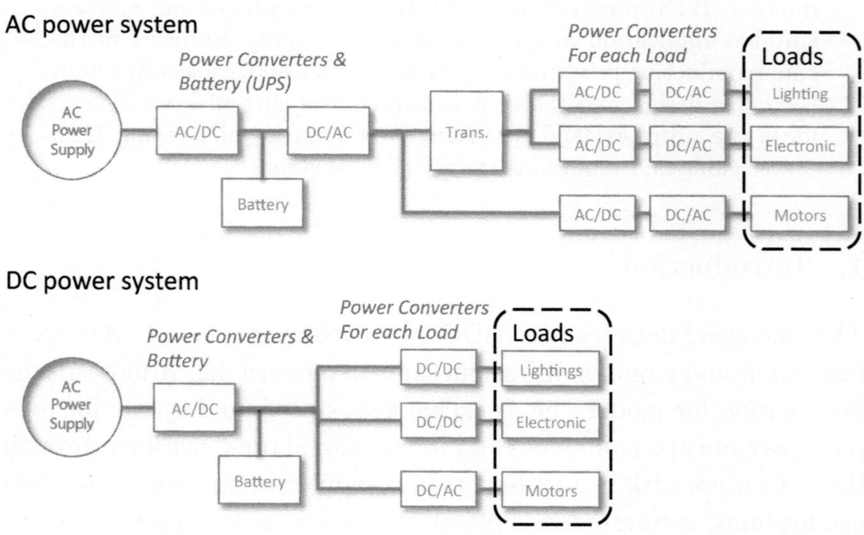

Fig. 1. Two schematic illustrations of iDC. AC power system is a major system in the present time, and it needs to use many power converters as compared with DC power system.

The result of the test is successful. The iDC power consumption was reduced by ~20%, and the instrument cost of the power supply was also reduced by ~20%. Therefore, the design is a viable commercial system, and can now be expanded to many iDCs. The voltage of the DC-iDC is around 380 V DC, and the total amount of current can be as large as 10 to 100 kA.

Superconducting technology has strong compatibility with a DC-iDC power system because a superconducting circuit is of low voltage and has large current system. We do not need to consider the AC power losses of the superconducting cable and can save the electric power for refrigeration.

In this Chapter the design of a future iDC is discussed. The characteristics and parameters of the iDC will be described in Sec. 2. The losses of the DC superconducting system depend mainly on heat leakage from the room temperature side. The structure of the cable and the cryogenic pipe are discussed in Sec. 3, together with the low heat leak current lead. Finally, I discuss the future of the iDC and my perspectives.

2. Characteristics and Parameters of an Internet Data Center

The DC electric power transmission line for GRID requires 100 kV to 800 kV in presently existing, non-superconducting copper cable systems. However, the superconducting transmission line does not need high voltage because the resistive loss of the cable is negligibly small. The present design parameter of the superconducting cable experiment is ~20 kV.[1] Unfortunately, this system is not yet suitable for commercial use. Although the DC voltage of an iDC is around 380 V, the DC power system is considered high voltage DC (HVDC) for several reasons.[1] The largest iDCs consume ~40 MW at

[1]Once telephone exchanger system is similar to the present iDC, and the common voltage of telephone exchanger system is DC48 V. But it was too low to operate iDC to save electric power because it consumes higher power much more, and high voltage system was needed and the DC380 V system is developed. The subjects

the present time, with a total current of ~100 kA. However, it is better that one superconducting DC cable in the DC-iDC should carry less than 5 kA because it must distribute electric power to many loads at low voltage. The insulation of the cable should also differ from the usual transmission line cable because of the low voltage. The cables should be easy to connect and their current density should be high because of the combination of many cables and cryostats and to reduce the heat leakage. The heat leak of the cryogenic pipe and the cryogen flow is not as simple as for the long transmission line. Thus, the structure of the cable can be significantly different if we find a good structure for the cable and its cryogenic pipe.

The cable length inside the iDC is shorter than the usual transmission line, around 200 m. Therefore the major loss of the superconducting system will be heat leakage from the current lead at the terminal cryostat. Since the electrode at the room temperature side is directly connected by copper wire to the electrode of the high temperature superconducting (HTS) tape at the cryogenic temperature, the superconducting cable system will not be economical and feasible if we cannot reduce the heat leak from the current lead. Therefore, the heat leakage of current leads are discussed in the next section.

3. R&D Subjects for an Internet Data Center

Research and development on many subjects will be required to realize an superconducting DC-iDC. The system of the DC cable and its cryogenic pipe is described here. The major power loss of a DC superconducting power cable is the heat load. If it is high, a large refrigerator is needed and may reduce the operating efficiency and increase the cost of the iDC. The major heat load is the heat leak from the current lead and the cryogenic pipe. For a conventional copper current lead, the heat leak is ~50 W/kA per lead. If the total current in the superconducting cable is 20 kA, its total heat leak is estimated to be 4 kW due to the four sets of current leads. The additional heat

to develop high voltage iDC are the dc switch and the electrification. It is long time to fix the voltage in the iDC.

leakage from the cryogenic pipe is assumed to ~1 W/m, which for a 200 m cable gives a total heat leak of 200 W. The remaining heat leaks and loads are estimated to be 50 W to 100 W. Therefore, the major heat leak comes from the current leads. In reference 2, the goal is a heat leak from the current lead of 25W/kA.[2] The cable and cryogenic pipe structure and the current lead are described below.

3.1. *Cable and cryogenic pipe structure*

Since the insulation voltage of the cable is lower than 1 kV and the total length of one cable is less than 200 m, the structure of the cable will be different from the power transmission cable in GRID. We proposed the stack conductor for a low voltage, large current transmission cable.[3] Similar structures have been proposed and tested by others.[4,5] A typical stacked conductor and its insulation are shown in Fig. 2. Since the current density of the stack conductor is ~10 times larger than the usual DC superconducting power cable[6] because it uses a thick electric insulator layers not needed. The small diameter of the cryogenic pipe has a low heat leakage rate, which reduces its cost significantly.

Two HTS tapes can be used to couple the conductors, as shown in Fig. 2, and this structure makes it easy to add terminals along the

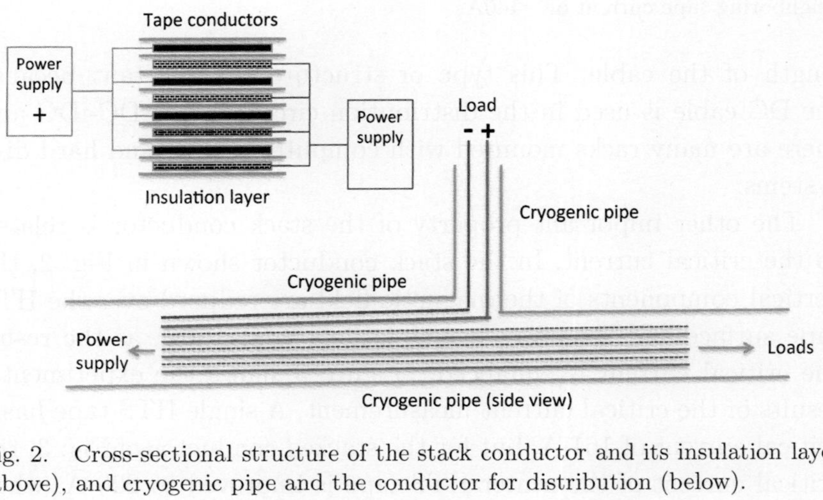

Fig. 2. Cross-sectional structure of the stack conductor and its insulation layer (above), and cryogenic pipe and the conductor for distribution (below).

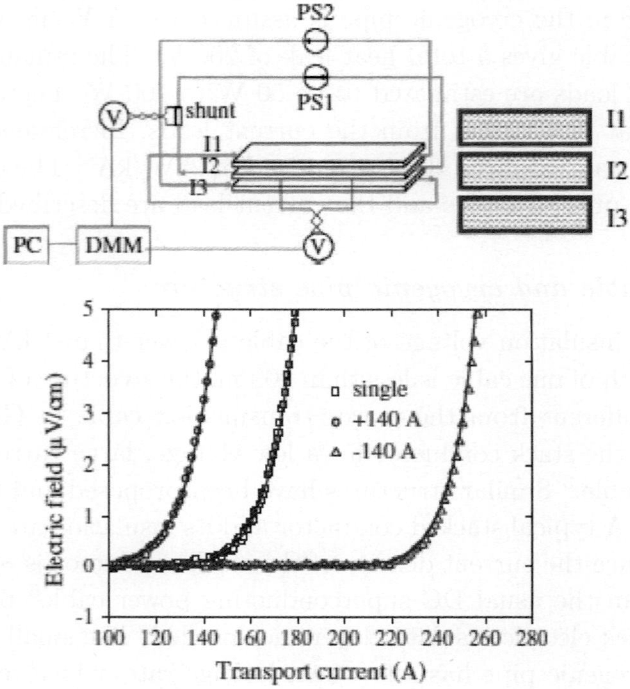

Fig. 3. Schematic set-up of the experiment to measure the critical current (above), and its experimental results for the critical current. In a single HTS tape the critical current is 161A, but here the critical current of the central tape is 220A when the neighboring tapes have currents of −140A, and +140A fora neighboring tape current of −140A.

length of the cable. This type of structure is important because the DC cable is used in the distribution circuit in the DC-iDC and there are many racks mounted with computer servers and hard disk systems.

The other important property of the stack conductor is related to the critical current. In the stack conductor shown in Fig. 2, the vertical components of the magnetic field are reduced over the HTS tape surface and the stray magnetic field is also low, as the result the critical current is enhanced.[7] Figure 3 shows the experimental results of the critical current measurement. A single HTS tape has a critical current of 161 A, but for the stacked conductor of Fig. 3, the critical current of the center HTS tape is improved to 220 A, when the current of the adjacent HTS tapes is −140 A (opposite direction to the center tape), and is a conductor. A high critical current density in the cable also provides a extra margin of safety for the short circuit current if the rated current can be kept low.

3.2. *Peltier current lead*

Reducing the heat leak from a current lead contradicts the natural behavior of the material because it requires a material that has high electrical conductivity and low thermal conductivity at the same time. Therefore, we seek a new mechanism that introduces a material that has different processes of heat and electrical transport. I proposed to use the Peltier effect, and it has been named the Peltier current lead (PCL).[8] The schematic structure and a typical temperature profile along the lead are shown in Fig. 4.[9]

The Peltier effect is a type of heat pump. The PCL is composed of *P*- and *N*-type thermoelectric semiconductors to make the pair of leads. Bismuth telluride alloy (BiTe) is used to make the PCL. Since the performance of the P-type and N-type BiTe is not so high to connect from room temperature to liquid nitrogen temperature, we use copper leads to connect with BiTe, as shown in Fig. 4. The COP of the Peltier material is not high, but since the COP of low

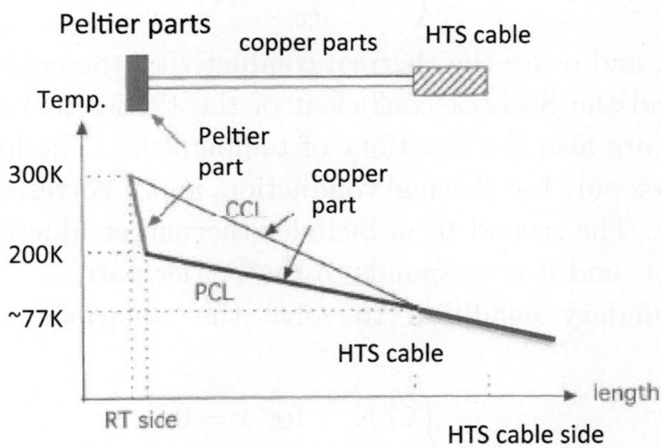

Fig. 4. Schematic structure of Peltier current lead (PCL) (above), and its temperature profile along the lead (lower). CCL is all copper current lead.

temperature refrigerator is lower, there is merit to using the PCL. Moreover, since the thermal conductivity of BiTe is ~0.3% of copper, the heat leak is lower even in the zero current phase.

To minimize the heat leakage, in the PCL we solve the one dimensional steady-state thermal conduction equation, given by

$$\frac{dQ}{dx} - fmC_P\frac{dT}{dx} + \eta\frac{I}{A}^2 = 0, \qquad (1)$$

where $Q, x, f, m, C_P, T, \eta, I, A$ are the heat flux, length along the lead, heat exchange ratio, mass flow rate of cryogen gas, constant pressure specific heat, temperature, electrical resistivity, current and cross-section of lead, respectively.

These parameters vary with temperature, so we must measure the parameters experimentally before solving Eq. (1) numerically. The solution of the equation shows an optimum ratio of the cross-section and length of the copper lead parts and the Peltier parts to minimize the heat leak per current. We first consider a system without gas-cooling of the current lead, then $f = 0$ and we can neglect the second term in Eq. (1).

The heat flux Q is defined as

$$Q = \begin{cases} -\kappa A\dfrac{dT}{dx} \\ -\kappa A\dfrac{dT}{dx} - \alpha T I \end{cases}, \qquad (2)$$

where κ, A, and α are the thermal conductivity, the cross-section of the lead, and the Seebeck coefficient of the Peltier material. These parameters are also the functions of temperature. The first term of Eq. (2) shows only the thermal conduction, and it corresponds to the copper part. The second term includes thermal conduction and the Peltier effect, and it corresponds to the Peltier part.

The boundary condition to solve the equations are usually given by

$$T = \begin{cases} 77\,\mathrm{K} & \text{for } x = 0, \\ 300\,\mathrm{K} & \text{for } x = 1. \end{cases} \qquad (3)$$

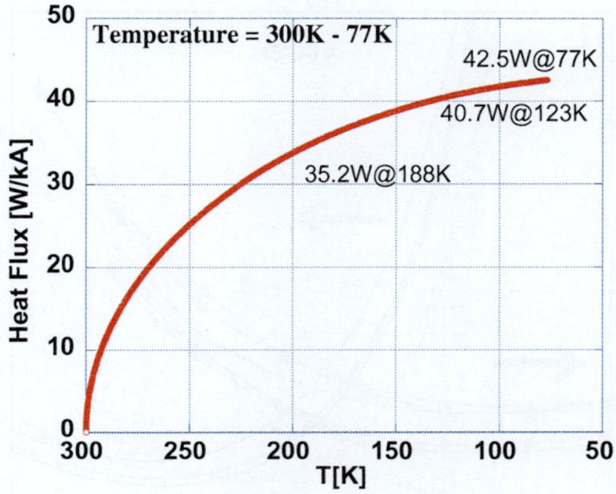

Fig. 5. Calculated heat flux of the copper lead for temperature.

One solution is shown in Fig. 5 for the heat flux in the copper current lead. The values of the heat flux at temperatures of 188 K, 123 K, and 77 K are shown in the figure.

PCLs were made using various kinds of Peltier materials and shapes for the copper leads and were tested in the laboratory. Figure 6 shows the heat leak and heat leak per current for different PCLs. The horizontal axis is current. We found there can be an optimum current to minimize heat leakage per current. The heat leak tends to increase with current, but its behavior depends on the design and the material characteristics of the PCL system.

The minimum value of the heat leak per current is ~30 W/kA at currents of around 100 A and 140 A. The temperature of PCL on the room temperature side increases with current because the PCL is acts like a heat pump that pumps the heat to the hot side. This property can be used to reduce the heat leak in future PCL designs especially for high figure of merit of Peltier materials in low temperature.

3.3. *Multi-stage current lead*

The idea of a multi-stage current lead is suggested by the temperature dependence of the COP in refrigerators. The COP of a refrigerator depends on the Carnot cycle efficiency and it is low for a refrigerator that operates at low temperature. If we can use a high

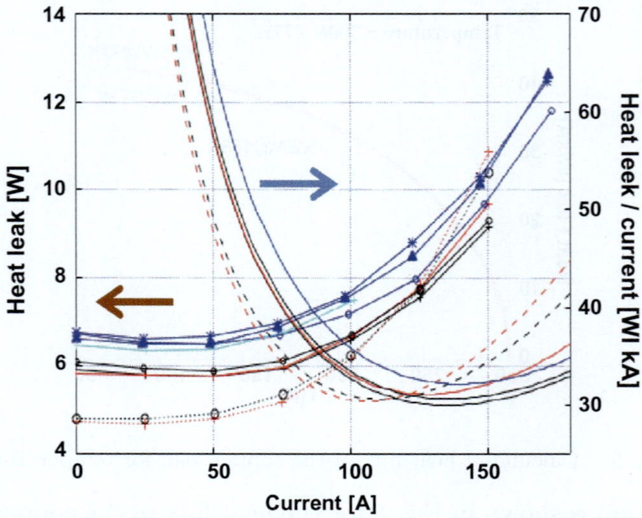

Fig. 6. Experimental heat leakage and heat leakage per current of different PCL designs as a function of current.

temperature refrigerator to keep the superconductor at low temperature and/or to reduce the heat leak of the superconductor system, we can save on the electric power consumption of the total system.

The multi-stage current lead uses two different temperature refrigerators.[11,12] Figure 7 shows its schematic structure. One refrigerator, labeled Ref. 1, keeps the superconductors at the temperature of 77 K. The second refrigerator, Ref. 2, has an outlet temperature of 188 K and cools the current lead only. The heat leak from 300 K to 188 K is 35.2 W for 100 A, and therefore the heat leak from 188 K to 77 K is 7.3 W. If the mechanical efficiency is the same in both refrigerators, the ratio of the COPs of the 77 K and 188 K refrigerators is 4.86. Therefore the ratio of the electric power consumption of two stage refrigeration system compared to the single refrigerator system is 0.342, a savings of 65% in electrical power consumption.

Another option for this type of design is to use liquid natural gas (LNG). The temperature of LNG is around 121 K, but usually its cold temperature is not used at the present time. Some of iDCs use fuel cells (FCs) to supply electric power and natural gas is the fuel of the FC. If we can use LNG to cool the current lead, we do not need

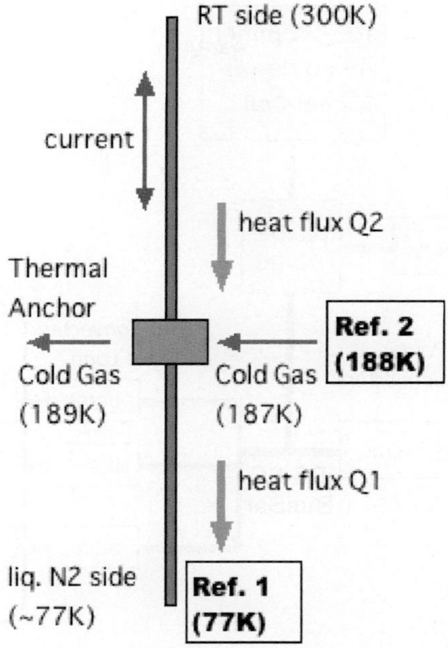

Fig. 7. Schematic structure of a two stage current lead. Outlet temperature of refrigerator 1 (Ref. 1) is 77K, and the other (Ref. 2) is 188 K.

to use a second refrigerator (Ref. 2), and we can save ~95% of the electric power. Moreover, the LNG can be used to cool the internet equipment in the iDC.

4. New Proposals for iDC Power System

An iDC needs a huge amount of electrical power for its operation and the cost of electricity is important. Fortunately, renewable energy systems (RE), such as photovoltaic cells (PV) and wind power generators are spreading and their electricity cost is falling. For example, many people insist that electricity from PV is the cheapest from 2014 in California, USA. However, the output power of RE is not stable, and we need to use some instruments to stabilize the electric power. Usually, a second battery system is used to stabilize the output power from the RE. The iDC has a huge secondary battery system to supply electric power because of safety concerns, but during most times

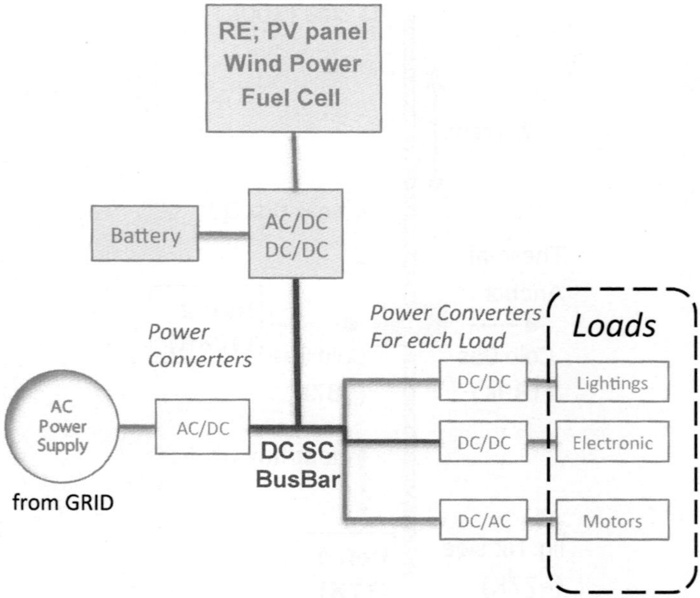

Fig. 8. A proposed electric power supply system including the RE and its stabilize battery system for iDC.

it is not used. If some of the secondary battery can be used to stabilize the electric power generated by the RE, it would become a good combination with the iDC.

An experiment using this type of operation with an RE will be started in 2015 in Japan. One of the proposed system is shown in Fig. 8. In order to get maximum power from the RE, an electric power converter system is needed to connect the RE to the battery. The DC HTSC power supply system is a good candidate for this system, and therefore DC-iDC will be major trend in technology near future.

5. Conclusions and Outlook

The superconducting power transmission cable is not used at the present time in iDCs. But it becomes feasible once the heat leakage from the current lead can be reduced below 25 W/kA and/or a comparable current lead, such as the multi-stage current lead and/or an improved PCL, are realized. Thus, once we succeed in establishing the technology to save electricity in iDCs, as discussed in this chapter, the superconducting power cable system including power converter, RE and backup system, is likely to be used widely, especially for DC-iDCs.

Acknowledgments

This work was supported partially by the Ministry of Economy, Trade and Industry (METI), and the Ministry of Education, Culture, Sport, Science and Technology (MEXT), Japan. The author would like to thank the members of Center of Applied Superconductivity and Sustainable Energy Research (CASER) at Chubu University and to acknowledge Prof. A. Iiyoshi, the Board Chairperson of Chubu University, Dr. Hiroshi Fujiwara, the President of Broad Band Tower (BBT), Japan and Mr. Kunihiro Tanaka, the President of Sakura Internet Corporation, Japan for their continuous support. He also thanks Prof. B. Coppi, Drs. L. Sugiyama and J. Minervini for their useful discussions in MIT.

References

1. S. Yamaguchi, T. Kawahara, M. Hamabe, H. Watanabe, Y. Ivanov, J. Sun, and A. Iiyoshi, *Physica C*, **471**, 1300 (2011).
2. Navigant Consulting, Costing predictions of Superconducting components in 2008–2012.
3. J. Sun, S. Yamauchi, H. Watanabe, M. Hamabe, T. Kawahara, and S. Yamaguchi, *Physics Procedia*, **36**, 1290 (2012).
4. M. Takayasu, L. Chiesa, L. Bromberg, and J. V. Mernivini, *IEEE Applied Supercond.* **17**, 3121 (2007).
5. A. Ballarino, L. Martini, S. Mathot, T. Taylor, and R. Brambilla, *IEEE Applied Supercond.* **21**, 2340 (2011).
6. S. Yamaguchi, S. Hamabe, I. Yamamoto, T. Famakinwa, A. Sasaki, A. Iiyoshi, J. Schultz, J. Minervini, T. Hoshino, Y. Ishiguro, and K. Kawamura, *J. Physics Conf. Series*, **97**, 012290 (2008).
7. J. Sun, H. Watanabe, M. Hamabe, N. Yamamoto, T. Kawahara, and S. Yamaguchi, *Physica C*, **494**, 297 (2013).
8. S. Yamaguchi, K. Takita, O. Motojima, Proc. 16th Int. Cryogenic Conf and Int. Cryogenic Mat. Conf., p. 1159 (1996).

9. S. Yamaguchi, Y. Hasegawa, H. Okumura, and K. Sato, *Rev. Sci. Instr.* **75**, 207 (2004).
10. S. Miyata, Y. Yoshiwara, H. Watanabe, K. Yamauchi, K. Makino, and S. Yamaguchi, 12th European Conference of Applied Superconductivity, **2A-LS-P-07.04**, 2015.
11. S. Yamaguchi, M. Emoto, T. Kawahara, M. Hamabe, H. Watanabe, Y. Ivanov, J. Sun, N. Yamamoto, and A. Iiyoshi, *Physics Procedia*, **27**, 448 (2012).
12. L. Bromberg, P. C. Michael, J. V. Minervini, and C. Miles, *AIP Conf. Proc.* **1218**, 577–584 (2010).

Chapter 3.9

DC Cables for Metropolitan Networks: St. Petersburg Project

Victor Sytnikov

R&D Center@FGC UES, Moscow, Russia

Superconducting cables have evident benefits when transmitting large power flows through electrical networks in comparison with traditional ones. However, the use of DC superconducting cable lines together with converters brings additional advantages like reduction of losses in cables and suitable lowering of refrigerating plant capacity, as well as the realization of the function of short-circuit currents limitation by means of the appropriate setting of converter equipment. The presence of the current limitation function makes it possible to use such lines for the connection of megalopolises' power network sectors at the medium-voltage side without increasing the short-circuit currents.

1. Introduction

The efficiency of electricity generation and its transportation to the consumer are becoming the priorities in the development of power industry in the 21st century. Simultaneously, the demands for environmental and resource-saving parameters in all phases of production and distribution of electricity are increasing. Meeting the growing demands is only possible by using the most progressive and advanced technologies, including superconductivity. These problems are appearing most acutely in megalopolises.

The load growth in metropolitan power systems makes it necessary to put into operation new generating and network equipment. At the same time in order to improve the reliability of consumer power supply the reservation of substations is implemented (new transmission lines are built and brought into service) that occurs in multiple connection network creation, so called an extensive network. These trends contribute to higher levels of short circuit currents in power systems. To reduce the levels of short circuit currents to safe values, current-limiting reactors and network sectioning are used. However current-limiting reactor application occurs in additional power losses and network sectioning occurs in the reduction of consumer power supply reliability. Moreover network development (based on XLPE cables) increases the complexity of power flow controllability and may lead to transmission lines overcurrent. HTS DC cable line commissioning (instead of XLPE cables with current-limiting reactors) is delivered from these drawbacks. When using HTS DC cable line, a new quality of transmission system is acquired. Cable line becomes an active controlled element of the network that is capable of power flow regulation and fault current limitation.

2. HTS Links between Substations in Metropolitan Areas

Power energy system of megalopolises is a dynamically developing structure, which has the following features:

- Rapid growth of energy consumption that, in general, exceeds the increase of consumption throughout the country.
- High density of energy consumption.
- Areas' deficiency and branching of distribution networks.
- Partitioning the urban area electrical network to reduce short-circuit currents.

All these factors ultimately determine the main problems in power grids of metropolitan areas, namely:

- A high level of power losses in distribution networks.
- High levels of short-circuit currents that in some cases exceed the breaking capacity.

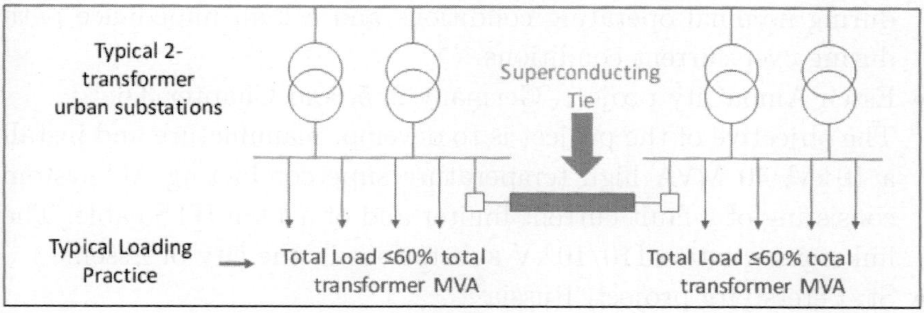

Fig. 1. Substation interconnection concept.

- Low level of controllability.

On the other hand, the loading of substations in the city is very uneven. In many cases, substation transformers are loaded only 40–60%. As a rule, the input substations in the cities feed from separate high voltage feeders. The connection of substations on the side of the medium voltage would ensure mutual redundancy of power network sectors, release backup transformer capacity that will ultimately lead to reduction of energy losses in the network. In addition, this type of connections, allow release capacity to connect additional load without additional transformers or new substations and power transmission lines.[1,2,3] Fig. 1, reproduced from EPRI,[1] illustrates the concept.

However, the direct connection of substations will lead to a substantial increase in short circuit current. There are three superconducting cable projects purposed to connect two substations without increasing the short-circuit currents.

- Project HYDRA, New York, USA[1,4]
  The project purpose is to install an HTS AC cable system in New York City that would provide a high capacity link (96 MVA) on the secondary side of the transformers in nearby urban substations. The cable system would have an inherent ability to limit fault current magnitude by using 2G conductors that are specially designed to quickly transition from the superconducting state to the resistive state, thereby providing a low impedance current path

during nominal operating conditions and a high impedance path during over current conditions.

- Essen AmpaCity project, Germany 2, 5, and Chapter 3.6.
  The objective of the project is to develop, manufacture and install a 10 kV, 40 MVA high temperature superconducting AC system consisting of a fault current limiter and of a 1 km HTS cable. The link connects two 110/10 kV substations in the city of Essen.
- St. Petersburg project, Russia[3,6]
  The intention of the project is to enhance the reliability of electric power supply and to limit fault current in metropolitan network of St. Petersburg. The project provides the installation of HTS DC cable lines between two downtown substations on the medium voltage side. Length of the line will be 2.5 km and the transmission power will be 50 MW.

All three projects have the objective of high power transmissions, at distribution voltage between the two substations, while limiting short circuit currents. Project HYDRA combines the function of transmitting a large capacity and current limitation in one device — superconducting cable of special design. This makes it extremely difficult to optimize the cable with regard to possible modes of the network, the conditions of cooling, and cable routing. Essen project divides these functions between the cable and the current limiter. This simplifies the task of designing each device; however, it requires coordination of their characteristics. The St. Petersburg project divides the two functions between the HTS DC power cable and conversion system which is adjusted accordingly. Unlike the above two projects, DC line is an active network element here and allows to manage power flows in adjacent lines in a wide range up to power reverse.

3. St. Petersburg Project

Investigations of the grid in more than ten scenarios showed that in the central region of the city there is the potential for a number of post fault system oscillation modes that result in power supply interruption. It was further shown that strengthening the system

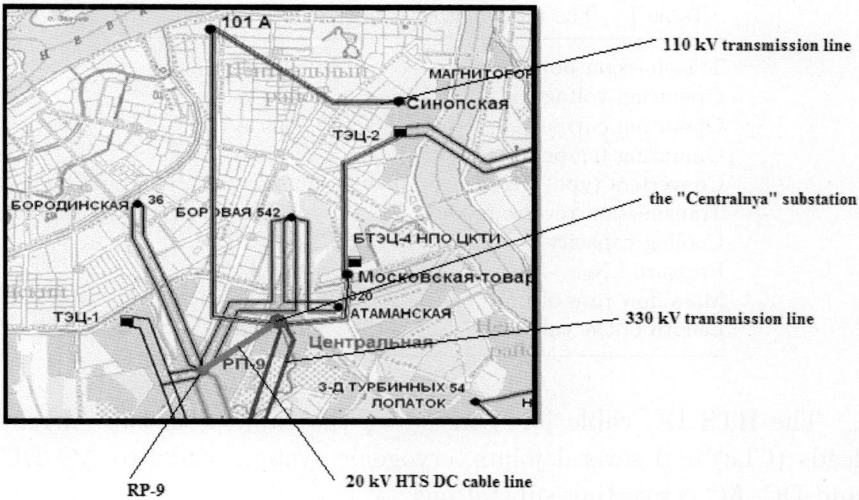

Fig. 2. Connection of HTSC DC cable line in St. Petersburg power system.

with a conventional XLPE cable or gas-insulated line (GIL) between two downtown substations worsened the post-fault modes because of additional current overloads in adjacent transmission lines and substations. The HTS DC cable option, with AC–DC converters, avoided these scenarios and furthermore strengthened the system.[7]

The Central part of St. Petersburg was selected as the project site.[3,6] HTS DC line will connect 330 kV substation "Tsentralnaya" and 220 kV substation "RP-9". The connection of these substations will create conditions for "deliverance" of reserve power network capacity in the power system to connect new consumers to the electric networks of the central district of St. Petersburg. Additionally, this link will increase the reliability of power supply to the existing consumers. Investigations have shown that the connection of two substations will increase the manageability of network and will not lead to the increase of short circuit currents.[3]

In the future, the creation of MV ring in metropolitan areas will significantly increase the reliability of power supply to consumers. Figure 2 shows the connection of HTSC DC cable line in the power system. The main characteristics of the cable line are presented in Table 1.

Table 1. The main characteristics of HTS DC CL.

Transmission power	50 MW
Operating voltage	20 kV
Operating current	2500 A
Operating temperature	65–80 K
Converters type	Twelve pulsed converters
Transmission reverse mode	Exists
Cooling capacity	12 kW@70 K
Pressure LN_2	Up to 1.4 MPa
Mass flow rate of LN_2	0.1–0.6 kg/s
Length of the cable line	2500 m

The HTS DC cable line consists of the cable with two current leads (CLs) and several joints, cryogenic system, and two AC–DC and DC–AC converting substations.

Russian Federal Grid Company financed the project, and its 100% affiliate R&D Center is the general contractor. One of the aims of the project is the creation of an effective scientific and industrial cooperation of various organizations which will be able to use the results obtained for other object of electric power engineering.[3]

4. HTS Cable

As a basic design, a unipolar cable was chosen with coaxial layout of the direct and reverse conductors. The direct conductor operates at 20 kV and the reverse conductor is grounded at one side. The electromagnetic field of this cable is only between two conductors, and the stray fields are negligible. The absence of stray fields and the use of the liquid nitrogen as a cooling agent make these cables environmental friendly and fire-safe. The cable consists of concentric layers containing the following elements:

— Former and stabilizing element.
— Superconducting central wire — the forward conductor (D = 19.8 mm).
— High voltage insulation.
— Superconducting reverse conductor (D = 31.8 mm).
— External stabilizer.

— External (screening) insulation.
— Electric (non-superconducting) screen.
— Cryostat (ID/OD: 64 mm/110 mm) consisting of two corrugated tubes with the vacuum thermal insulation and protecting layer.

The estimated external cable diameter is 39 mm. As a basic superconducting material for the cable, the Bi-2223 HTS tape produced by SEI (Japan), type HT-CA is used. Forward conductor include 22 tapes with $I_c = 160$ A, placed in two layers. Reverse conductor include 19 tapes with $I_c = 180$ A, placed in one layer. Directions and twist pitches of superconducting layers were designed from the ground up to ensure a uniform distribution of currents between the two layers of forward conductor.[8,9] The cable schematic diagram is shown in Fig. 3.

The cabling technology was developed for the serial cable works "Irkutskkable". Some special equipment and tools were made for the cable manufacture. Two 30 m cable samples were made for the preliminary testing and cabling technology verification. Full scale cable length in six sections was made after successful test of two 30 m samples. Cables were inserted into cryostats in the same cable works. All cable sections with a total length of about 2500 m were delivered to the test site of R&D Center@FGC UES, Moscow.

5. CLs and Joints

The CL box consists of pair of brass CLs in individual insulators, stainless steel double-wall cryostat evacuated to 10^{-3} Torr, high pressure N_2 vessel and two cryogen interfaces — for flexible cable cryostat connection and for the sub-cooled nitrogen inlet (Fig. 4).[10] The thickness of flanges was calculated in the model of circular plate with uniformly distributed load. The side flanges have the OD of 466 mm and at pressure of 0.1 MPa, its minimum thickness appeared to be equal to 5 mm corresponding to mechanical stress up to 130 MPa at the flange edge. The maximum deflection in the center of the flange during the evacuation is 2 mm. The N_2 vessel is much more complicated junction as it has to withstand an overpressure of sub-cooled

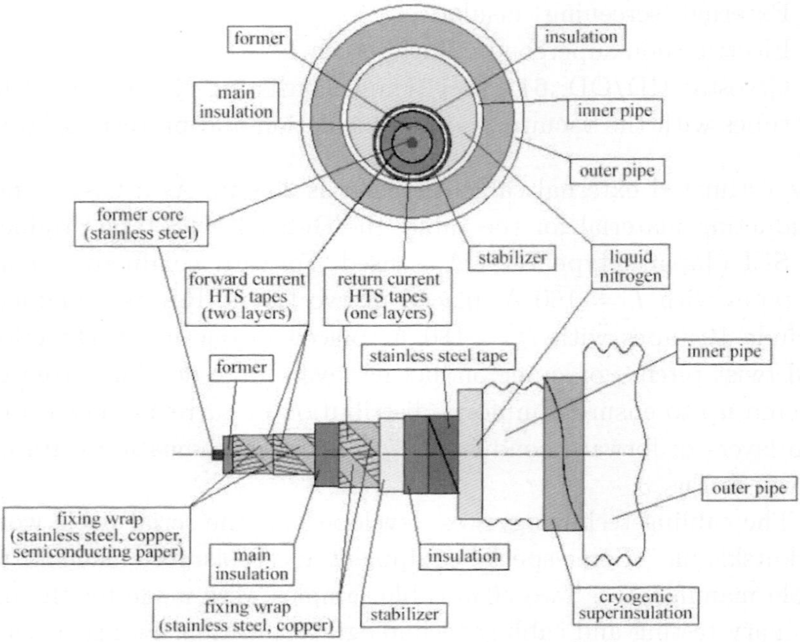

Fig. 3. Schematic diagram of HTS DC cable.

nitrogen up to 1.4 MPa. To decrease the mechanical load in this junction, an additional reinforcing ring was used. The same method was used for reinforcement of N_2 vessel in joint box.

After assembling, the N_2 vessel was subjected to hydraulic tests. Room temperature water was pumped into the N_2 vessel up to pressure of 1.5 MPa (for 30 minutes). After pressure tests, the vessel remained hermetic. It is well known that during the transition to lower temperatures, steel is strengthened; (77 K = 370 MPa, 300 K = 170 MPa). So, the design of high pressure vessel has some safety margin.

The CLs were made up of brass rods. Effective length is 0.45 m. Lateral surface is thermal insulated. An optimal cross section was calculated[10] for the working current of 2.5 kA, which appeared to be equal to 16 cm^2 (OD 46 mm).

The calculated dependencies of the temperature *vs.* the CLs' length are presented in Fig. 5 for three regimes: Without current,

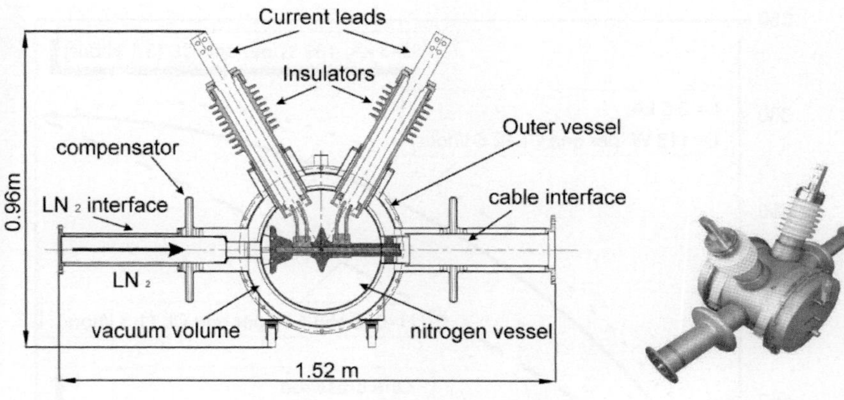

Fig. 4. CLs: Schematic diagram and photo.

on rated current of 2.5 kA, and in overloaded regime with $I = 3\,\text{kA}$. The liquid nitrogen evaporation velocities due to the thermal flux through the CLs are also given there. Joint box was designed for the connection of separate cable segments and for monitoring the cable joint properties. It is a cylindrical stainless steel vessel with a double-wall for high vacuum insulation[10,11]. The inner wall was covered with several layers of thermal insulation to decrease the radiation. The box is equipped with vacuum, pumping, and thermometry gauges.

6. Cable and Cable Fitting Test

Preliminary test was carried out by cryogenic scheme shown in Fig. 6. Voltage taps were placed on each conductors and joints in cold area.

The main goals of the first test were: Checking of cabling technology and measurement of electrical resistance and vacuum tightness of CLs and coupling box. Current source "plus" pole was connected to the direct conductor and "minus" pole to the reverse conductor on one side of the test line. CLs on another side of the line were shortened. The outlet temperature of the experiments was about 80 K. Electrical resistances of the CLs and joints remained stable in all range of currents. The resistance of two conductors in the coupling box was less than 1 μOhm. Resistance to all four CLs was within 20–22 μOhm that corresponds to electrical heat generation not more

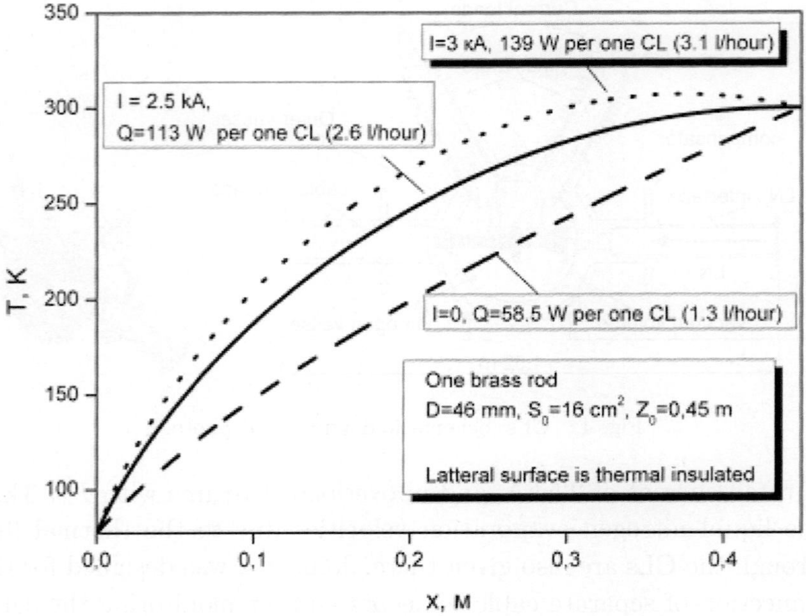

Fig. 5. Temperature distribution along the brass CL and heat leakage into the cold zone of box (through one CL).

than 140 W at the rated current. The critical current of HTS conductors was within 3420–3550 A.[6] The critical current of the cable is practically equal to the sum of critical currents of used superconducting tapes that indicates the development of a reliable cabling technology.

The voltage taps were removed from the cold zone for the next testing stage and 60 m line was cooled in circulation mode through Stirling cryogenic machine. The $V-I$ curves of the line at different temperatures are shown in the Fig. 7(a). Voltage taps were placed in warm ends of the CLs.

As it can be seen, resistance of the CLs remains stable up to current 5000 A and resistance varies between 37–39 μOhm (for two CLs) depending upon temperature. Critical current dependence upon liquid nitrogen input and output temperature is shown in Fig. 7(b). The full scale testing was made on two 430 m cables with two CLs

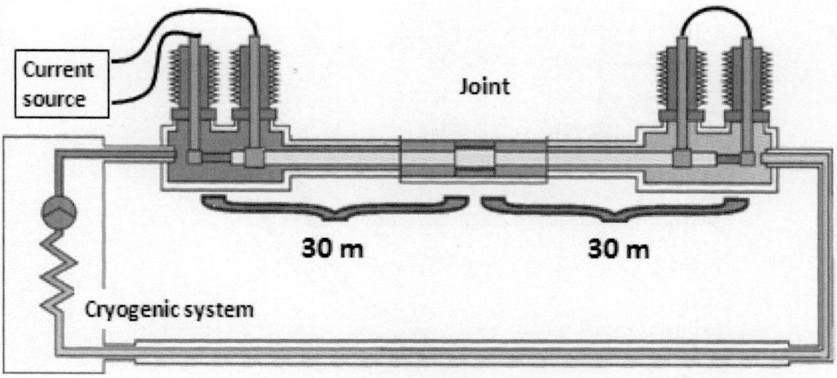

Fig. 6. Cryogenic scheme for testing 60 m line.

and one joint. Common view of testing cables and $V-I$ curve for 860 m line is shown in Fig. 8.

The mark "star" on the Fig. 7(b) corresponds to the critical current of the 860 meters line; defined as a beginning of $V-I$ curve deviation from linear part. Comparison of $V-I$ curves with Figures 7 and 8 at 78 K shows their identity, which testifies to high quality manufacturing of long length cables.

All cables withstood high-voltage testing of rectified voltage of 50 kV for primary isolation and 10 kV for the screen isolation during 30 min.

7. Converter Stations for HTS DC Cable Line

For our HTS DC cable line, twelve-pulse line commutated converter technology based on thyristor valves should be used. Both are connected to converter transformers 63 MVA 110/8.27/8.27 kV with winding group Y/Y0/D11. Direction of power flow can be changed from +50 MW to −50 MW by changing a polarity of converter voltage.

On DC site of converter, LC filters for 12th harmonic 600 Hz have to be connected. Converter is equipped with water cooling system. Heat generation of converters will be used for heating of substation

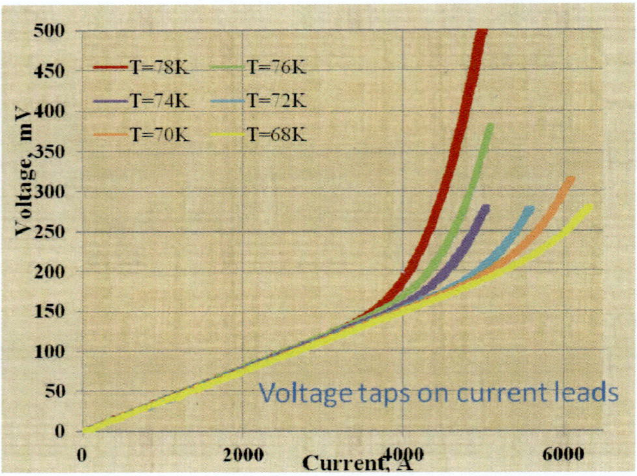

(a)

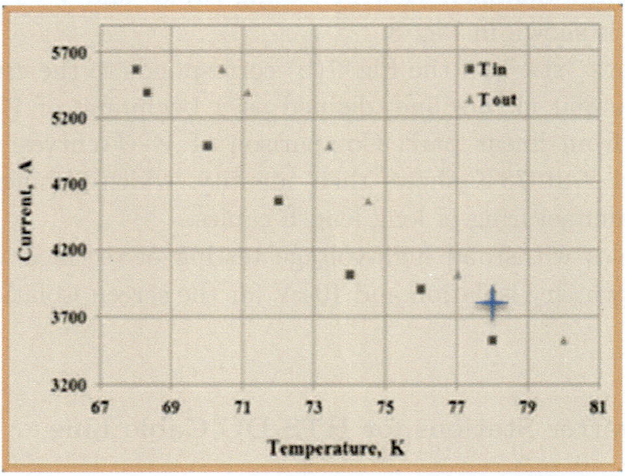

(b)

Fig. 7. $V - I$ curves of the 60 m line at different temperatures (a) and critical current dependence upon input and output temperatures (b).

buildings, using a heat pump system. The control system of converter has capability of over current protection, converter unit faults protection, capacitor unbalance protection, and commutation failure protection, etc.

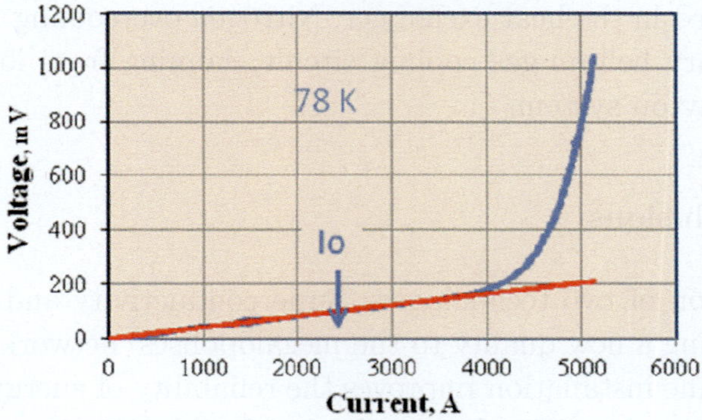

Fig. 8. Common view of testing cables and $V - I$ curve for 860 m line.

8. Outline of Cryogenic Cooling System

The cooling of the HTS cable is performed by the liquid nitrogen pumping through the space between the outer cable surface and internal cryostat surface. Cooling problem is typical for long cable lines associated with increasing the value of heat inflows and also the heat dissipation due to friction, with length. As a result of calculating the pressure drop and temperature gap in the flexible cryostat with the cable placed inside,[12] the next accepted parameters were $\Delta\ Tmax \leq 12\,\mathrm{K}$ over the pipe length 5.0 km and $\Delta\,Tmax \leq 6\,\mathrm{K}$ over the cable length (2.5 km). The operating parameters given above may be reached when using the Nexans cryostats with the ID of the corrugated tube 64 mm and the external tube diameter 110 mm. Closed loop cryogenic machine has been developed to provide the required parameters. The nitrogen circulating over the cryostat with the cable is sub-cooled in the heat exchanger. Nitrogen overcooling is provided by secondary helium gas cooling circuit, running from low pressure Turbo–Brayton system.

9. Conclusions

Combination of two technologies: Superconductivity and DC transmission bring a new quality to the megalopolises' network. The HTS DC cable line installation improves the reliability of energy supply to the consumers by mutual redundancy of grid sectors and enhancement of controllability. Along with this, it does not increase short circuit currents.

HTS tape Bi-2223 is currently the best material for the manufacture of DC cable due to the high current-carrying capacity and acceptable mechanical characteristics that allow its processing on the cable equipment.

St. Petersburg Project is carried out in accordance with the schedule. All units of equipment have been developed and will be produced in the year 2015. The introduction of this HTS DC CL into the St. Petersburg power system will allow checking up of the basic technical solutions and get an experience for the commercial application. It will be first step toward further building of circular DC electric power chain in megalopolises.

References

1. EPRI, *Superconducting Power Equipment Technology Watch 2012*. Palo Alto, CA, USA, 1024190 (2012).
2. M. Stemmle, F. Merschel, and M. Noe, *Phys. Procedia* **36**, 884 (2012).
3. V. E. Sytnikov *et al.*, *IEEE Trans. Appl. Supercond.* **23**, 5401904 (2013).
4. J. Maguire *et al.*, *IEEE Trans. Appl. Supercond.* **19**(3), 1740 (2009).
5. M. Stemmle *et al.*, *Proc. IEEE Int. Conf. Appl. Supercond. Electromag. Devices*. Beijing, China, ID3122, October (2013).
6. S. I. Kopylov *et al.*, *J. Phys.: Conf. Ser.* **507**, 032047 (2014).
7. V. Sytnikov, Superconductivity news forum, Category 6, Ref. STP 377 (2013).
8. V. E. Sytnikov and P. I. Dolgosheev, *Inst. Phys. Conf. Ser.* **158**, 1199.
9. V. E. Sytnikov *et al.*, *Physica C* **310**, 357 (1998).

10. A. V. Naumov and V. E. Sytnykov, *et al.*, *Adv. Cryog. Eng.* **1573**, 1692 (2014).
11. V. V. Lobyntsev, D. N. Durakov, V. E. Sytnikov, and A. S. Dinisilov, *ASC Charlotte Report* **3**, $LP_0$2L-01 (2014).
12. Y. Ivanov, M. Romashov, and V. Sytnikov, *Adv. Cryog. Eng.* **1573**, 887 (2014).

Chapter 3.10

The Development and Demonstration of a 360 m/10 kA HTS DC Power Cable

Liye Xiao
Institute of Electrical Engineering,
Chinese Academy of Sciences
P. O. Box 2703, Beijing 100190, China

With the quick development of renewable energy, it is expected that the electric power from renewable energy would be the dominant one for the future power grid. Due to the specialty of the renewable energy, the HVDC power transmission would be very useful for the transmission of electric power from renewable energy. DC power cable made of High T_c Superconductor (HTS) would be a possible alternative for the construction of HVDC power transmission system. In this chapter, we report the development and demonstration of a 360 m/10 kA HTS DC power cable and the test results.

1. Introduction

With the growing pressure on the environment by traditional energy use, it is necessary that the mankind should increase the use of the renewable energy and decrease the use of fossil energy gradually. This significant change on the energy structure will have a revolutionary impact on the future power grid, leading to critical challenges to the development of future power grid.[1] In order to deal with the challenges from the renewable energy and make full uses of the spatial and temporal complementarities of the renewable energy in wide area, it has been suggested that the DC power grid should be developed instead of the current AC power grid,[2] because the DC power grid has many advantages over the AC system, such as that the DC transmission grid is not subjected to the stability problem, and its transmission distance and power are not restricted by the synchronous operation of the power system; and the VSC-HVDC system is easier and quicker to regulate and control the power and current of DC transmission lines.

In China, most of the power sources are located at the west and northwest of the country while most of the load centers are located in the south and east. Therefore, the possible DC transmission power grid would transport large-capacity of energy for long distances, and thus leading to a potential application for the superconducting applications for DC power cable in the future, because the superconducting power cable is very low in losses, high in efficiency, and compact in structure.

In last few years, High T_c Superconductor (HTS) DC power cables have been developed in Japan, Korea, Russian, and China. For instance, a group headed by Prof. Yamaguchi at Chubu University, Japan, has successfully fabricated and demonstrated a HTS DC power cable with length of 200 m, rated voltage of ±10 kV and rated current of 2 kA in 2010.[3] In Russia, a 2.5 km long HTS DC cable of 20 kV/50 MW was planned to connect two electric power substations and now is still under production in 2015 in St. Petersburg.[4] In this chapter, we will report the development and demonstration of a 360 m/10 kA HTS DC power cable in China.

2. The Development of the 360 m/10 kA HTS DC Power Cable[5,6]

In order to demonstrate the possible application of a superconductor in DC power transmission, a 360 m/10 kA HTS DC power cable project was started in 2007. This project is supported by China Ministry of Science and Technology (MOST), Chinese Academy of Sciences (CAS) and the Zhongfu Group, an aluminum production company, and sponsored by the Institute of Electrical Engineering, CAS.

Table 1. Main parameters of Bi-2223 tapes for the conductor.

Manufacturer	Tapes Specifications
InnoST, China (Standard HTS Tape)	Width: 4.2–4.4 mm; Thickness: 0.26–0.28 mm; Critical tensile stress (77 K): 100 MPa; Minimum bending radius: 30 mm; Critical current (77 K, self field): 90 A; Piece length: 400 m (with one joint at maximum)
SEI, Japan (DI-BSCCO Tape)	Width: 4.2–4.8 mm; Thickness: 0.32–0.40 mm; Critical tensile stress (77 K): 250 MPa; Minimum bending radius: 30 mm; Critical current (77 K, self field): 150 A; Piece length: 400 m (with one joint at maximum)

2.1. *The cable conductor*

The cable conductor consists of the former, Bi-2223 tapes, and the inter-layer insulation tapes. In the design of this cable conductor, different Bi-2223 tapes from China's Innova Superconductor Technology Company (InnoST) and Japan's Sumitomo Electric Industries (SEI) are employed. Table 1 lists some parameters of the Bi-2223 tapes. The cable conductor consists of five layers of Bi-2223 tapes, in which one layer is fabricated with the Bi-2223 tapes designed by InnoST, and four layers are fabricated with the Bi-2223 tapes designed by SEI. The inner and outer diameters of the cable conductor are 41.0 mm and 46.2 mm, respectively. The design of the cable conductor is listed in Table 2.

2.2. *The cryogenic envelope and insulation*

The cryogenic envelope for the cable conductor consists of two concentric corrugated stainless steel tubes. The super-insulating materials are placed between the corrugated tubes, reducing the heat losses. Table 3 lists the design parameters of the cryogenic envelope. Test shows that heat loss of the cryogenic envelope is about 1.0 W/m.

Table 2. Design parameters of the cable conductors.

Items	Value
Outer diameter of the former (mm)	41.0
Critical current of the HTS tapes used (A)	150/90
Minimum tensile stress of the HTS tapes (MPa)	250/100
Outer diameter of the conductor (mm)	46.2
Layer of conductors	5
Pieces of HTS tapes	115
Total length of HTS tapes used (km)	46

Table 3. Design parameters of the cryogenic envelope.

Parameter	Cryogenic Envelope
Inner diameter (mm)	62
Outer diameter without steel textile (mm)	131
Outer diameter with steel textile (mm)	133
Outer diameter with insulation (mm)	151
Operational pressure with (bar)	1–5
Minimum bending radius (m)	1.2
Operational temperature (K)	70–77
Vacuum (pa)	$\leq 10^{-2}$

In order to fabricate the cryogenic envelope whose length reaches up to 360 meters, eight segments are fabricated at first, and then the segments are joined while each joint is not placed at the bending part of the cable. Joint between two segments is made by means of a dual vacuum layer insertion, and the heat loss of the joint is 3W. This joint technique for cryogenic envelope would be useful in the development of long distance HTS cable.

The cable is fabricated in a warm-dielectric type, and the insulation of the segments is made by the same technique as the conventional power cable, making the insulation of the HTS cable easier.

2.3. *The terminator and current lead*

The terminator of the cable consists of a main body, a current lead (CL), and a chamber which are designed and manufactured separately, so that it can be easily integrated. For the CL, the ratio of length to cross section is 318.5 m^{-1} and the length is 2.9 m, this design can reduce the heat losses of lead to 43 W/kA.

3. Installation and Test

3.1. *The refrigeration system*

The HTS cable is cooled by a sub-cooled liquid nitrogen (LN_2) system in which the LN_2 can be cooled down to 70 K. In order to obtain the temperature difference between 70 K and 77 K, the pressure of the system is kept in the range of 1–5 bar, and then the backward LN_2 is cooled down by a Sterling refrigerator whose cooling power is 4 kW at 77 K and about 3.3 kW at 70 K. Figure 1 shows the refrigeration system for the power cable. In order to keep the consistent operation of the cable when the refrigerator breaks down, an open-loop cooling system *via* evaporation of LN_2 is adopted as backup.

3.2. *The installation of the power cable*

Actually in the power grid, there is not suitable DC power transmission line which can transport DC current as high as 10 kA. In order to test the HTS DC power cable for operation in large transport

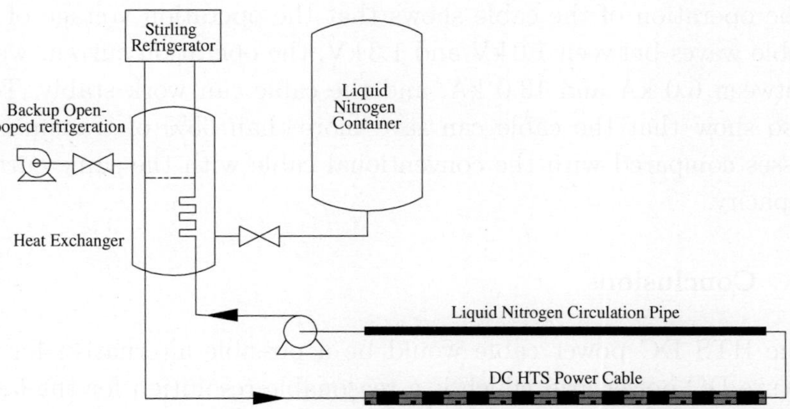

Fig. 1. Refrigeration system of the power cable.

Fig. 2. A view of the 360 m/10 kA HTS DC cable after installation.

current, the power cable is installed at Henan Zhongfu Industrial Co. Ltd., which is a company producing aluminum. The HTS power cable is then used to connect the substation and the bus bar of an aluminum electrolyzing workshop, serving for the power supply of the factory together with the conventional current conductor. A view of the cable after installation is shown in Fig. 2.

In order to test the bending performance of the cable, the HTS DC cable was bent at nine separate locations during installation, and three of them are in the vertical direction and six of them are in the horizontal direction, and the minimum bending radius is 3 m.

3.3. *Test of the power cable*

After installation, the cable was tested at 77 K. Tests show that the critical current of the cable by 1 μVm/cm was more than 12.5 kA. The operation of the cable shows that the operation voltage of the cable waves between 1.0 kV and 1.3 kV, the operation current waves between 6.0 kA and 13.0 kA, and the cable can work stably. Tests also show that the cable can save more than 65% of transmission losses compared with the conventional cable with the same current capacity.

4. Conclusions

The HTS DC power cable would be a possible alternative for the future DC power grid which is a reasonable resolution for the large-scale acceptance of renewable energy power. In order to develop the HTS DC power cable, a 360 m/10 kA HTS DC power cable has been fabricated and tested, and it has been successfully demonstrated at aluminum factory. The test also shows the cable can work stably when the current and voltage waves are optimized, and it can save about 65% of transmission losses compared with the traditional power cable of the same capacity.

References

1. L. Xiao and L. Lin, *Adv. Tech. Electr. Eng. Energy* **30**, 56 (2011).
2. L. Xiao, L. Lin, and Y. Liu, *Energies* **4**, 1025 (2011).
3. S. Yamaguchi, Y. Ivanov, J. Sun, H. Watanabe, M. Hamabe, T. Kawahara, A. Liyoshi, M. Sugino, and H. Yamada, *Phys. Procedia* **36**, 1131 (2012).
4. V. E. Sytnikov, S. E. Bermert, S. I. Kopylov, M. A. Romashov, T. V. Ryabin, and Y. G. Shakaryan, *IEEE Trans. Appl. Supercond.* **24**, 2373814 (2014).
5. L. Xiao, S. Dai, and L. Lin *et al.*, *IEEE Trans. Appl. Supercond.* **22**, 5800404 (2012).
6. S. Dai, L. Xiao, and H. Zhang *et al.*, *IEEE Trans. Appl. Supercond.* **24**, 5400104 (2014).

Chapter 3.11

DC Cable for Railway

Masaru Tomita
Railway Technical Research Institute, Japan
tomita@rtri.or.jp

The development of a superconducting cable for railways has commenced, assuming that a DC transmission cable will be used for electric trains. The cable has been fabricated based on the results of current testing of a superconducting wire, and various evaluation tests have been performed to determine the characteristics of the cable. A superconducting transmission cable having zero electrical resistance and suitable for railway use is expected to enhance regeneration efficiency, reduce power losses, achieve load leveling and integration of sub-stations, and reduce rail potential.

1. Introduction

The substation interval in the DC electric railway differs depending on the electricity generation system, railroad track conditions, electric current, driving conditions, power supply, etc. The voltage fluctuates according to the electric current drawn off because there is impedance for the line feeder to send electricity to the train.[1] A ministerial ordinance on railway technology prescribes "To secure proper movement of trains, the voltage of the overhead trolley must be kept sufficiently high," so the voltage drop of the overhead trolley must remain within the permitted range. For example, for a DC 1500 V electricity generation system, the voltage must not fall below 1000 V. Therefore, the substation interval is typically 10 to 15 km at the line connecting cities and 3 to 5 km in a large metropolitan area which

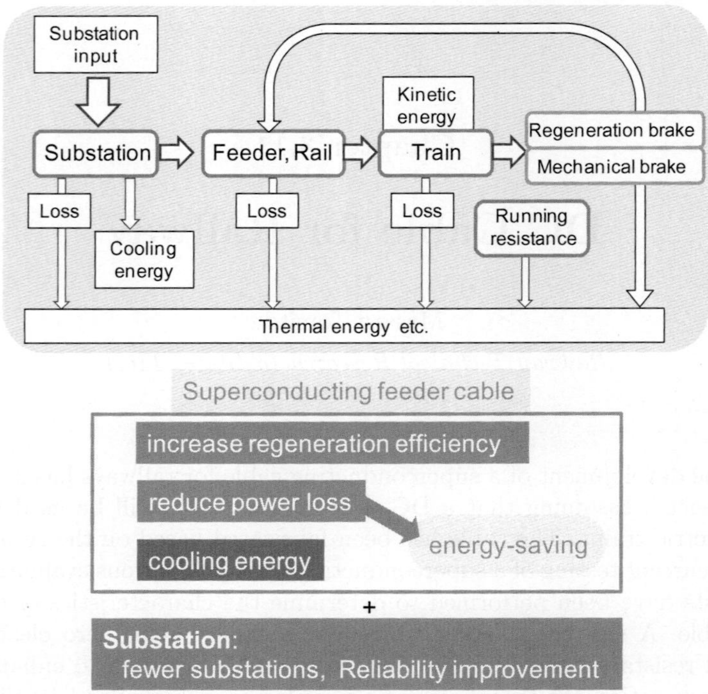

Fig. 1. Benefits of superconducting feeder cable.

requires an enormous electric current capacity and the installation of many substations.

We assume that a superconducting cable is applied to a railway feeder system, and are fabricating a prototype superconducting cable by using superconducting tapes and performing characteristic evaluation tests. A superconducting transmission cable having zero electrical resistance and suitable for railway use is expected to enhance regeneration efficiency, reduce power losses,[2] achieve load leveling and integration of sub-stations, and reduce rail potential (Fig. 1). It will be bidirectional; by flowing a retrace current into the superconducting cable, the leak magnetic field can be reduced. The leakage of current from the rail to earth also disappears, the rail electric potential can be reduced, and the problem of electrolytic corrosion can be resolved.

The characteristic evaluation results of a material required when introducing superconducting cable into railways are described. Based on the results, a superconducting cable was set up at the testing track of the Railway Technical Research Institute. After confirming its soundness from the results of cooling and current tests, a train running test was conducted on a test line.

2. Development of Superconducting Feeder Cable

2.1. *Superconducting material for railway feeding cable*

2.1.1. *Bending test on superconducting tape material*

The superconducting feeder cable was made by wrapping superconducting tape, protective copper sheet, insulating paper, etc. around a former in a spiral pattern. The superconducting tape is subjected to mechanical stress as it is wrapped around the former, which can degrade the tape's current properties. Therefore, to evaluate the allowable bending properties, bending stress was applied to the material of the superconducting tape, as shown in Fig. 2.

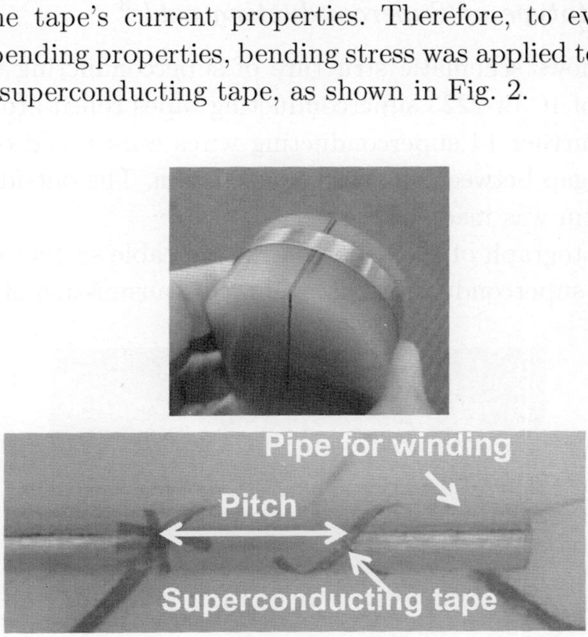

Fig. 2. Bending test on superconducting tape material.

With the superconducting tape subjected to spiral bending stress and dipped in liquid nitrogen, critical current (Ic) were measured. The allowable bending radius of the material was thus determined, which was then used as a reference when winding the material into a cable.

2.1.2. *Overcurrent test on superconducting tape material*

In the case of a short circuit or other event, the current capacity of the superconducting feeder cable can be far exceeded. If the current in a superconducting tape exceeds its critical current, part of the current branches out to the silver and copper around the tape, but the heat generated by the current may eventually burn the tape. To evaluate the overcurrent properties, the superconducting tape material was subjected to typical rail short-circuit currents, and the results were used to design the allowable current levels for the superconducting feeder cable (Fig. 3).

2.2. *Prototype of Superconducting cable*[3]

Figure 4 shows schematic structure of superconducting cable. It is composed of 10 Bi-2223 superconducting tapes rolled around a copper core. Further 14 superconducting wires were rolled onto the 10 wires. The gap between the wires was 0.5 mm. The outside diameter of the system was made of Former.

The photograph of the superconducting cable system is shown in Fig. 5. The superconducting cable allowed transmission of 1.5 kA DC

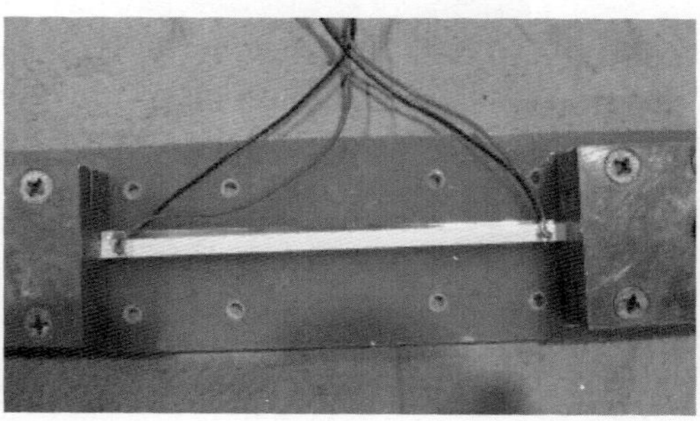

Fig. 3. Overcurrent test on superconducting tape material.

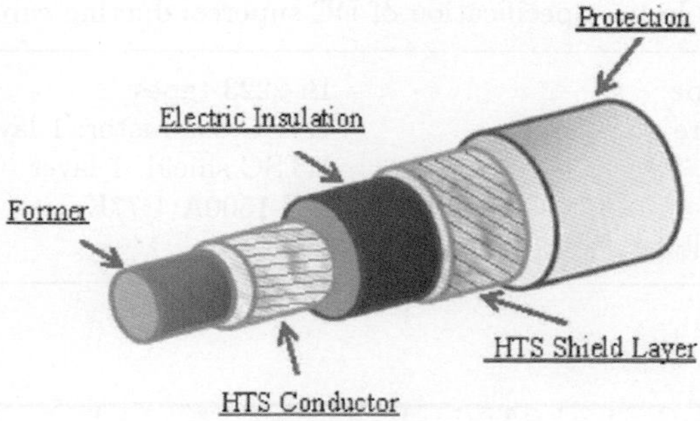

Fig. 4. Structure of superconducting cable.

Fig. 5. Prototype 2 m long Bi-2223 superconducting cable for railway system.

current under the cooling by liquid nitrogen at the ambient atmospheric pressure as shown in Table 1.

When using only the inner sheet wire, the critical current (defined by the criterion of 1 uV/cm) was 1720 A as shown in Fig. 6. In the outer sheet wire the current reached 2430 A. In this design, the nominal Ic of each superconducting wire at liquid nitrogen temperature was 160 A. Therefore, the total nominal current carrying capacities of the inner and outer layers were 1600 A and 2240 A, respectively.

The experimental data confirmed these nominal values and showed even higher currents, namely 172 A and 173 A per each wire

Table 1. Specification of DC superconducting cable.

Tape	Bi-2223 tapes
Core	HTSC conductor: 1 layer
	HTSC shield: 1 layer
Ic	DC 1500A @77K
Voltage	1500 V

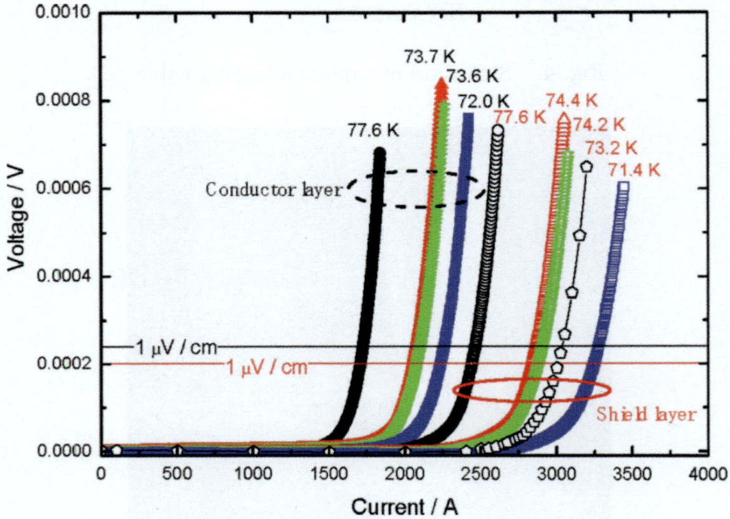

Fig. 6. Measurement result of critical current Ic of 2 m long Bi-2223 superconductor tapes at liquid nitrogen temperature.

of the inner and outer sheets, respectively. The influence of magnetic self-field was negligible. Moreover, then n-values were very high, 17 and 18, respectively.

In Japan, the railway tracks are very close to the residential areas. Therefore, the effect of leaked magnetic field on the society should be evaluated. For example, due to leaked magnetic field picture of electron microscope may get distorted, the image of TV screens may appear blurring, or electrocardiogram data may get disturbed. The stray field may also affect patients with heart stimulator. The magnetic field profile around the superconducting cable was checked using a Hall probe. The Hall probe was attached to a measuring arm that

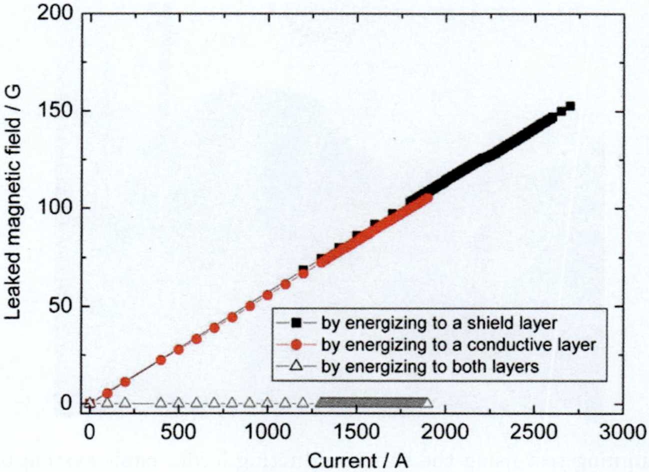

Fig. 7. Leaked magnetic field result of shield layer and conductive layer at liquid nitrogen temperature.

could move in three directions of translation and rotation around the axis. Eventually, the point by point field mapping was done. First, we applied the DC current of 0–1800 A at 77.3 K in the seal layer of the superconducting cable and mapped the leaked magnetic field at 45 mm radial distance around the cable. A similar experiment was repeated for the conductor layer. Finally, we sent the current through the both layers in the opposite direction and measured the leaked magnetic field value at the same, 45 mm radial distance. The leaking magnetic field results are presented in Fig. 7. It is clear that the individual layers show a negligible leaked magnetic field. When the same current was sent through both layers in opposite directions, the leaked magnetic field was zero. The results suggest that the DC superconducting cable can be used for the railway system without any hazard for public.

2.3. *Verification of superconducting feeder cable system*

A superconducting cable was introduced to the test line of the Railway Technical Research Institute, based upon the performance

Fig. 8. Running test using the superconducting feeder cable system on the test track at the Railway Technical Research Institute.

characteristics of cable system determined thus far.[4] The cooling test and the current test were examined using the superconducting cable and, we succeeded in the train running test using the cable. Additionally, in parallel in the evaluation trials, actual line simulations have been progressively verified.[5] We will aim to clarify the results of the demonstration trials to determine the consequences of installing the superconducting cable, all with the aim of realizing a cable applicable for railway use.

References

1. S. Hase, *Railway Research Review* **60**, 22 (2003). (in Japanese)
2. H. Ohsaki *et al.*, Superconducting power cable application in DC electric railway systems, *IEEE Trans. Appl. Supercond.* **23**, 3600705 (2013).
3. M. Tomita *et al.*, Next generation of prototype direct current superconducting cable for railway system, *J. Appl. Phys.* **109**(6), 63909 (2011).
4. M. Tomita *et al.*, Research and development of superconducting cables for railway system, *Abstracts CSSJ Conf.* **87**, 20 (2013). (in Japanese)
5. M. Tomita *et al.*, Development of superconducting feeder cable system for railway, *Abstracts CSSJ Conf.* **90**, 25 (2014). (in Japanese)

Chapter 3.12

Economical Aspects of Superconducting Cable

Masayoshi Ohya

Sumitomo Electric Industries, Ltd.
1-1-3 Shimaya, Konohana-ku, Osaka, Japan

High-temperature superconducting (HTS) cables are expected to resolve technical problems with power grids because they put large-capacity, low-loss power transmission into a compact package. One problem is replacing old 275-kV oil filled (OF) cables with cross-linked polyethylene insulated vinyl sheath cables (XLPE cables). This is difficult because XLPE cable has a lower transmission capacity than OF cable. In addition, the high concentration of public infrastructure underground makes it extremely difficult to build new ones. However, if 66-kV HTS cables can be installed inside existing underground conduits and can achieve a power capacity equivalent to conventional 275-kV cables, construction costs could be significantly reduced. Moreover, if XLPE cables are used for a 1,000 MVA-class transmission line, then three circuits of nine 275-kV single-core cables would be required, which would incur a transmission loss of 90 W/m/cct. Three circuits of three 66-kV Three-in-One HTS cables, however, with an AC loss of 1 W/m/ph@3 kA, heat invasion of 2 W/m, and cooling system efficiency of 0.1, would reduce transmission loss to less than three-fifths that of XLPE cables.

1. Introduction

Installing high-voltage power transmission systems has been widely promoted in Japan over the last 40 years since their full-fledged

Type	Specifications	Photo
OF Cable	Fluid-filled cable Oil-impregnated paper insulation with the oil under pressure	
XLPE cable	Dry-type cable Cross-linked polyethylene insulation	
HTS Cable	Fluid-filled cable LN2-impregnated paper insulation with the forcibly circulated LN2 under pressure	

Fig. 1. Conventional and HTS cables.

introduction in the 1970s, in concert with large-scale power source development. Although underground oil filled (OF) cables (see Fig. 1) are widely used to supply electric power to urban areas, cables installed during the years of rapid economic growth have been in service for more than 30 years now and are functionally deteriorating due to the degradation of various components.[1] Enormous civil engineering and construction spending is required to replace these old underground power cables with cross-linked polyethylene insulated vinyl sheath cables (XLPE cables) (see Fig. 1) and to install new conduit lines or cable tunnels to increase transmission capacity in preparation for future increases in power demand.

Installing high-temperature superconducting (HTS) cables is drawing attention as a way to solve this problem economically. HTS cables have many advantages (see Fig. 2).[2] First, they can carry more power than conventional cables of the same diameter, one way to sharply reduce construction costs.[3] Second, transmission loss from HTS cables is much lower than conventional cables, making HTS cables more economical. HTS cables also have environmental benefits. They are EMI-free [the superconducting shield layer prevents

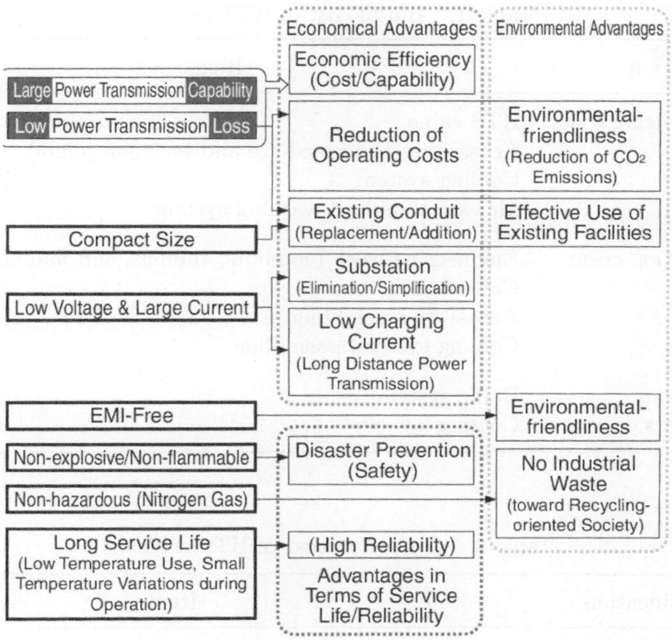

Fig. 2. Economical and environmental advantages of HTS cables.

electromagnetic (EM) wave leaks], more environmental friendly (less transmission loss leads less CO_2 emission into the atmosphere) and non-flammable (the oil in OF cables is flammable).[4] Discussions of the costs of HTS cables usually focus on two major issues. One is that an HTS cable system can sometimes be *less expensive* than a conventional cable system; the second is that an HTS cable can sometimes be *more expensive*. In the former case, when the economical advantage of HTS cables is clear they are used in place of conventional cables. However, in the latter case, it is necessary to consider both the operational and environmental costs.

2. Economic Efficiency of HTS System

2.1. *Initial cost*

Table 1 shows the items included in the initial cost of an HTS cable system. There is no substantial difference in these items without

Table 1. Initial cost of HTS cables.

Classification	Items
Manufacturing costs	HTS cable Accessories (intermediate and terminal joints) Cooling system Measurement and warning system
Construction costs	Engineering work (conduits, tunnels, and manholes, etc.) Cable laying Accessory assembling Cooling system assembling
Qualification test costs	Cable system Cooling system

Table 2. Operation cost of HTS cables.

Classification	Items
Power transmission loss	Power transmission loss × electric power unit price
Cable maintenance	Vacuum maintenance
Cooling system maintenance	Refrigerator maintenance Circulation pump maintenance, etc.
Miscellaneous costs	Measurement and warning system, etc.

cooling system compared with a conventional cable system. If HTS cables become more compact, the cost for engineering work is lower. Therefore, even if their manufacturing cost is higher than conventional cables, there is an economical advantage for the total initial cost.

2.2. *Power transmission loss and operation cost*

Table 2 shows the items included in the operation cost of an HTS cable system. There is not only the AC loss of the HTS wires, Q_{ac} but also the dielectric loss of the electric insulation layer, Q_e and the invasion heat of the thermal insulation pipe from outside Q_h. In addition, it is necessary to examine the invasion heat and the joule heat of the terminal and intermediate joints, since they use normal conducting metals as the current loading material. Here, these losses are collectively defined as the accessory loss, Q_a. It is necessary to cool down the coolant [liquid nitrogen (LN_2)] with refrigerators. However, since there is also heat invasion into the refrigerators' Q_c, it is necessary to count Q_c within the total loss from an HTS cable system. Finally, the power transmission loss is calculated using Eq. (1) when considering the efficiency η of the refrigerators.

$$\text{Power transmission loss} = (Q_{ac} + Q_e + Q_h + Q_a + Q_c)/\eta, \ldots, \quad (1)$$

The operation cost is obtained by multiplying the electric power unit price for the power transmission loss. In addition, when the power transmission loss of HTS cables is lower than that of conventional cables, the effect of the CO_2 emission decrease can be considered in the economical discussion. The conversion formula of the economical effect of the CO_2 emission decrease is shown in Eq. (2).

$$\text{Economical effect} = \Delta W \times E \times F, \ldots, \quad (2)$$

where ΔW: Difference of the power transmission loss.

E: Amount of CO_2 emission for unit electric power.

F: Trading amount of CO_2 emission right for unit CO_2 weight.

2.3. *Economical advantages of HTS cables*

In this section, the results of an economical examination are described using the example of replacing an underground power line in an urban area. The three existing lines are old 275-kV OF cables, and these cables are in conduits with an inner diameter of 150 mm. Each line length is 10 km, and electric power of 1,050 MVA (350 MVA × 3 lines) is being transmitted. Because the transmission capacity of XLPE cables is lower than that of OF cables, it is necessary to build a new tunnel if XLPE cables are used. However, if high-capacity, compact HTS cables are used, construction costs would be significantly reduced: Power capacity equivalent to that of conventional 275-kV cables can be achieved if 66-kV HTS cables are installed within the

Table 3. Specifications for transmission lines.

Items	XLPE cable	HTS cable
Capacity and number of lines	350 MVA × 3 lines	350 MVA × 3 lines
Rated voltage and current	275 kV/0.75 kA	66 kV/3 kA
Length of lines	10 km	10 km
Construction form	Building new tunnel	Reusing existing conduit

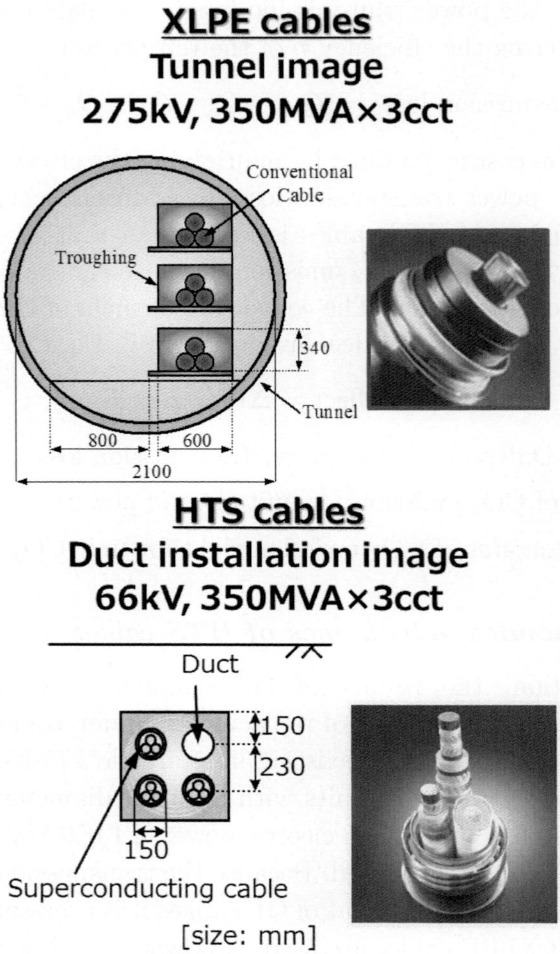

Fig. 3. Replacement of old of cables using XLPE and HTS cables.

Table 4. Examination conditions of construction cost.

Items	XLPE cable	HTS cable (current state)	HTS cable (2020)
Cable	200 k JPY/m	—	—
HTS wire	—	10 JPY/Am	5JPY/Am
Cooling system	—	30 MJPY/kW	15 MJPY/kW
Engineering	3 MJPY/m	50 kJPY/m	50 kJPY/m

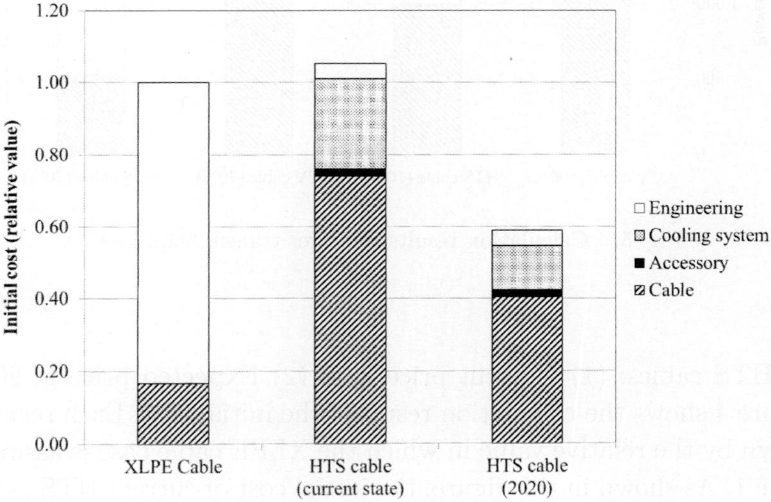

Fig. 4. Calculation result of initial costs (relative value).

Table 5. Calculation conditions of HTS cable.

Items	Conditions
Cable (AC loss)	1 W/m/ph@3 kA
Invasion heat of thermal insulation pipe	2.0 W/m
Efficiency of refrigerators (COP)	0.1

existing underground conduits.[5] The specifications for the transmission lines are shown in Table 3, and the application image is shown in Fig. 3.

The details of the construction costs, which are used for the initial cost calculations, are shown in Table 4. Two cases are assumed

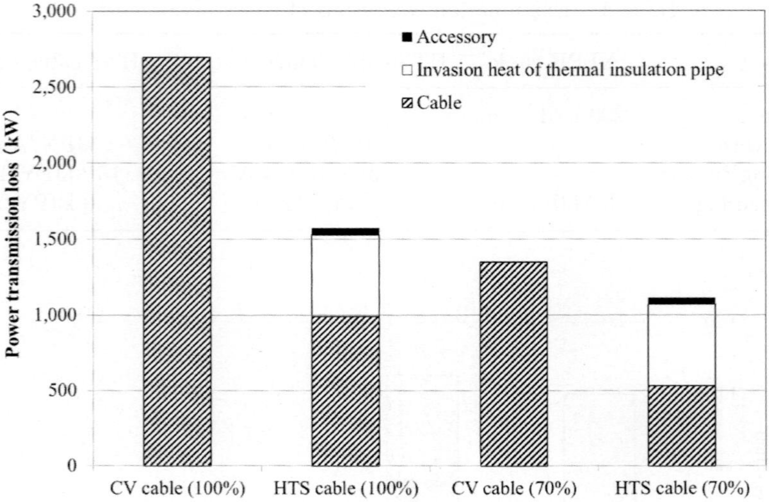

Fig. 5. Calculation results of power transmission loss.

for HTS cables: (1) Current price, and (2) Expected price in 2020. Figure 4 shows the calculation result of the initial cost. Each result is shown by the relative value in which the XLPE cable case is assumed to be 1. As shown in this figure, the initial cost of current HTS cables is comparable to that of XLPE cables; this is because the cost of the HTS wire and cooling system cancels out the advantage of less costly engineering work. If the cost of the HTS wire and cooling system is reduced by half, the total initial cost of HTS cables decreases to two-thirds that of XLPE cables.

The power transmission loss was also calculated. The calculation conditions are given in Table 5.[5] This calculation is executed [made] considering the current performance of HTS cables. Figure 5 shows each calculation result of the power transmission loss when the load factor is 100% and 70%. In this calculation, the efficiency of the refrigerators η (COP) is assumed to be 0.1. As shown in this figure, power transmission loss of HTS cables is less than three-fifths that of XLPE cables when the load factor is 100% and four-fifths when the load factor is 70%. It is anticipated that the advantages of HTS cable will further improve by decreasing AC loss and/or the invasion heat of thermal insulation pipe.

References

1. A. Toya, Overview of present technology of power cable, *IEEJ Trans. Power Energy* **126**, 396 (2006).
2. Y. Yamada *et al.*, Examples and future prospects of high-temperature superconductor products, *SEI Tech. Rev.* **65**, 51 (2007).
3. T. Masuda *et al.*, High-temperature superconducting cable technology and development trends, *SEI Tech. Rev.* **59**, 8 (2005).
4. M. Hirose *et al.*, Study on commercialization of high-temperature superconductor, *SEI Tech. Rev.* **62**, 15 (2006).
5. M. Ohya *et al.*, Japan's first operation of high-temperature superconducting cable systems in live grids (high-temperature superconducting cable demonstration project), *SEI Tech. Rev.* **70**, 84 (2010).

Chapter 3.13

MRI Magnet for Human Brain

Shin-ichi Urayama

Human Brain Research Center, Graduate School of Medicine, Kyoto University, Shogoin-Kawahara-cho 54, Sakyo-ku, Kyoto, Kyoto, Japan

Recent rapid demand increase and supply decrease for helium has raised the price in these years. Superconducting magnetic resonance imaging (MRI) magnets, which consume 20% of global production of helium as the cryogen, are therefore expected to be helium-free and high-temperature superconducting (HTS) materials are potent candidates to realize this. Because of the reason, we developed a cryogen-free 3T-MRI scanner for human brain research using Bi-2223 tapes. For scanning a subject in sitting position, a vertical bore was adopted. The magnet was designed for operating temperature of 20 K and for driven mode. Both target homogeneity and stability of the magnetic field in field of view (FOV) region were within 1 ppm. Not only the magnet but also the other important hard/softwares were produced by us. After the assembly, adjustments and imaging experiments with the scanner were carried out at 1.5 T successfully. Although ramp-up to 3 T succeeded three times, successive abnormal events happened for longer than ten minutes during the third ramp-down time, and finally the magnet got fatal damages. Here, we introduce the system and discuss on problems and potentials of HTS-MRI magnets.

1. Introduction

Although helium is the second most abundant atom in the universe, it is difficult to collect because it spreads thinly and broadly in the air and does not react chemically with any of the atoms or molecules.

Therefore, helium is obtained only as a by-product of selected natural gas deposits and has been collected and reserved mainly in US since the finding of helium in natural gas fields in the US in 1903. However, recent shale gas revolution decreases production of not only natural gas but its by-product helium, and the helium reserves in US is running out. In addition, demand for it has been increasing rapidly in a decade, especially from raising nations. Because of these, the price has doubled in recent years. Magnetic resonance imaging (MRI) magnets consume 20% of the global helium supply as a cryogen to maintain superconductivity in the magnet wires. MRI is one of major consumers of helium and so its rising cost is passed on to users of MRI systems.

High-temperature superconducting (HTS) materials show great potential for realizing helium-free magnets. Although studies on HTS material was started in 1986, when Bednorz and Muller found new superconducting material, $La_{2-x}Ba_xCuO_4$, which exhibits superconductivity below approximately 30 K,[1] there have been several problems in applying these HTS materials to an MRI magnet. First, HTS materials are generally ceramic and are not able to be wound as an electromagnetic coil. Second, it is not possible to make persistent current operation with HTS wires at present. Third, the combination of the electric current in the HTS tape and the resultant magnetic field generates strong Lorentz forces and stresses on the tape which further hinders the production of MRI magnets for human imaging.

Bi-2223 HTS tapes, which are now industrial products, have excellent performance[2–4] to overcome the above problems. Therefore, in 2008, we started a project to produce a cryogen-free, 3T-MRI scanner for human brain using Bi-2223 HTS tapes. Here, we describe the details of our MRI system and discuss the feasibility of HTS magnets.

2. Design and Fabrication

Bi-2223 HTS tapes (DI-BSCCO Type HT, Sumitomo Electric Industries, Ltd., Japan) were used for the magnet. In general, HTS material is ceramic and brittle, making winding impossible. However, Bi-2223 tape, a flattened HTS multi-filamentary-core covered

Fig. 1. Developed HTS-MRI system. Conceptual design (left), the HTS magnet just after the winding (middle) and integrated HTS-MRI system (right).

with Ag and Ag-alloy, is flexible and can be bent to enable coil winding. At the conditions of 20 K and 3 T, its critical currents are around 573/291 A (parallel/perpendicular to the tape surface),[5] which provide a considerable margin above the necessary electric current (184 A) required to generate a magnetic field of 3 T. The critical temperature at 184 A and 3 T is approximately 30 K.

Figure 1 shows the conceptual design of the MRI system.[5] The scanner is built for human brain research and the magnet bore is vertical. The magnet consists of five main coils chilled with only GM cryo-cooler and thermal conduction. Because this project is the first trial, no self-shielding of the magnet was incorporated and the five-gauss line is 6.5 m from the magnet's center. The magnet is designed for driven mode using super-stable power supply (Model 854, Danfysik, Denmark) in order to avoid reduction of electric current due to tiny resistance of soldered joints of Bi-2223 tapes. The inductance of all coils is 137.5 H, the maximum axial/hoop stresses are 4.7/137 MPa and the stored energy is 2.3 MJ, at 3 T.

In total, 44.6 km of tape was wound and fixed by painting stycast (2850FT, National Starch & Chemical Co., USA) to form the five main coils. The maximum piece-length of a single tape was about 400 m and 132 tapes were joined with soldering. The five coils were connected in series with copper lead wires and, as a protection circuit for quench; a diode clipping circuit was connected to each coil to avoid over-current through the coil. After windings, the outside surface of each coil was covered with cold plates and they were connected to the cold heads of the cryo-cooler for thermal conduction. Finally, the coils were fixed in the cryo-stat. The magnet vessel was vacuumized to 2.7×10 Pa and the coils were chilled using the cryo-cooler to a steady state of 6–8 K in 570 h.

In this project, not only the magnet but a gradient coil, a spectrometer, RF coils and system software were also developed by us. The gradient coil wire paths were designed for maximum gradient strength, less than 5% gradient field error and zero net torque whilst optimizing the maximum current density, slew rate, power dissipation, and eddy currents *via* active screening.[6] Based on the wire paths, their copper elements for the x/y-gradient and shim coils were produced by chemical etching and bent with a numerically-controlled bending machine. After assembling the elements and cooling pipes on a fiberglass/resin composite bobbin, they were cast with epoxy resin. The gradient coil efficiencies were 0.06/0.06/0.15 mT/m/A for the $x/y/z$ gradients, respectively. An FPGA (programmable IC chip) based spectrometer with gradient coil control functions was developed based on the "OPENCORE NMR" technology.[7,8] For transmitter/receiver RF coil, a birdcage type coil was produced.

Finally all hardware and software were assembled as an MRI system. To isolate the magnet from gradient coil vibration, the coil was fixed not to the magnet but to an aluminum frame surrounding the magnet.

3. Operation at 1.5 T

As an intermediate step to 3 T, adjustments and experiments were carried out at 1.5 T.

Magnetization to 1.5 T was carried out more than 60 times without problems. Target electric current was 92 A and it took about 2/1.5 h to ramp-up/down with constant speed of 0.75/1.0 A/min, respectively. After the ramp-up process, it took about 20 minutes before the ripple current decayed to a negligible level. Temperatures of the five coils increased during ramp-up by 1–3° and then decreased by 0–1° in two h before reaching the steady states. During ramp-down, they again increased by 0–3° and then decrease to the initial states in nine h.

Magnetic field homogeneity of the un-shimmed magnet was 889 ppm in the targeted field of view (FOV) region; a $25 \times 25 \times 20$ cm spheroid. To increase the homogeneity, two passive shim layers are arranged in the space between the magnet and the gradient coil. The main layer is placed just inside of the magnet bore and is for primary shimming before installation of the gradient coil, and the secondary layer is placed just outside of the gradient coil and is for fine shimming after the gradient coil installation. After the sixth primary shimming trial, the homogeneity was improved to 8 ppm. After the gradient coil installation, it was worsened to 9 ppm, but after the twice fine-shimming trials, it was improved to 5 ppm, the target homogeneity of this project.

The magnetic field strength at the magnet center was monitored for 24 h after the magnetization to evaluate the stability. In order to examine effect of shield current to the stability, the monitoring was carried out two times with/without 1%-overshoot at the end of the ramp-up process.[9,10] Without the overshoot, field strength increased monotonically by 0.85 ppm/h for the measurement duration. However, with the overshoot, the field strength decreased rapidly by 14 ppm in the first five h and then reached a steady state within $\pm$ 0.5 ppm. This result shows that, although there is shield current which causes field change for more than 24 h, it can be suppressed well with the overshoot technique. The resultant stability was much higher than that of the power supply (nominal stability is 0.3 ppm/30 min), probably because of the large inductance of the magnet.

Figure 2 shows cross-sectional images of an extracted brain of a Macaca monkey with 2D spin echo sequence (TR/TE: 600/13 msec, slice thickness: 2 mm, pixel size: 1×1 mm). Although still weak zipper-artifacts at the center area and non-uniform pattern in the immersed liquid area are observed, brain structures are well depicted.

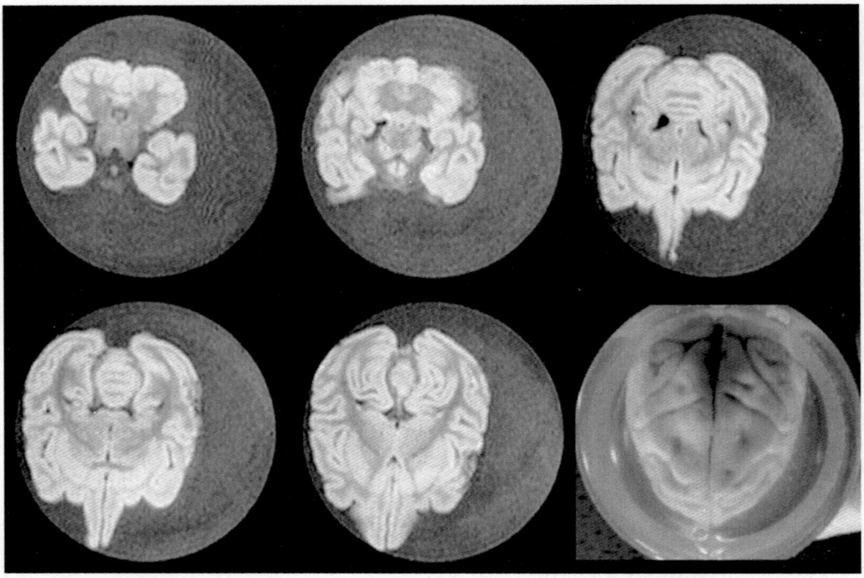

Fig. 2. Cross-sectional images of an extracted Macaca brain obtained with the developed HTS-MRI system.

4. Magnetization to 3 T

After we confirmed the feasibility of our scanner at 1.5 T, we tried to magnetize to 3 T.

First magnetization was successfully carried out in five h. Temperatures of the five coils increased during ramp-up by 4–6 degrees and then increased another 1 degree in four h before reaching the steady states. During ramp-down, they changed by −1–2.5 degrees in 2.5 h and then decreased to the initial states in 12 h.

Magnetic field distribution was measured at the second and third magnetization for passive shimming, and field inhomogeneity could be decreased from 845 to 64 ppm.

However, during the third ramp-down, suddenly loud noise and abnormal voltage change happened about one minute after ramp-down started from 2.5 T (intermission stage). They continued for longer than 10 minutes and finally the power supply shut down. Since temperature of the coils was increasing after the start of the

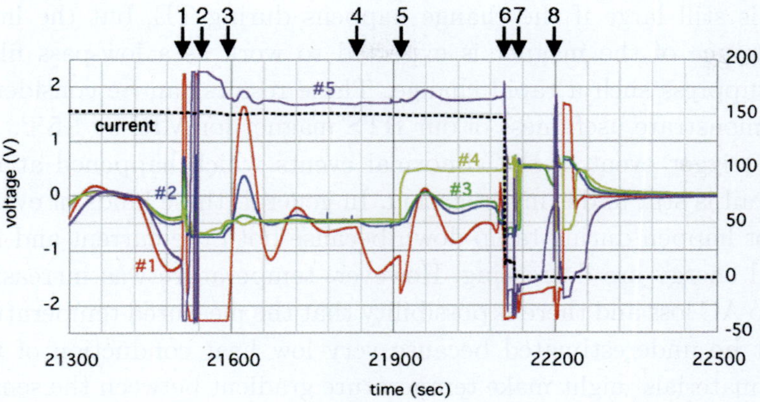

Fig. 3. Recorded voltage and current data during the abnormal events.

abnormal events, we switched off the cryo-cooler and decided to start investigation of the events.

Inspection of the decomposed magnet revealed three major damages, a gap between coil #2 and the flange, melted electrodes at coil #1 and #2, and burned HTS tapes in coil #5. Recorded data of electric current, voltages, and temperature indicates sequential critical events shown with eight arrows in Fig. 3, that is, (1) Abnormal events start (probably normal transition in coil #5), (2) Thermal run away causing short circuits expanding in coil #5, (3) Ramp-down halt by the operator, (4) Ramp-down restart, (5) Quench in coil #4, (6) Power supply shutdown, (7) Spark at the electrode of coil #2, and (8) Spark at the electrode of coil #1.

5. Discussion

In this study, we developed an MRI scanner with a HTS magnet for human brain imaging and demonstrated its feasibility with phantom experiments at 1.5 T.

Although a magnet using HTS tapes might be expected to have problems with magnetic field stability (because it must be power supply-driven) and homogeneity (because this was the first trial to develop a large-scale HTS magnet for human MRI), both of them were not problematic in our study. The stability of less than 0.5

ppm is still large if the change happens during TE, but the huge inductance of the magnet is expected to work as a low-pass filter and suppress such a rapid change. These results can be considered to demonstrate usefulness of our HTS magnet for MRI at 1.5 T.

A trigger event of the abnormal events which happened at the 3 T trail is still under investigation. In general, these kinds of events do not happen during ramp-down because both the current and the stored energy are decreasing. However, temperature was increasing due to AC loss and there is possibility that the measured temperature might be underestimated because very low heat conduction of the HTS materials might make temperature gradient between the sensor position (surface of the coil) and inside of the coil. Therefore, thermal run away caused by temperature increase due to AC loss might be the trigger event.

The sequential abnormal events show problems of HTS-MRI magnets. In conventional LTS MRI magnet, quench propagates to whole magnet immediately. This means it shares stored energy with whole coils and the magnet is difficult to be damaged, resulting in unnecessity of quench detection and protection. While, normal transition which happened in HTS magnet is difficult to propagate because of large temperature margin and high specific heat capacity of the HTS materials. If there is no cryogen which can suppress rapid temperature increase, local normal transition easily causes thermal run away at the local points, resulting in fatal damages. Therefore, for HTS-MRI magnets, detection of a normal transition and magnet protection are very important.

In contrast, there are potentials in HTS-MRI magnets. Because they have sufficient margins to their limits, there must be various possibilities in applications, like a helium-free ultra high field HTS-MRI, a LN_2 HTS-MRI or a compact and inexpensive HTS-MRI with helium. Application trials on HTS magnet to MRI have been started in these years and we believe that HTS magnet is a possible and practical option for future MRIs.

Acknowledgments

First of all, we would like to thank all other forty team members of this project for their great contribution. We are deeply grateful to Dr. R. Turner at MaxPlank Institute, Leipzig (Germany), Dr. D. Le Bihan at NeuroSpin (France) and Dr. G. Sorensen at MGH (USA) for the valuable discussion and information about MRI production.

This study is supported by JST-SENTAN, System Development Program for Advanced Measurement and Analysis.

References

1. J. G. Bednorz and K. A. Müller, *Z. Physik* **B 64**, 189 (1986).
2. N. Ayai, K. Yamazaki, M. Kikuchi, G. Osabe, H. Takaaze, H., Takayama, S. Kobayashi, J. Fujikami, K. Hayashi, K. Sato, K. Osamura, H. Kitaguchi, S. Matsumoto, T. Kiyoshi, and J. Shimoyama, *IEEE Trans. Appl. Supercond.*, 3014, (2009).
3. N. Ayai, S. Kobayashi, M. Kikuchi, T. Ishida, J. Fujikami, K. Yamazaki, S. Yamade, K. Tatamidani, K. Hayashi, K. Sato, H. Kitaguchi, H. Kumakura, K. Osamura, J. Shimoyama, H. Kamijyo, and Y. Fukumoto, *Physica C*, 1747, (2008).
4. N. Ayai, M. Kikuchi, K. Yamazaki, S. Kobayashi, S. Yamade, E. Ueno, J. Fujikami, T. Kato, K. Hayashi, K. Sato, R. Hata, J. Iihara, K. Yamaguchi, and J. Shimoyama, *IEEE Trans. Appl. Supercond.*, 3075 (2007).
5. H. Kitaguchi, O. Ozaki, T. Miyazaki, N. Ayai, K. Sato, S. Urayama, and H. Fukuyama, *IEEE Trans. Appl. Supercond.*, 710, (2010).
6. M. Poole, H. S. Lopez, O. Ozaki, H. Kitaguchi, I. Nakajima, S. Urayama, K. Sato, H. Fukuyama, and S. Crozier, *J Phys. D: Appl. Phys.*, 095001 (2010).
7. K. Takeda, *Ann. Rep. NMR Spectros.*, 355 (2011).
8. http://kuchem.kyoto-u.ac.jp/bun/projects/OPENCORE/opencore.html.
9. Y. Koyama, T. Takao, Y. Yanagisawa, H. Nakagome, M. Hamada, T. Kiyoshi, M. Takahashi, and H. Maeda, *Physica C*, 694 (2009).
10. Y. Yanagisawa, H. Nakagome, Y. Koyama, R. Hu, T. Takao, M. Hamada, T. Kiyoshi, M. Takahashi, and H. Maeda, *Physica C*, 1996 (2009).

Chapter 3.14

LTS/Bi-2223 NMR Magnets Operated Beyond 23.5 T (1 GHz)

H. Maeda, S. Iguchi, M. Takahashi, T. Yamazaki,*, H. Nakagome,† T. Kiyoshi, K. Hashi, T. Shimizu, S. Ohki, S. Matsumoto, G. Nishijima, A. Goto, K. Deguchi, K. Yamada, T. Noguchi, S. Sakai,‡ M. Hamada,§ K. Saito, M. Yoshikawa, T. Miki, A. Otsuka,¶ R. Tanaka, T. Nemoto, H. Suematsu,‖ and Y. Yanagisawa*

*RIKEN-CLST, Yokohama, Japan

†Graduate School of Engineering, Chiba University, Japan

‡National Institute for Materials Science, Tsukuba, Japan

§Japan Superconductor Technology, Inc., Kobe, Japan

¶Kobe Steel Ltd., Kobe, Japan

‖JEOL RESONANCE Inc., Akishima, Japan

The use of a Bi-2223 magnet provides a high current density above 23 T, enabling a nuclear magnetic resonance (NMR) magnet to exceed a magnetic field of 23.5 T (1 GHz). The development and perspectives of super high field (LTS)/Bi-2223 NMR spectrometers that can operate at fields as high as 1.2 GHz NMR are described.

1. Introduction

The nuclear magnetic resonance (NMR) spectrometer is widely used for studies in life science, pharmaceutical research, organic chemistry, materials science and so forth. The principle difficulty of the NMR measurement is its inherently low sensitivity. The sensitivity is proportional to the 1.5^{th} power of the magnetic field strength, while the resolution of the n-dimensional NMR experiment is proportional to the nth power of the magnetic field strength. Hence, a higher magnetic field is a goal pursued by NMR spectroscopists.

Conventional low-temperature superconducting (LTS) NMR magnets cannot exceed 23.5 T, corresponding to a ^{1}H NMR frequency of 1 GHz, as the critical current density of a typical LTS conductor decreases steeply beyond 23 T. The use of a high-temperature superconducting (HTS) magnet provides a high current density above 23 T, enabling an NMR magnet to exceed an operating magnetic field of 23.5 T. In fact, a field of 33.8 T,[1] corresponding to a ^{1}H NMR frequency of 1.44 GHz, has been achieved by using a Bi-2212 insert coil in a background field of 31.2 T, while 35.4 T (1.5 GHz) was attained[2] with a REBCO insert coil. Thus, super-high field NMR magnets, with field that substantially exceeds 1 GHz (23.5 T) such as 1.3 GHz (30.55 T) are made possible if we employ low-field LTS outer coils and high-field HTS inner coils, an arrangement which is called in this case "an LTS/HTS NMR magnet".

Few LTS/HTS NMR magnets have been developed so far. MIT has developed a 350 MHz (8.2 T) LTS/Bi-2223 NMR magnet and a 700 MHz (16.45 T) LTS/Bi-2223 NMR magnet.[3] As the project emphasized developing LTS/HTS NMR magnets, the obtained NMR spectrum resolution, 5 ppm, was insufficient for high resolution NMR measurements. MIT is now developing a 1.3 GHz (30.55 T) LTS/Bi-2223/REBCO NMR magnet.

On the other hand, Japan started a project to develop an LTS/Bi-2223 NMR spectrometer operating beyond 1 GHz (23.5 T)[4]: The project emphasized its use for high resolution NMR measurements. The first stage resulted in a 500 MHz LTS/Bi-2223 NMR spectrometer.[5] The second stage was to construct a 1.03 GHz (24.2 T) LTS/Bi-2223 NMR spectrometer. Japan is now considering the development of a 1.3 GHz LTS/HTS NMR spectrometer.

In this chapter, the development of the 500 MHz LTS/Bi-2223 NMR spectrometer[5] and the 1.02 GHz LTS/Bi-2223 NMR spectrometer are described. A perspective on the potential of super high field NMR spectrometers, such as a 1.2 GHz NMR will be presented.

2. Development of LTS/Bi-2223 NMR Magnets

2.1. *Specifications of NMR spectrometers*

The specifications for high resolution NMR spectrometers are as follows:

(1) High central magnetic field strength.
(2) Temporal magnetic field stability; 0.1 ppb.
(3) Spatial magnetic field homogeneity over an NMR sample; <1 ppb.

A conventional LTS NMR magnet is usually operated in persistent current mode with a temporal magnet field drift rate of <0.01 ppm/h. It is stabilized to 0.1 ppb using a field-frequency lock system installed in an NMR console. The field-frequency lock system continuously measures the NMR frequency of ^{2}H in a sample solution, which is proportional to the central magnetic field intensity; it compensates for temporal magnetic field fluctuations using a compensation coil to hold the axial central magnetic field intensity constant.[6] The spatial distribution of the axial magnetic field intensity at the magnet center can be expressed as a series of Legendre spherical harmonics.[7] Spherical harmonics that are symmetrical about the coil axis (i.e. the z-axis) are indicated by z, z^2, z^3, z^4, ... On the other hand, those that are asymmetric about the z-axis are indicated by x, y, xy, $x^2 - y^2$, zx, zy, ...; here x and y are the radial coordinates of the NMR magnet.

The NMR main coil comprises numbers of thin solenoids as can be seen in Fig. 1; the outermost coils are field correction coils, concealing error harmonics such as the z^2 and z^4 harmonics generated by the main coil. Eight superconducting shim coils, located outside the main coil, and several tens of room-temperature shim coils, located inside the NMR magnet bore tube, are installed to correct the spatial inhomogeneity of the magnetic field; each individual shim-coil compensates a corresponding spherical harmonic.

Due to the low n-index value of the Bi-2223 multi-filamentary tape conductor and the impossibility of preparing a superconducting

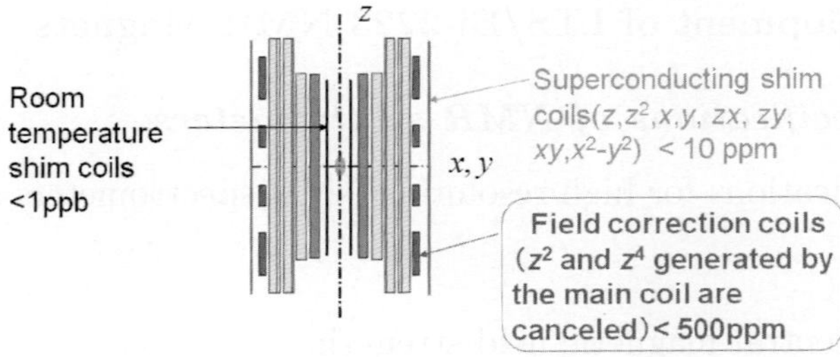

Fig. 1. Cross-sectional view of an NMR magnet. Field correction coils, superconducting shim coils, and room-temperature shim coils are usually installed in the NMR magnet system.

persistent joint, it is unlikely that a Bi-2223 inner coil can provide a persistent current sufficient for NMR measurements. Therefore, the Bi-2223 high field inner coil is usually connected in series to the LTS low-field outer coils and charged simultaneously by an external ultra-stabilized DC power supply; i.e. it is operated in a driven mode. The current fluctuations of the ultra-stabilized DC power supply, typically 0.3 ppm to 1 ppm, causes a temporal field fluctuation, generating noticeable sidebands on the NMR spectrum and a deviation in the peak frequency. The field-frequency lock system stabilizes the temporal magnetic field fluctuations caused by the DC power supply, as long as the magnetic field fluctuations do not exceed 3 ppm.[5]

2.2. *500 MHz LTS/Bi-2223 NMR magnet*

2.2.1. *Design and fabrication of the LTS/Bi-2223 NMR magnet*

The innermost Nb_3Sn coil of a 600 MHz LTS NMR magnet was replaced by a Bi-2223 coil, 81.2 mm in inner diameter, 121 mm in outer diameter, and 375 mm in length.[5] A copper alloy reinforced Bi-2223 tape conductor, 4.6 mm in width and 0.36 mm in thickness, was used for coil fabrication. From the view point of magnetic field homogeneity, the solenoid winding method is preferable to the pancake winding method as the solenoid coil is closely spaced by conductors; therefore, the solenoid winding method was employed. The total length of the Bi-2223 tape conductor was 1.34 km.

In the driven mode of the LTS/Bi-2223 NMR magnet, the current leads become the major source of heat leak into the 4 K helium bath. Thus, a two stage pulse-tube cryo-cooler was installed on top of the cryostat, to re-liquefy the vaporized helium gas[5]: A pulse-tube cryo-cooler offering a cooling power of 0.5 W was used as it has low levels of mechanical vibration. Long term NMR operation over several months were made possible without refilling liquid helium which was sufficient for NMR measurements.

2.2.2. *Temporal magnetic field drift and fluctuations/spatial magnetic field correction*

After the LTS/Bi-2223 NMR magnet was charged to the operating current (145 A, 11.856 T, and 499.971 MHz), it was operated in driven mode. The initial change in the central magnetic field intensity is shown in Fig. 2(a).[5] The magnetic field intensity steeply increased initially at a rate of 20 ppm/h (2×10^{-5}/h), and then it tended to saturate at 0.05 ppm/h (5×10^{-8}/h) at t = 110 h. The latter drift allows for the continuous operation of a field-frequency lock system for 40 h, sufficient for high resolution NMR measurements.

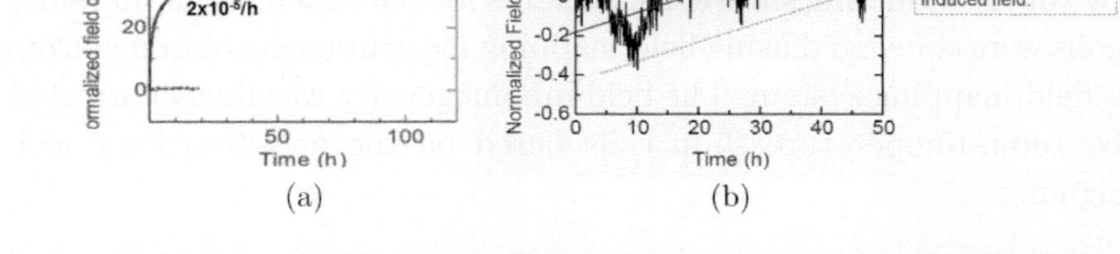

Fig. 2. (a) Positive drift of the central magnetic field intensity with time after charging the 500 MHz LTS/Bi-2223 NMR magnet. (b) Temporal magnetic field drift and fluctuations on the 20th day after the magnet charge. Figure modified from Ref. 5.

The positive magnetic field drift seen in Fig. 2(a) is explained as follows: When the NMR magnet is charged, a screening current is induced in the Bi-2223 tape conductor by the radial component of the local magnetic field, generating a negative magnetic field at the magnet center. Relaxation of the screening current resulted in temporal positive drift of the magnetic field intensity with time as seen in Fig. 2(a).[8]

Figure 2(b) shows a magnetic field fluctuation on the 20th day after the magnet charge. The temporal magnetic field fluctuation consisted of three components[5]:

(1) Long-term positive drift of the average magnetic field at a rate of 0.01 ppm/h (10^{-8}/h), which is due to the relaxation of the screening current-induced magnetic field.
(2) Daily variations in the magnetic field of 0.5 ppm (5×10^{-7}), which were due to daily changes in the DC power supply current.
(3) Short-term field ripples with a peak–peak amplitude <0.2 ppm (2×10^{-7}), which were due to short-term current ripples caused by the DC power supply.

The magnetic field fluctuations seen in Fig. 2(b) are stabilized by the operation of the field-frequency lock system.[5,6] Long term field stability over 74 h under field-frequency lock operation is shown in Fig. 3. The deviation in peak frequency over the running time was as small as 0.5 ppb (5×10^{-10}), nearly of the same quality as the persistent current LTS NMR spectrometer, which is sufficient for high resolution NMR measurements.

The spatial magnetic field inhomogeneity was coarsely corrected by superconducting shim coils; currents for the superconducting shim coils were converged using field mapping measurements obtained from a field mapping system. The field inhomogeneity was finely corrected by room-temperature shim coils based on the field-frequency lock signal.

2.2.3. *Quality of the NMR spectra*

The ^{1}H NMR resolution is determined by the half-height line width of the NMR peak for 1% $CHCl_3$ in deuterated acetone[5]: It was

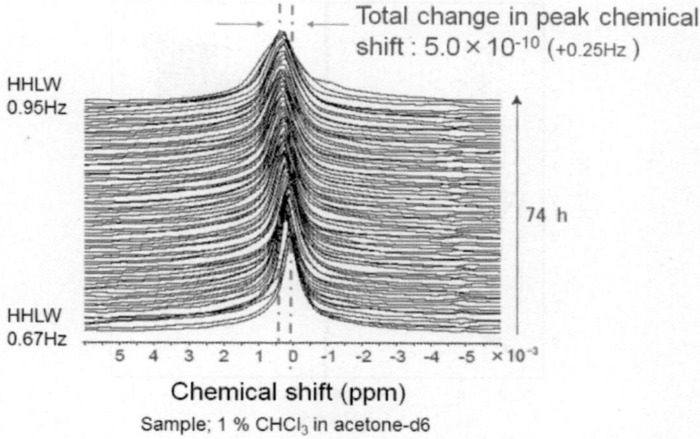

Fig. 3. Deviation in the NMR peak frequency of ^{1}H for 1% $CHCl_3$ in deuterated acetone over the running time of 74 h. Figure adapted from Ref. 5.

1 ppb (10^{-9}) for the 500 MHz LTS/Bi-2223 NMR magnet, equivalent to that obtained in a persistent LTS NMR magnet. Side bands, appearing due to mechanical vibrations generated by the pulse-tube cryo-cooler, were negligibly small. On the other hand, the ^{1}H NMR sensitivity is determined as the ratio of the central peak height of the ethylbenzene quartet peak to the base line noise for 0.1% ethylbenzene in $CDCl_3$. It was 512 and very close to that (599) obtained by a 500 MHz LTS NMR. Thus, both the NMR resolution and the NMR sensitivity obtained by the 500 MHz LTS/Bi-2223 NMR spectrometer are equivalent to those achieved by a conventional LTS NMR.

Figure 4 illustrates the two-dimensional Nuclear Overhauser effect correlated spectroscopy (NOESY) spectrum for a protein (chlorella ubiquitin) obtained using the 500 MHz LTS/Bi-2223 NMR magnet. The NOESY spectrum provides information about the distance between proton nuclei based on the intensity of cross peaks.[6] The NOESY experiment on a protein solution is one of the most precise experiments. If there is magnetic field instability with time, noticeable noise will appear along the vertical axis, which is not observed in Fig. 4. If the NMR spectrum is not sharp enough, the water peak at 4.7 ppm is not suppressed and signals around 4.7 ppm becomes

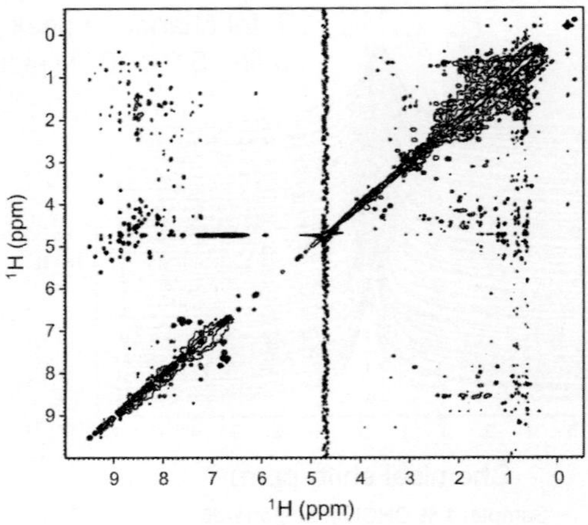

Fig. 4. Two-dimensional NOESY signal for a protein aqueous solution sample (chlorella ubiquitin), which is equivalent to that obtained by a conventional LTS NMR. Figure adapted from Ref. 5.

unobservable. The quality of the multi-dimensional NMR spectra obtained by the 500 MHz LTS/Bi-2223 NMR magnet are equivalent to those achieved by a conventional LTS NMR.

2.3. *1.02 GHz LTS/Bi-2223 NMR magnet*

The innermost Nb_3Sn coil of a 920 MHz (21.6 T) NMR, having been installed in the National Institute of Materials Science (NIMS), was replaced by a Bi-2223 coil,[4] 78.4 mm in inner diameter, 127 mm in outer diameter, and 840 mm in length. The Bi-2223 tape conductor, having been used for the 500 MHz LTS/Bi-2223 magnet, was used for the NMR magnet; the total conductor length was 3.2 km. Both LTS outer coils and the Bi-2223 inner coil are connected in series and charged by the ultra-stabilized DC power supply which is used for the 500 MHz LTS/Bi-2223 NMR magnet; the NMR magnet is operated in driven mode. The magnetic field strength is 1.02 GHz (24.0 T); the Bi-2223 innermost coil generates 3.6 T, while the LTS outer coils provide 20.4 T. The hoop-stress due to electromagnetic force is 200 MPa, which is close to the critical tensile stress for conductor degradation. The magnet is cooled by pressurized superfluid helium below 1.8 K. The NMR magnet was charged in NIMS and attained 1.02 GHz (24.0 T) in 2014.[9]

Daily magnetic field variation, without the field-frequency lock operation, was <1.2 ppm in peak-peak amplitude which was more than double that of the 500 MHz NMR (Fig. 2(b)). On the other hand, short term field ripple was <0.3 ppm, nearly the same as the 500 MHz NMR. The magnetic field was then stabilized to <1 ppb/10 h (10^{-9}/10 h) with the field-frequency internal lock operation, sufficient for high resolution solution NMR.[6]

Magnetic field inhomogeneity was corrected by superconducting shim coils, ferromagnetic shims, and room temperature shims, respectively, based on the field distribution measured by a shim robot. In case there is a large error component that is uncorrectable with superconducting shim coils, it is necessary to provide a retrofitted strong shim device, such as ferromagnetic shims. An NMR spectrum resolution, i.e. field homogeneity, of 0.7 ppb (7×10^{-10}) was finally achieved.

A field-frequency external lock using a laboratory-made microprobe was operated to stabilize the magnetic field fluctuation for solid-state NMR measurements.[9] The temporal stability was 0.17 ppm/10 h (1.7×10^{-7}/10 h), sufficient for high resolution solid-state NMR measurements.

High resolution 2D solid-state NMR spectra were measured for a membrane protein. The NMR spectrum resolution was more than double that achieved by a conventional 700 MHz LTS NMR; thus solid-state NMR spectra achieved with the 1.02 GHz LTS/Bi-2223 NMR were of a sufficiently high quality.

3. Perspectives on Super-High Field LTS/Bi-2223 NMR Magnets beyond 1 GHz

A design concept for the 1.02 GHz LTS/Bi-2223 NMR magnet so far described is shown as follows:

(i) The outer LTS coil generates 20.4 T, which is close to the upper limit of the magnetic field strength achievable with a conventional LTS NMR magnet. The magnetic field strength is enhanced to 24.0 T by the Bi-2223 booster coil.

(ii) The Bi-2223 conductor hoop-stress due to the electromagnetic force is as low as 200 MPa, resulting in lower current density.

(iii) The magnet is cooled by super-fluid helium below 1.8 K, resulting in a large cryostat.

A 1.2 GHz (28.2 T) LTS/Bi-2223 NMR magnet can be designed based on the similar criteria; the result is shown in Fig. 5(b). It is clear that the magnet becomes impractically large. A stronger Bi-2223 conductor has been developed by Sumitomo Electric Industries, Ltd.; it endures the tensile stress of 400 MPa, while the conventional Bi-2223 conductor endures ~250 MPa. If we use this type of new Bi-2223 conductor and assume the following new design criteria, the 1.2 GHz NMR magnet becomes much smaller, as seen in Fig. 5(c):

(i) The Bi-2223 inner coil generates ~21 T, while the NbTi outer coil generates ~7 T. Removal of the Nb_3Sn coil reduces the magnet volume.

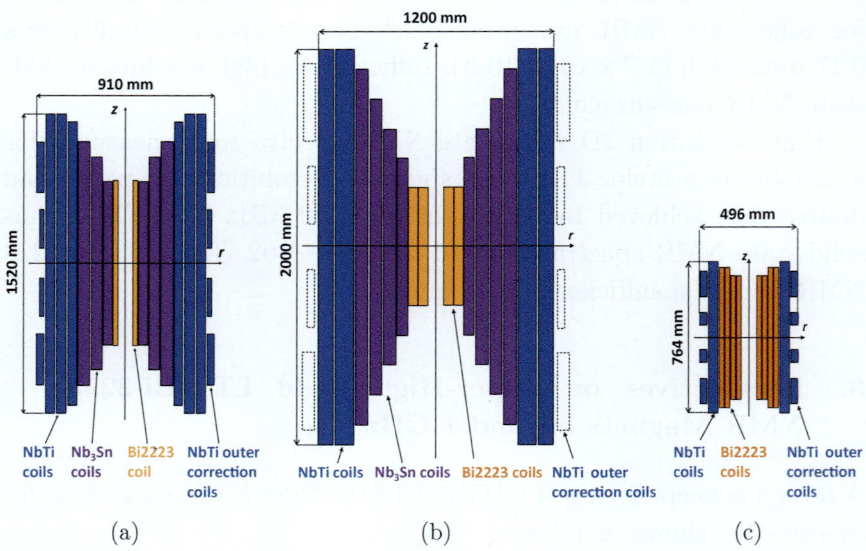

Fig. 5. Cross-sectional view of the super high field LTS/Bi-2223 NMR magnet design. (a) 1.02 MHz NMR. (b) 1.2 GHz LTS/Bi-2223 NMR magnet. (c) Optimal design of the 1.2 GHz LTS/Bi-2223 NMR magnet. Figure modified from Ref. 10.

(ii) The Bi-2223 hoop-stress due to the electromagnetic force is as high as 400 MPa, increasing the overall current density of the NMR magnet.
(iii) The magnet is cooled by liquid helium, which is recondensed by cryocoolers, resulting in the reduction of the size of the cryostat.

Similar results may be obtained if we use REBCO HTS-coils, as the permissible tensile stress for this conductor is as large as 700 MPa.[10] However, according to NMR experiments using a 400 MHz LTS/REBCO NMR magnet,[10] the effect of screening current is extremely larger than that for a LTS/Bi-2223 NMR magnet. Therefore, from the view point of magnetic field homogeneity and magnetic field stability, the LTS/Bi-2223 NMR magnet is preferred to the LTS/REBCO NMR magnet as seen in Fig. 5(c).

4. Summary

Conventional LTS NMR magnets cannot exceed 23.5 T, 1 GHz, as the critical current density of the LTS conductor decreases steeply beyond 23 T. The use of a Bi-2223 magnet provides a high current density above 23 T, enabling an NMR magnet to generate a magnetic field that exceeds 23.5 T (1 GHz). Development of the 500 MHz LTS/Bi-2223 NMR magnet and the 1.02 GHz LTS/Bi-2223 NMR spectrometer that has been constructed, and a perspective on super-high field LTS/Bi-2223 NMR magnets, such as a 1.2 GHz NMR, are described.

References

1. H. W. Weijers, U. P. Trociewitz, W. D. Markiewicz, J. Jiang, D. Myers, E. E. Hellstrom, A. Xu, J. Jaroszynski, P. Noyes, Y. Viouchkov, and D. Larbalestier, High field magnets with HTS conductors, *IEEE Trans. Appl. Supercond.* **20**, 576–582 (2010).
2. U. P. Trociewitz, M. D. Canassy, M. Hannion, D. Hilton, J. Jaroszynski, P. Noyes, Y. Viouchkov, H. W. Weijers, and D. C. Larbalestier, 35.4 T field generated using a layer-wound superconducting coil made of (RE) $Ba_2Cu_3O_{7-x}$ (RE = rare earth) coated conductor, *Appl. Phys. Lett.* **99**, 202506 (2011).

3. S. Hahn, J. Bascunan, H. Lee, E. S. Bobrov, W. Kim, M. C. Ahn, and Y. Iwasa, Operation and performance analyses of 350 and 700 MHz low-/high-temperature superconductor nuclear magnetic resonance magnets: A march toward operating frequencies above 1 GHz, *J. Appl. Phys.* **105**, 024501 (2009).
4. T. Kiyoshi, S. Choi, S. Matsummoto, K. Zaitsu, T. Hase, T. Miyazaki, A. Otsuka, M. Yoshikawa, M. Hamada, M. Hosono, Y. Yanagisawa, H. Nakagome, M. Takahashi, T. Yamazaki, and H. Maeda, HTS NMR: Present status and future plan, *IEEE Trans. Appl. Supercond.* **20**, 714–717 (2010).
5. Y. Yanagisawa, H. Nakagome, K. Tennmei, M. Hamada, M. Yoshikawa, A. Otsuka, M. Hosono, T. Kiyoshi, M. Takahashi1, T. Yamazaki1, and H. Maeda, Operation of a 500 MHz high temperature superconducting NMR: Towards an NMR spectrometer operating beyond 1 GHz, *J. Mag. Res.* **203**, 274–282 (2010).
6. A. E. Derome, *Modern NMR Techniques for Chemistry Research*. Oxford, UK: Pergamon Press/Elsevier (1987).
7. Y. Iwasa, *Case Studies in Superconducting Magnets, 2nd edition*. New York: Springer (2009).
8. Y. Koyama, T. Takao, Y. Yanagisawa, H. Nakagome, M. Hamada, T. Kiyoshi, M. Takahashi, and H. Maeda, Towards beyond 1 GHz NMR: Mechanism of the long term drift of screening current-induced magnetic field in a Bi-2223 coil, *Physica C* **469**, 694–701 (2009).
9. K. Hashi, S. Ohki, S. Matsumoto, G. Nishijima, A. Goto, K. Deguchi, K. Yamada, T. Noguchi, S. Sakai, M. Takahashi, Y. Yanagisawa, S. Iguchi, T. Yamazaki, H. Maeda, R. tanaka, T. Nemoto, H. Suematsu, T. Miki, K. Saito, and T. Shimizu, Achievement of 1020 MHz NMR, *J. Mag. Res.* **256**, 30–33 (2015).
10. Y. Yanagisawa, R. Piao, S. Iguchi, H. Nakagome, T. Takao, K. Kominato, M. Hamada, S. Matsumoto, H. Suematsu, X. Jin, M. Takahashi, T. Yamazaki, and H. Maeda, Operation of a 400 MHz NMR magnet using a (RE = rare earth) $Ba_2Cu_3O_{7-x}$ high temperature superconducting coil: Towards an ultra-compact super-high field NMR spectrometer operated beyond 1 GHz, *J. Mag. Res.* **249**, 38–48 (2014).

Chapter 3.15

Compact HTS Beamline and Magnetic Resonance Magnets

D. Pooke, M. Fee, T. Huang, V. Chamritski, and M. Christian
HTS-110 Ltd.,
69 Gracefield Road, Lower Hutt, New Zealand

Over the last decade significant progress has been made in the commercialization of HTS magnets, paralleling the improvements in wire performance and quality. In this chapter, we review the development at HTS-110 of magnets for synchrotron and neutron beamlines which leverage the benefits of high current density relative to copper and a relatively high operating temperature compared to the low-temperature superconductors. New products extending this development into the demanding realm of magnetic resonance are also discussed.

1. Introduction

At the time of this review, HTS magnets have been manufactured commercially for over a decade. Earlier efforts were mostly concerned with developing and validating coil manufacturing methods, but at least by the turn of the new century, the wire technology had advanced sufficiently that useful scientific magnets could be realized. One of the earliest magnets to be put into a demanding scientific use was an ion-beam steering magnet[1] co-developed by (the former) Industrial Research Ltd., American Superconductor Corporation (AMSC), Alphatech Systems (Auckland, New Zealand) and ISYS (USA). That magnet was used to direct an ion beam to one of three target beam lines in a tandem Van de Graaff accelerator facility mainly employed for carbon dating and beryllium archeological and geophysical studies.

The magnetic circuit chosen employed an H-shape, room temperature iron yoke and two Bi-2223 wire coils each of 9200 A-turns generating a uniform magnetic field of 0.72 T in a 30 mm air gap between the two poles. This provided an ion deflection angle of ±25 degrees and a deflection radius of 1654 mm. Several HTS tapes were required to be wound-in-hand to carry the design current of 100 A at temperatures up to 50 K, with the HTS coils, manufactured by AMSC, configured as race-track shaped solenoids. The magnet (Fig. 1) operated successfully for over five years at which point failure of the cryocooler (a single-stage Gifford-McMahon "GM" unit) led the users to replace the HTS coils with copper. Nevertheless, the unit was the first large-scale HTS physics magnet in a demanding working environment, and demonstrated a then-unproved capability of the wire to perform reliably over the long term.

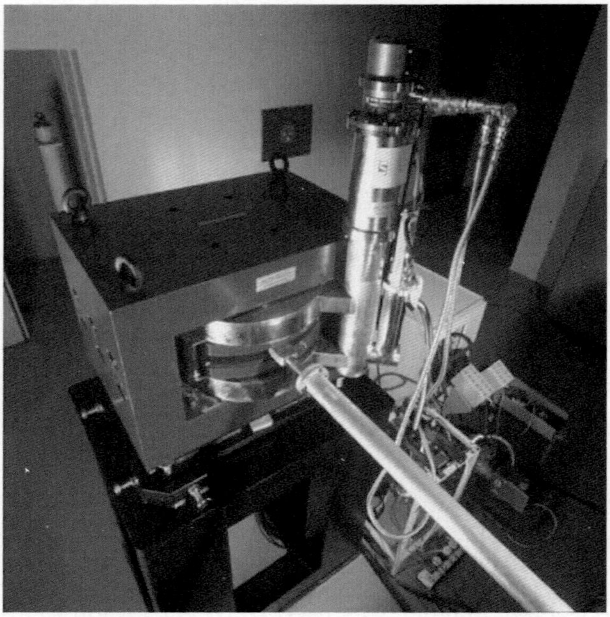

Fig. 1. Ion-beam steering magnet installed in 1995.

Following this effort, a 3T-class magnet was developed with the initial target being to upgrade a vibrating sample magnetometer (VSM) from an existing 1 T copper electromagnet.[2] The design requirement was for a 50 mm pole-gap and 120 mm pole-face diameter to house the variable-temperature VSM sample tube and detection coil set; a 3 T field strength over this sample volume is difficult to achieve with Cu-coiled electromagnets so the HTS conductor provided a solution without undue increase in size or system power consumption. The target field was achieved using an assembly of four HTS double pancake coils with an operating current of 130 A; this project employed the "high-strength" type Bi-2223 conductor laminated on both sides with thin stainless-steel strips (AMSC). Refrigeration was provided by a small single-stage cryo-cooler (Cryomech AL63) providing 25 W of cooling capacity at the operating temperature of 35 K (Fig. 2). Other early commercial magnets were built around the same time, including magnets for an ECR ion source[3] and gyrotron.[4]

Fig. 2. A 3 T H-frame laboratory electromagnet (2003).

Over the last decade of HTS magnet manufacture, the performance of Bi-2223 wire has improved at least a factor of two-fold in critical current and, just as importantly for utility, the uniformity with length has increased dramatically. Defects in early production wire dictated the employment of quality assurance measurements on incoming wire, which could be carried out with non-contact inductive methods on a continuous basis. Identified defects of sufficient severity would be cut from the wire and a low-resistance splice added. Now, however, defects in the wire are the exception and both aspects of quality improvement have greatly aided the production of a broadening range of HTS magnets.

2. Magnet Construction

The flat tape aspect of HTS wire lends itself to a modular pancake style of coil manufacture, either with a splice at the inner diameter or a continuous helical winding. This modularity provides a great deal of flexibility in terms of inner and outer diameters of the individual coils, which aids magnetic design, but also allows individual testing of the coil modules prior to assembly, and in principle provides for an easier repair pathway, should the magnet be damaged in operation. A typical magnet construction then consists of a one or more stacks of double pancakes interleaved with conductive plates which are connected to a thermal bus linked to a cold head. The double pancakes can be interconnected *via* copper or superconducting links.

The use of iron in the magnetic circuit can serve three purposes, firstly to reduce the perpendicular component of magnetic field being applied to the Bi-2223 conductor in order to maximize its performance; this was the primary focus of earlier magnet design efforts simply because the performance drop with perpendicular fields can be two to three times higher than for parallel fields over the target range of operating temperatures. Secondly, the iron contributes to the central field through its own magnetization even though parts of the iron circuit may well be fully saturated. Thirdly, the iron circuit contributes to a reduced fringe field without the use of active shielding coils. Many of the magnets produced at HTS-110 have employed iron circuits for these reasons, and often the iron is also employed to form at least part of the vacuum cryostat for the coil environment.

Typically HTS magnets are operated at temperatures ranging from 20 K to 40 K where the heat capacity of the materials is around 10-fold higher than at 4.2 K, rendering an HTS coil very stable against thermal perturbations which would quench an LTS magnet. Combined with a comparatively low power-law index for the superconducting-normal transition, the development of any significant coil voltage can be detected with sufficient time for the coil to be safely de-energized.

The high operating temperature also means that for a given cryocooler input power there is much more cooling power available at the target operating temperature, typically also by a factor of 10. The outcome is that very compact cryogenic cryostats can be designed for the magnets and that the coil suspension system can be made very robust. A solid thermal radiation shield is not required, with multilayer insulation alone allowing the physical gap to room temperature to be as small as 4 or 5 mm.

3. Beamline Magnets

The very compact magnet envelope possible with HTS conductors lends itself well to magnets for synchrotron and neutron beamlines, where space is generally at a premium and more specifically where other end-station instrumentation imposes particular space or weight constraints.

3.1. *X-ray diffraction, resonant magnetic scattering*

A good example of this was the first 5 T HTS magnet built by HTS-110 expressly designed for resonant magnetic scattering and X-ray diffraction at the MAGS beamline of the Berlin Electron Synchrotron (BESSY).[5]

This was a split-pair magnet with a 90° scattering slot, and a 40 mm room temperature bore for the sample which was designed to be compatible with standard sample cryo-coolers. The magnet

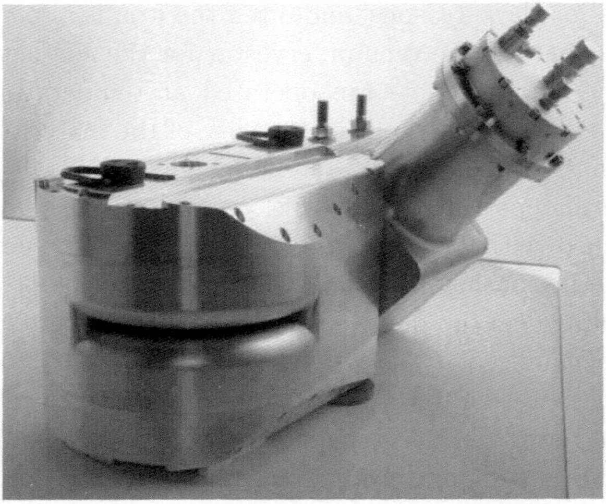

Fig. 3. A 5 T magnet for X-ray diffraction and resonant magnetic scattering.

was designed to operate with the diffraction slot in either vertical or horizontal orientations providing either vertical scattering with horizontal field at the sample, or horizontal scattering with vertical field at the sample; the rotation could be achieved by direct mounting of the magnet into the Euler cradle of a six-circle Huber goniometer. In addition to restricting the physical dimensions, this mounting imposed a weight constraint of 100 kg on the magnet assembly; the final magnet assembly employed an all-alloy cryostat and had a mass of 95 kg (Fig. 3). The radial diffraction slit width was a minimum of 10 mm; this and the span determined the minimum separation of the two HTS coil packs. These consisted each of six double-pancakes and in selected layers employed the highest-performing commercially available stainless-steel laminated HTS wire at the time with 160 A self-field critical current (T = 77 K). Ti-alloy plates were employed to cope with the 12 t compressive force generated between the two coil packs at full field. Cryo-cooling was by a two-stage pulse-tube refrigerator (Cryomech PT-810) mounted at 45° to the diffraction slot so that operation in either of the required orientations could be achieved. Cooling performance of the cryo-cooler was reduced by about 10% for an operational angle of 45° from vertical[6] but the system had more than sufficient cooling capacity, to the extent that the magnet was stable to beyond 220 A operation against a design operating current of 200 A. A remote valve-motor for the cold head was also employed as this contributed to weight reduction of the magnet assembly and also offered a lower level of mechanical vibration.

A similar design of magnet was later produced featuring a central field over 6 T; this resulted from the use of a partial iron circuit (allowable since it did not have to fit within a rotation stage), along with a higher winding density through the use of a single co-wound stainless-steel strip replacing the previous double-sided stainless steel laminated wire. Use was also made of a modest increase in wire performance between the two dates of manufacture, with self-field I_c (77 K) then exceeding 170 A.

3.2. *UHV operation*

HTS magnets have been constructed for scattering experiments where ultra-high vacuum (UHV) operation is a requirement. In the first case a 2.2 T wide-bore dipole was constructed around a customer-supplied UHV chamber for X-ray magnetic circular dichroism (XMCD) experiments (Fig. 4).[7] The cryo-cooler used was

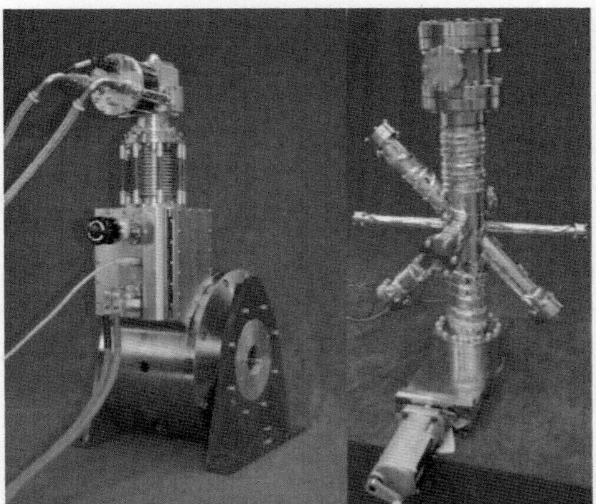

Fig. 4. A 2.2 T magnet (left), housing a UHV chamber (right) for XMCD.

a two-stage type GM-type for a total thermal load estimated at 6 W and 20 W at the first and second stage respectively. The pole gap of the magnet was 90 mm and the access bore through the pole direction was 44 mm to accommodate the XMCD chamber with nine access ports for sample manipulation and XMCD analysis. For bake-out, the chamber itself was wound with nichrome ribbon element and thermally insulated from the body of the magnet with a high temperature insulating blanket. A water cooling circuit was added to the magnet poles to limit the temperature rise; in practice the cryostat was stable at 30°C with the UHV chamber at 100°C.

The XMCD experiment measures the difference between two X-ray absorption spectra taken either by switching the helicity of the circularly polarized light or of the magnetization direction of the sample. High reversing rate for the magnet is thus a distinct advantage; in this case the ramp rate of the magnet was limited by the peak voltage of the bipolar power supply, but provided a full field reversal in around 40 secs. Over the range of ± 1 T, the field ramp rate is equivalent to 10 T/min. In such a ramp the maximum measured temperature rise of the HTS coils was less than 1 K. With continuous ramping at 1 A/s the coil temperatures stabilized at around 20 K, equating to an average ramping loss (AC loss) of 1.7 W using the manufacturer-provided cooling power capacity of the refrigerator. This loss is close to that predicted (1.3 W) from models for frequency-independent magnetic hysteresis loss.[8] Extrapolating this to a full-field reversal in a time of 1 sec indicated a likely rise in coil temperature of 3 K.

The second magnet was designed for operation *within* a UHV chamber on the BOREAS beamline for magnetic spectroscopy and scattering at the ALBA synchrotron light facility in Spain.[9] Achieving the desired field and field uniformity with large optical and sample access (50 mm pole gap) with flexible sample manipulation, were the main design challenges for the system, in addition to the stringent UHV requirements. Additionally, a low fringe field was desired. The magnetic structure is again presented as a dipole with room temperature iron circuit and poles surrounding the HTS coils. The magnet is shown in Fig. 5; it is able to be rotated in the chamber in order to

Fig. 5. UHV compatible magnet with wide optical access angles for magnetic spectroscopy and scattering.

provide magnetic field both along and perpendicular to the incident beam direction, with ±46° and ±10° of rotation about its vertical axis in the longitudinal and perpendicular field orientations, respectively. Additionally, the sample holder can rotate independently of the magnet, and also translate by ±10 mm in the direction perpendicular to the ±46° access. Employing a two-stage GM cryo-cooler (Sumitomo Heavy Industries CH208L), the operating temperature of the coils was 21 K at the 2 T maximum central field with an operating current of 150 A. An additional requirement on the magnet was for a sweeping field performance of 1.6 T/min for a succession of full field cycles (±2 T) for up to 10 cycles. Under these conditions, the operating temperature rose to 25 K. The UHV compatibility with the magnet mounted inside the chamber was achieved by retaining a separate vacuum for the HTS coil space, and allowed for a bake-out temperature of 100°C on the exterior surface of the magnet with the cryo-cooler operating to protect the HTS coil assembly. Vented Viton O-ring seals were used on all magnet-to-UHV boundaries and a metal seal CF flange on the UHV-to-atmosphere interface. In addition to the main HTS field coils, additional copper coils were installed to provide superimposed DC fields of up to 3 mT vertically and 13.3 mT horizontally, and at up to 50 kHz AC excitation.

3.3. *NEXAFS*

One of the largest HTS-110 magnet systems to date has been for a near-edge absorption fine structure (NEXAFS) imaging electron spectrometer developed by Synchrotron Research Inc. (USA). The Large Area Rapid Imaging Analysis Tool "LARIAT" provides 3D chemical imaging capability, wherein photo-emitted electrons follow, in helical orbits, the lines of magnetic field from sample to detector, forming a parallel imaging system with high collection efficiency and large depth of field.[10] It provides a non-destructive, depth-selective measurement of electronic structure, chemistry and bond orientation. Data is acquired as a series of images at each primary X-ray energy (providing the chemical selectivity), and can be displayed as a linear combination of images or as extracted spectra. The electron cyclotron radius determines to first order the imaging resolution, so high sample fields are advantageous, and variation of the detector field provides an image zoom capability.[11] The first LARIAT was based on a 1 T electromagnet; the HTS magnet upgrade in LARIAT II, with an 8.5 T sample field (110 mm RT bore) and 0.6 T detector field (110 mm RT bore) improved resolution 10-fold to approximately 7 μm with an image magnification of up to 16 times. The large bore provides for imaging of samples up to 20 mm $\times$ 20 mm with a collection speed of 725/h employing a 4000 $\times$ 4000 pixel CCD camera for LARIAT-II. The LARIAT instruments were commissioned on National Institute of Standards and Technology beamlines at the National Synchrotron Light Source (NSLS) (Fig. 6), and are currently being transferred to the new NSLS-II facility.[12]

3.4. *Neutron beamlines*

Similar magnets have also been supplied into neutron beamline facilities. The first produced at HTS-110 was for neutron reflectometry, small-angle neutron scattering and time-of-flight in elastic neutron scattering at the OPAL neutron facility of the Australian Nuclear Science and Technology Organization (ANSTO).

A 5 T split-pair magnet (Fig. 7) with horizontal field and large accessibility angles ($\pm$8–12° in the transverse direction and $\pm$16° in

Fig. 6. The LARIAT- II instrument *in situ* at the NSLS beamline.

the axial direction) was developed,[5] with capability for mounting samples of up to 50 × 50 mm in size. The vertical sample entry port was compatible with a closed-cycle cryo-furnace (Janis SHI 4-800). Sapphire windows of 170 mm diameter were employed at each of four optical ports, sealing a beam tube constructed from carbon-fiber. This magnet did not employ a soft iron circuit for weight considerations; internal forces between the coil pack were calculated at 55 t at full field, and this was supported by a combination of Ti and composite plates. The magnet had an operating temperature of 20 K with the design field achieved at 192 A.

A more recent magnet for time-of-flight experiments again used an iron circuit but with the aim of very large optical access angles, in this case 150° in the scattering plane; a room temperature bore of 80 mm and a transverse split of 80 mm provided 40° optical access in both vertical and transverse directions (Fig. 8). The very compact dimensions required a high engineering current density which was achieved with Bi-2223 wire of 190 A (77 K self-field) performance.

Fig. 7. A 5 T split-pair magnet (centre) with large sample volume at ANSTO.

3.5. *The opportunity for energy savings in storage rings*

Generally, energy efficiency has not been a major selling attribute for HTS-110 magnets, with the key advantages being instead the system compactness or robustness in operation. However in the case of the bending magnets for a synchrotron storage ring, which together might consume over 1 MW of power along with substantial cooling water requirements, there is a compelling energy efficiency argument in favor of an HTS-based ring architecture. Responding to growing concerns over resource usage, especially for planned new rings, HTS-110 in collaboration with Brookhaven National Laboratory (BNL), undertook a retrofit demonstration in which the copper coils of a

Fig. 8. A 2.2 T magnet for neutron time-of-flight experiments.

standard vacuum-ultraviolet (VUV) storage ring bending dipole from NSLS at BNL were replaced with equivalent ampere-turn HTS coils. The bending dipole copper electromagnets were of *c*-type construction, providing a central field of 1.41 T. A spare iron yoke was shipped to HTS-110 facilities for the coil replacement. In this case, the two coils were mounted in o-ring sealed cryostats consisting of aluminum shells with stainless-steel lids. Given the length of the coil, two GM-type cold heads were employed (Cryomech AL125), one at each end of the magnet and each linked to both coil packs; the two cold heads were operated from one compressor (Cryomech CP970). In assembly, each individual coil pack was fitted separately into the yoke then interconnected within manifolds at each end. The wire used in this magnet was the high-strength type of 145 A performance (77 K, self-field); the operating temperature at full field was 43 K, with an operating current of 111.2 A. The completed magnet was shipped back to BNL for testing, where the magnetic field profile was mapped, and this compared favorably with the original copper coiled system.[13] In comparison to the Cu-coiled original version, the HTS magnet consumed approximately half the electrical power (7.5 *vs.* 13.1 kW). In addition, in this design the HTS coils and cryostat were of smaller cross-section than the original copper coil envelope and this widened the aperture for beamline equipment. The original and retrofitted magnets are shown in Fig. 9. Even in this case of an isolated one-magnet retrofit there were thus tangible benefits in energy consumption and utility; greater efficiency gains would be possible with a ring of magnets operating from a centralized cryogenic plant. The success in replacement of the copper coils with HTS coils in an existing magnet demonstrated that retrofitting of an existing synchrotron ring would be viable.

Following the demonstration at BNL, HTS-110 supplied a second HTS dipole system for the then-planned new synchrotron ring at the National Synchrotron Radiation Research Center (NSRRC) in Taiwan.

In this project a new iron *c*-type yoke was first tested with Cu coils in Taiwan prior to fitting with HTS coil packs at HTS-110. The yoke was made from 1 mm low-carbon steel laminations, with a pole gap of 46 mm and a pole-face width of 172 mm; the central field was designed to be 1.191 T providing a bend angle of 7.5° and bend radius of 8.4 m for the beam energy of 3 GeV.

Originally the design target was for coil refrigeration with sub-cooled nitrogen but the wire performance at the time of manufacture was insufficient to realize this target in a cost-effective manner. The cryogenic design then was similar to that in the BNL project, except that the yoke could be split in the center-plane allowing the cryostat to be fully assembled before integration with the yoke. The cold heads selected in this case were pulse-tube types of lower nominal cooling power (Cryomech PT-60), but providing advantages of low maintenance and low vibration compared to GM coolers, though with

Fig. 9. The bending dipole magnet with Cu coils (top); retrofitted with HTS coils (bottom).

some penalty in efficiency. Again connected to a common compressor (Cryomech CP289C) which consumed 7.5 kW (50 Hz) input power, this provided an operating temperature at the maximum field tested (1.4 T) of 54 K at an operating current of 132 A; cool-down of the system was achieved within 20 h, with a base temperature of around

45 K. Higher performance wire was available for the coils, (160 A at 77 K, self-field) and this together with flux-deflector plates that were designed into the coil packs were responsible for the shift upward in both operating current and temperature.

Field measurements were performed at NSRRC, with field distributions and integrated field deviations comparable to the Cu-coiled system.[14] Estimates made at the time of the development indicated that if the 48 dipole magnets forming the ring of the Taiwan Photon Source were replaced with HTS versions, the energy consumption when operated *via* a closed liquid nitrogen (LN_2) circuit would be decreased by a factor of 3.6 relative to copper-coil dipole magnets, and there would be an almost 20-fold reduction in cooling water requirements. Operation with a sub-cooled LN_2 circuit remains an attractive target for the technology and subsequent improvements in wire performance now make this a more likely possibility.

4. High Uniformity Magnets

A similar design approach to beamline magnets has been used to construct HTS magnets for high-performance metrology applications, including a 3 T dipole magnet for NMR relaxometry and solenoid magnets for NMR spectroscopy at 200 MHz (4.7 T), and with a 400 MHz (9.4 T) system and a 3 T small-mammal-sized MRI system currently in construction. The key parameter for the MR-class magnets is field uniformity and stability, with HTS technology able to meet these demands whilst bringing advantages of compactness, freedom from liquid cryogens, the ability to be turned on and off as required, plus potential system mobility. These are new and very attractive attributes for NMR users.

4.1. *Relaxometry/metrology*

A variable field dipole magnet with iron yoke and poles was designed for application in NMR relaxometry experiments, in association with Stelar s.r.l. of Italy. Fast Field Cycling (FFC) relaxometry gives a measurement of the nuclear spin-lattice relaxation rate as a function

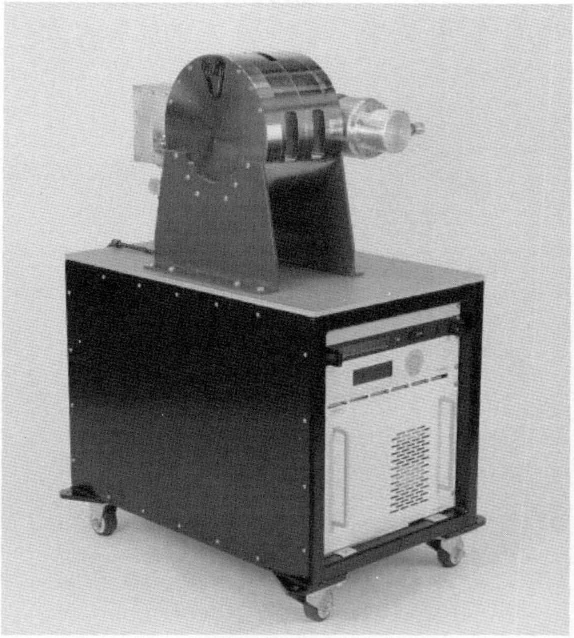

Fig. 10. A 0–3 T variable field magnet for NMR relaxometry.

of the applied magnetic field strength, providing a unique characterization of the local dynamics of molecules over a wide frequency range.[15,16] The 3 T magnet extends the range of field-cycling relaxometry to higher fields which is necessary in applications such as the acquisition of dispersion (NMRD) profiles of MRI contrast agents being developed for the latest generation of 3 T MRI systems.

The NMRD measurement requires field uniformity across the sample of better than approximately 50 parts per million (ppm) over the applied field range. The present magnet design (Fig. 10) achieves this through a combination of pole design and coil placement optimized for uniformity even as the iron yoke transitions through saturation. Sample access consists of a slot through the magnet yoke of dimensions 25 × 150 mm, with the smaller dimension matching the pole gap of the iron-dominated design. In operation, the NMR probe is housed in a removable cassette sized to fit accurately in the access slot. The magnet is driven by a high-stability power supply

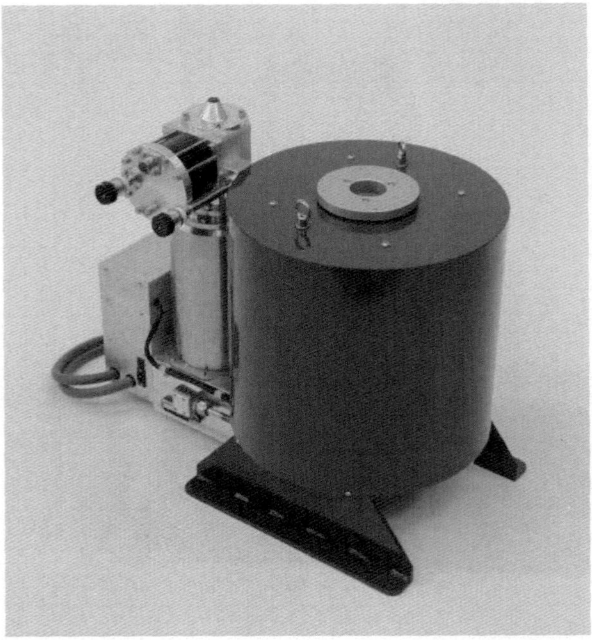

Fig. 11. A 200 MHz (4.7 T) solenoid with 54 mm clear bore for NMR spectroscopy.

(0.5 ppm/h, Danfysik, Denmark), which has proven more than sufficiently stable for the NMRD measurement. The cryo-cooler selected is a two-stage GM unit (SHI CH208L) providing coil operating temperatures in the mid-20 K region for an operating current of 160 A. Room temperature shims have also been designed to further increase field uniformity across small samples (to order 1 ppm), and although these are not required for the relaxometry experiments, they have been used for limited spectroscopy experiments. This magnet could also be operated from a bipolar power supply and is ideal for other applications in metrology such as in Hall sensor calibration since the field can be specified to NMR levels of accuracy.

4.2. *HTS magnets for high resolution NMR spectroscopy*

There is considerable interest in developing HTS magnets for applications in magnetic resonance, either because of the capability for very high fields that enable NMR beyond 1 GHz, or because of the possibility of compact systems with simplified cryogenics compared to LTS technology.[18,19] The driver behind development of an HTS magnet-based NMR system at HTS-110 is the potential for deployment of NMR in non-traditional environments, such as in chemistry laboratories, pilot plants, and industrial settings. In such situations, direct access to NMR as a measurement and quality assurance tool is highly desirable, especially to address a growing interest in in-line or at-line reaction monitoring for chemical industries, however conventional LTS magnets have proven too cumbersome for such application, given their need for a low vibration environment, freedom from nearby ferromagnetic objects, and facilities for re-filling of liquid helium and nitrogen, along with exhaust ducting and oxygen monitoring in case of a magnet quench.

The attractive attributes of an HTS solution center on the one hand on a very compact magnet envelope, and on the other with the high operating temperature affording a cryogenic mechanical structure which can be made largely immune to local vibration, and which in combination provide the opportunity for a mobile system. Effectively, the HTS NMR magnet can be considered as having the convenience of existing low-field permanent magnet benchtop systems but with the power of high-field systems. In contrast to the relatively stiff cold-mass support system in the HTS magnet, along with the absence of a solid radiation shield, the traditional LTS NMR magnet has the superconducting coil carried by low-cross-section supports, with surrounding helium and nitrogen containment vessels, and a solid, high reflectivity radiation shield, all contained within an outer vacuum chamber. There, the design imperative is to minimize helium boil-off to ensure long periods of operation of the magnet between helium refills, however a direct outcome is a sensitivity to vibration, especially floor-borne mechanical noise; this can induce relative movement between the solid radiation shield, for example, and the superconducting coil leading to currents being induced in these components which in turn produce unwanted contributions to the magnetic field at the sample. In many cases buildings can require structural modification to provide the necessary level of vibration isolation, and it would not be tenable to consider a *mobile* LTS system.

For maximum compatibility with existing NMR spectroscopy hardware, HTS-110 has developed a series of solenoid NMR magnets suitable for high resolution NMR. This is perhaps the most demanding application of HTS magnets, arising through the combined requirements of both spatial field uniformity and temporal stability. The concerns regarding the ability to deliver spatial uniformity arise from the nature of the conductor itself, being a relatively broad tape compared to the more typical LTS round wires, and also the pancake style construction compared with a solenoidal layer wind. The concern for temporal stability primarily arises from the lack of a truly persistent-current mode of operation of the magnet, at least for Bi-2223 magnets. This is a combination of both a lack of adequate jointing technology for Bi-2223 wire (whereas there have been some concrete developments for YBCO)[20] and the relatively large flux creep observed with these materials, evidenced in low-index power-law transitions of conductance to the normal state, especially in elevated magnetic fields. The lack of a persistent mode also means that superconducting shim coils, which are usually employed for coarse adjustment of field uniformity in the as-built LTS magnet, are not an ideal option for the HTS magnet, since each shim circuit would need a dedicated and continuously energized power supply.

Treating the issue of spatial uniformity, the approach taken for coarse shimming in the HTS magnet is to replicate the passive ferromagnetic shimming of MRI magnets. Here, small pieces of ferromagnetic material are placed around the bore of the magnet to correct for the non-uniformity to the level of about 1 ppm, beyond which a standard set of room-temperature electrical coils can be used to deliver the remaining desired homogeneity. Passive shimming requires the use of field-mapping to establish the as-built field profile, and then efficient correction algorithms lead to a rapid reduction in field variation such that acceptable field quality is achieved within a few iterations. It is an advantage that the proximity of the ferromagnetic material to the sample zone provides for an efficient correction means, so that relatively large levels of inhomogeneity can be corrected compared with electrical shim sets, which are anyway limited in the maximum power that can be dissipated within the bore of a magnet.

The lack of a persistent mode forces the HTS magnet to be operated with a continuously connected power supply, so the requirement in part converges on the stability of the supply. Suitable high-stability units are available, with current stabilization provided by feedback from a high-quality current transducer. For these spectroscopy magnets, we have again employed units with specified drift rates of order ±0.5 ppm over 1 h and ±2 ppm over 8 h (Danfysik). This level of stability is sufficient for a standard NMR field-lock to function, enabling high resolution NMR, at least after an initial settling period.

Immediately following magnet energization, decay of magnetization currents can lead to a large initial field drifts. The wide conductor with untwisted filaments allows the establishment of magnetization super-currents which circulate in each tape (with somewhat undefined closure paths) in response to locally increasing applied magnetic field, and which sum to produce a central field contribution in opposition to the main field.[21] Unlike an LTS magnet, where the magnetic field drift is downward in response to the decay of transport current through residual resistances in the coil and persistent joints, decay of magnetization currents in this case yield an increasing background field. The phenomenology of these currents is still being investigated, particularly in the context of insert coils for NMR beyond 1 GHz,[21] as well as mitigation strategies.[19] The currents act in concert with environmental factors (principally temperature) and any settling behavior of the power supply after turn-on to produce the observed time dependence of the central field; at long times this comprises any residual monotonic increase in field from the exponentially decaying magnetization currents, with some shorter-time structure arising from environmental changes (temperature).

For an example, a 200 MHz system has been supplied for reaction monitoring; this consists of the 4.7 T HTS magnet, a new compact FPGA-based digital spectrometer from RS2D (France), Bruker room temperature electric shim stack and Protasis (USA) capilliary flow probe.[22] The system was installed in a mobile cart, and has met required key performance specifications which include a spectral line width of less than 1.5 Hz for chloroform in deuterated acetone, and a splitting of the anomeric proton in sucrose of greater than 20%. The system is successfully being employed for reaction monitoring; in this case the NMR spectra can be acquired with the field unlocked, and realigned with post-processing in software. In more recent work, we have established a chloroform line width of <0.5 Hz for a spinning sample with multiple-average acquisition, and <10 Hz width at a height of 0.55%. In all cases there is no evidence of mechanical vibration in the NMR spectra. The field is able to be locked *via* a deuterium lock channel for long periods and these magnets appear compatible with auto-shimming algorithms. Currently HTS-110 has a 9.4 T magnet in construction (400 MHz), extending the potential range of applications for this new class of NMR magnet; maintaining a target operating temperature of 20 K, this will employ Bi-2223 wire of class-leading critical current performance, 200 A at 77 K, self-field.

5. Summary

Several application areas have proven attractive for HTS magnets; while there is significant focus on utilizing the very high upper critical field of these conductors for magnetic fields beyond 20 T (especially for >1 GHz NMR), there are nevertheless useful applications that instead benefit from the relatively high operating temperature which can be leveraged either by specification of a very compact cryogenic envelope, where the physical separation between the cold mass and ambient environment is very small, or through the implementation of robust coil mechanical supports, which then provide immunity to environmental vibration (including that of possible cryo-cooler action) and the ability to be oriented or moved at will. Many of the magnets manufactured by HTS-110 have been employed in synchrotron or neutron beamlines, mostly with the benefit of compact dimensions and wide optical access, but an energy-conservation role has also been identified at the facility level for synchrotrons. This development in magnets has occurred largely on the basis of the utility of Bi-2223 conductor, and has benefitted from the improvement in critical current and long-length uniformity of this conductor over the years.

References

1. D. M. Pooke, L. L. Tallon, R. G. Buckley, G. Todd, S. S. Kalsi, A. Szczepanowski, A. Sidi-Yekhlef, G. Snitchler, H. Picard, R. E. Schwall, B. MacKinnon, and R. Neale, Cryo-Integrated HTS Ion-Beam Switching Magnet, In *Proc. 15th Int. Conf. Magnet Technol. 1997(MT-15)*, L. Lin, G. Shen, and L. Yan (eds.). Beijing: Science Press, 789–792 (1998).
2. D. M. Pooke, V. Chamritski, S. Gibson, M. P. Staines, M. Fee, and R. G. Buckley, A versatile laboratory electromagnet with HTS coils, *IEEE Trans. Appl. Supercond.* **14**(2), 1202–1205 (2004).
3. R. McMahon, S. Harrison, S. Milward, J. Ross, R. Stafford Allen, C. Bieth, S. Kantas, and G. Rodrigues, Design and manufacture of high temperature superconducting magnets for an electron cyclotron resonance ion source, *IEEE Trans. Appl. Supercond.* **14**(2), 608–611 (2004).
4. R. W. McGhee, E. E. Burkhardt, A. Berryhill, and D. M. Coffey, Design and test results of a BSCCO-2223 magnet for gyrotron application, *IEEE Trans. Appl. Supercond.* **15**(2), 1189–1191 (2005).
5. D. M. Pooke, V. Chamritski, M. Fee, S. Gibson, B. T. King, J. L. Tallon, M. Meissner, R. Fayerherm, S. R. Olsen, S. J. Kennedy, and R. A. Robinson, HTS 5 tesla synchrotron and neutron beamline magnets, *IEEE Trans. Appl. Supercond.* **19**(3), 1372–1375 (2009).
6. C. Wang, Efficient 10 K pulse tube cryocoolers, *Cryocoolers* **13**, 132–140 (2005).
7. T. Huang, X. Gao, D. Pooke, V. Chamritski, N. Briggs, M. Christian, S. Gibson, J. Mitchell, M. Miles, and J. deFeijter, A commercial HTS dipole magnet for X-ray magnetic circular dichroism (XMCD) experiments, *IEEE Trans. Appl. Supercond.* **22**(3), 4203504 (2012).
8. W. T. Norris, Calculation of hysteresis losses in hard superconductors carrying AC: Isolated conductors and edges of thin sheets, *J. Phys. D: Appl. Phys.* **3**, 489–507 (1970).
9. Z. Lazić, V. Chamritski, D. Pooke, S. M. Valvidares, E. Pellegrin, S. Ferrer, X. Granados, and X. Obradors, Integrating UHV (ultra high vacuum) and HTS (high temperature superconducting) magnets for X-ray synchrotron based experiments, *J. Phys.: Conf. Ser.* **425**, 102003 (2013).
10. P. Kruit and F. H. Read, Magnetic field paralleliser for 2pi electron-spectrometer and electron-image magnifier, *J. Phys. E: Sci. Instrum.* **16**, 313–324 (1982).
11. D. A. Fischer, C. Jaye, K. Scammon, C. Weiland, P. Sobol, and E. L. Principe, A novel large area imaging NEXAFS spectrometer enabling rapid combinatorial chemical and structural analysis, Presented at *2014 MRS Spring Meeting & Exhibits*, San Francisco, CA (Apr. 2014).

12. R. Reininger, J. C. Woicik, S. L. Hulbert, and D. A. Fischer, NIST NSLS-II spectroscopy beamline optical plan for soft and tender X-ray spectroscopy and microscopy (100 eV to 7.5 keV), *Nucl. Instrum. Methods Phys. Res. A* **649**, 49–51 (2011).
13. J. Muratore, J. Escallier, G. Ganetis, A. K. Ghosh, R. C. Gupta, P. He, A. Jain, P. Joshi, P. Wanderer, M. Fee, and M. Christian, Magnetic field measurements of an HTS retrofit synchrotron dipole, *IEEE Trans. Appl. Supercond.* **21**(3), 1653–1656 (2011).
14. C. K. Yang, C. S. Hwang, J. C. Jan, F. Y. Lin, C. H. Chang, M. Fee, and M. Christian, Design, fabrication, and performance tests of a HTS superconducting dipole magnet, *IEEE Trans. Appl. Supercond.* **22**(3), 4000804 (2012).
15. F. Noack, NMR field cycling spectroscopy: Principles and applications, *Prog. NMR Spectrosc.* **18**, 171–276 (1986).
16. E. Anoardo, G. Galli, and G. Ferrante, Fast field cycling NMR: Applications and instrumentation, *Appl. Magn. Reson.* **20**, 365–404 (2001).
17. Y. Iwasa, J. Bascuñán, S. Hahn, M. Tomita, and W. Yao, High-temperature superconducting magnets for NMR and MRI: R&D activities at the MIT Francis Bitter Magnet Laboratory, *IEEE Trans. Appl. Supercond.* **20**(3), 718–721 (2010).
18. T. Kiyoshi, S. Choi, S. Matsumoto, K. Zaitsu, T. Hase, T. Miyazaki, A. Otsuka, M. Yoshikawa, M. Hamada, M. Hosono, Y. Yanagisawa, H. Nakagome, M. Takahashi, T. Yamazaki, and H. Maeda, HTS-NMR: Present status and future plan, *IEEE Trans. Appl. Supercond.* **20**(3), 714–717 (2010).
19. H. Maeda and Y. Yanagisawa, Recent Developments in high-temperature superconducting magnet technology (review), *IEEE Trans. Appl. Supercond.* **24**(3), 4602412 (2014).
20. Y. Park, M. Lee, H. Ann, Y. H. Choi, and H. Lee, A superconducting joint for $GdBa_2Cu_3O_{7-\delta}$-coated conductors, *NPG Asia Mater.* **6**, e98 (2014).
21. Y. Yanagisawa, H. Nakagome, K. Tenmei, M. Hamada, M, Yoshikawa, A. Otsuka, M. Hosono, T. Kiyoshi, M. Takahashi, T. Yamazaki, and H. Maeda, Operation of a 500 MHz high temperature superconducting NMR: Towards an NMR spectrometer operating beyond 1 GHz, *J. Mag. Res.* **203**, 274–282 (2010).
22. B. D. Nguyen, M. T. Zell, D. A. Foley, B. L. Marquez, D. Strand, C. Milling, J. Norcross, J. Voccio, D. Pooke, M. Mallett, and J. Muller, A compact, portable 4.7 T driven NMR system for reaction monitoring, Presented at the *54th Experimental NMR Conference*, Pacific Grove, CA (2013).

Chapter 3.16

Conduction-Cooled Magnets for Industrial Applications

Takeshi Kato
Sumitomo Electric Industries, Ltd.,
1-1-3, Shimaya, Konohana-ku, Osaka 554-0024, Japan

Conduction-cooled magnets are useful and easy to operate. Their advantages and their configuration and application are explained. The technique required to protect the magnet is also described.

1. Introduction

A conduction-cooled magnet needs no liquid refrigerant such as liquid helium and has various advantages.

1. Easy to operate
   Liquid refrigerant can cause serious accidents when warmed, and handling liquid helium requires special knowledge and skills. The pressure in a containment vessel can increase dangerously when liquid helium flash-vaporizes. A conduction-cooled magnet requires only electricity, needing neither a refrigerant nor special skills.
2. No insecurity of refrigerant supply
   Helium's price is rising because its production is decreasing, while demand in developing countries is rapidly increasing. At worst, liquid helium may become unavailable, simply terminating any operation that requires it.

Magnets using Bi-2223 HTS wire have additional features beyond those that use metallic superconducting wire.

3. Quenchless
   A slight temperature rise leads to a very significant deterioration in the critical current of metallic superconducting wires, triggering a quench. In contrast, Bi-2223 HTS wire has a much higher critical temperature than metallic superconductors. If the temperature of the Bi-2223 HTS wire rises by several Kelvin from its operating temperature around 20 K, the critical current does not decrease significantly, due to its wider margin of safety. Furthermore, any temperature rise due to thermal disturbances is very small in the case of 20 K operation because heat capacities at 20 K are much higher than at 4 K. Therefore, quenchless operation is possible.
4. Higher ramping rate
   The wider margin between its operational and critical temperatures makes a Bi-2223 HTS magnet dramatically more stable, as described above. Thanks to this, a Bi-2223 HTS magnet can generate a magnetic field about 10 times faster than the conduction-cooled conventional superconducting magnets.
5. Wide variety of operating patterns
   A Bi-2223 HTS magnet can be operated across a wider temperature range. For example, a Bi-2223 HTS magnet can be operated until it heats up as far as 30 K. The Bi-2223 HTS magnet can generate a magnetic field with turning off refrigerator for some predetermined time. This is very useful as it allows the complete elimination of any vibration from a refrigerator.

2. Configurations

2.1. *Pancake Coil*

Since Bi-2223 wire is configured as a thin tape, a pancake coil structure is preferable because it is easier to wind coils.[1–3] Figure 1 shows an example of a double-pancake coil, which has two layers of pancake coil with a transition portion at the innermost turn. The double-pancake coil structure makes electrical connection of the coils easy because the ends of the wire are at the outermost turns.

Fig. 1. A double pancake coil (left) and a stacked coil (right).

The wire is insulated with a wrapping of polyimide tape. Glass fiber reinforced plastic (GFRP) sheets are placed on the layers to strengthen the electrical insulation between the wire and cooling plates. Figure 1 shows the appearance of a stacked coil. Cooling plates are properly inserted between double pancake coils.

2.2. *Magnet*

The coil is enclosed in a cryostat, as shown in Figs. 2 and 3. The cryostat is usually made of stainless steel, and employs an internal vacuum for thermal insulation, using multi-layer thermal insulation on a thermal shield. In some cases, the multi-layer thermal insulation can be placed directly on the coil for compactness, not needing a thermal shield because the cooling capacity is high enough to maintain the coil at around 20 K. The coil is fixed with a low thermal conductivity material such as GFRP. HTS current leads are usually employed to reduce the thermal heat leakage across the final step from around 60 K to 20 K. A Bi-2223 wire that includes a small amount of gold in the silver sheath shows excellent properties: Both low heat leakage and low heat generation. The coil is conduction-cooled through a highly thermo-conductive material such as copper

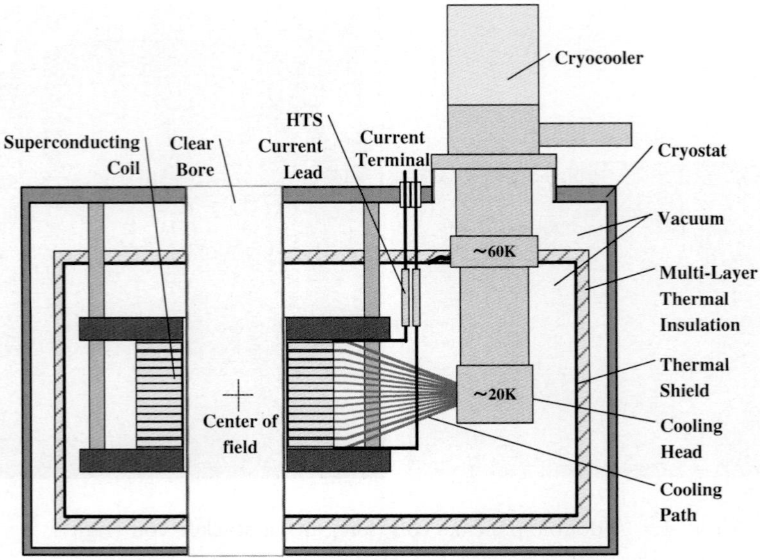

Fig. 2. Schematic view of a conduction-cooled magnet.

Fig. 3. Conduction-cooled 5 T magnet.

or aluminum. In the case of pancake coils, cooling plates between the coils can directly cool the coils, which have very low thermal resistance. In particular, the cooling structure is very effective for pancake coils placed at the upper or lower end of stacked coils because they generate a larger amount of DC and AC losses due to the normal field applied to the wire. From the point of view of conduction-cooling, the pancake coil structure is better than a solenoid coil.

3. Conduction-Cooled Magnets

A 20.1 T magnet utilizing an insert magnet made with Bi-2223 HTS wire was developed by Toshiba Corporation for Tohoku University in 2013.[4] The room-temperature bore was 52 mm. The backup field generated by NbTi and Nb_3Sn superconductors was 15.48 T, and the Bi-2223 insert magnet generated 4.45 T. Three GM cryo-coolers cooled the magnet and radiation shield.

Many applications have been published, including deflection magnets, flywheels, MRI, SMES, induction heaters, crystal pulling, and so on.[5–11] Some other examples are listed in Table 1.[12] These magnets were built using only Bi-2223 HTS wire. Their magnetic fields range from 5 T to 10 T. The very fast sweep rate of 5 T per 30 seconds was achieved by a 5T-100 mm magnet. High field and large-bore magnets used a high-strength type wire (commercial name Type HT-SS) to deal with the high hoop stresses.

Table 1. Examples of conduction-cooled magnets built by Sumitomo Electric Industries, Ltd.

Center field	+/−5 T	+/−10 T	+/−5.7 T	+/−5 T
R.T. bore	100 mm	100 mm	150 mm	300 mm
Current	175A	250A	250A	250A
Inductance	4 H	11 H	5 H	20 H
Stored Energy	61 kJ	344 kJ	156 kJ	625 kJ
Sweep rate	5 T/30s	10 T/10 min	5 T/50s	5 T/180s
DI-BSCCO Wire type	Type H	Type HT-SS Type H	Type HT-SS Type H	Type HT-SS Type H

4. Conduction-Cooled Magnet for Magnetic Property Measurement

Permanent magnets using rare-earth elements have become essential in a variety of applications, thanks to their very strong magnetic fields. Their magnetic properties have been measured with a B–H curve tracer or vibrating sample magnetometer (VSM). A conventional copper magnet can generate only a few Tesla. A pulse magnet can generate over 10 T, but it induces eddy currents in specimens because of its rapid ramping speed, which degrades the accuracy of the measurement.

A Bi-2223 HTS magnet can generate a strong magnetic field at an appropriate ramping rate. The magnet can improve measurement accuracy over 2 T and increase the measurement throughput compared to a conventional superconducting magnet.

Figure 4 shows the conduction-cooled 6T-70 mm magnet used for B–H curve measurement. The specifications are listed in Table 2. The sweep rate is 6T per 30 seconds. The time required to provide 1.5 cycles of a triangular AC wave form to measure a B–H curve is only three minutes (Fig. 5).

The temperature profile during this operation is shown in Fig. 6. A magnetic field of 1.5 cycles in three minutes with an interval of one

Fig. 4. Conduction-cooled magnet for B–H curve measurement.

Table 2. Conduction-cooled magnet for B–H curve measurement.

Center field	+/−6 T
R.T. bore	70 mm
Sweep rate	6 T/30sec.
Current	250A
Inductance	1 H
Homogeneity	0.3 %/10 mm DSV
Size	0.8 × 0.3 × 0.3 m
Weight	100 kg

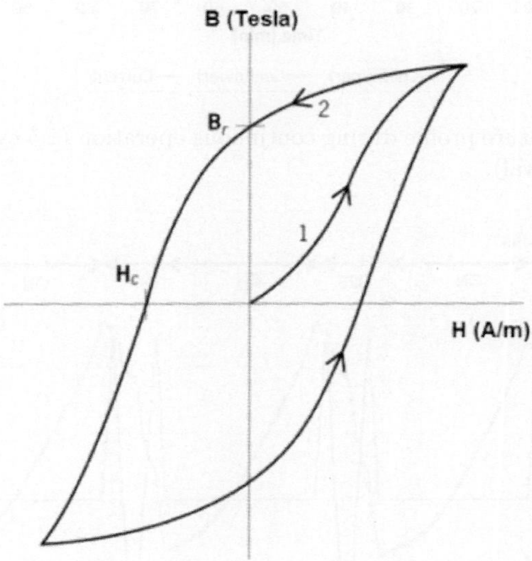

Fig. 5. Example of a B–H curve (a measurement requires 1.5 cycles of an AC magnetic field).

minute is generated repeatedly for 1.5 h. The magnet is very stable, although the maximum temperature of the coils reaches about 28 K. The magnet can also operate in a quite unique mode, as shown in Fig. 7. The cryo-cooler can be turned off during the excitation, yielding the great advantage of zero vibration. A zero-vibration operating mode is useful for various precision measurements.

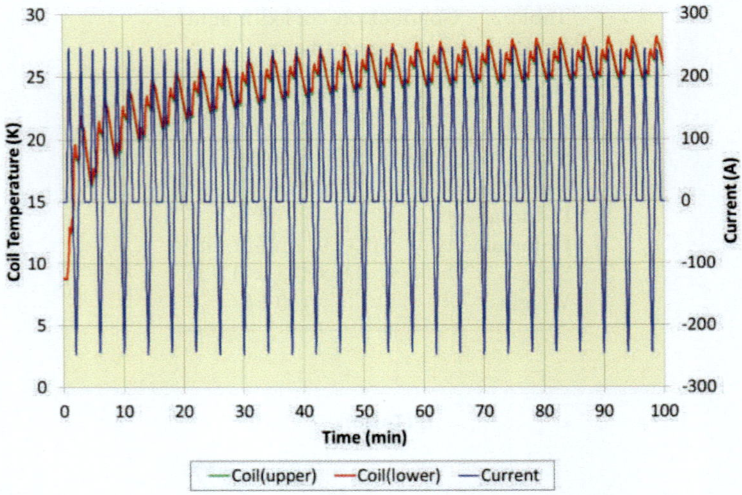

Fig. 6. Temperature profile during continuous operation (1.5 cycles in 3 minutes + 1-minute interval).

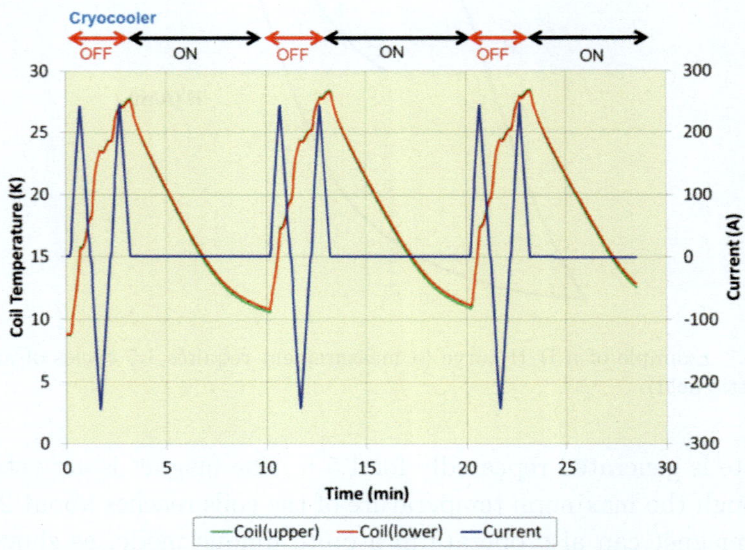

Fig. 7. Temperature profile operated with cryo-cooler turned off and on (1.5 cycles in 3 minutes + 7-minutes interval).

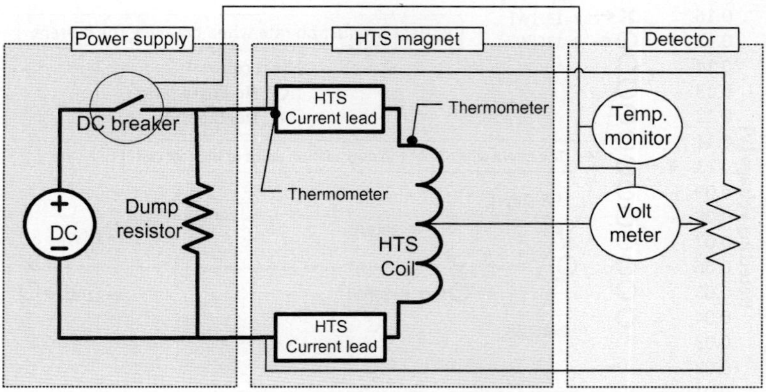

Fig. 8. An example of a basic circuit to protect an HTS conduction-cooled magnet.

5. Protection of Bi-2223 HTS Conduction-Cooled Magnets

Protection of a superconducting magnet is very important. Figure 8 shows an example of a basic protective circuit for an HTS conduction-cooled magnet. The detector monitors the temperatures and the voltage of the balance circuit, and opens the DC breaker in the event of abnormal readings, to protect the coil by rapidly shunting the current through a dump resistor. A balance circuit has commonly been used for conventional superconducting magnets because it can cancel the voltage induced during the excitation of the magnet.

The effectiveness of quench detection and protection by a balance circuit has been confirmed for actual Bi-2223 HTS coils.[13] The test coil consisted of four double-pancake coils with five cooling plates on each of their surfaces. The each coil size was inner diameter (ID) 70 mm, outer diameter (OD) 214 mm and height 9.7 mm. The center field was about 4T at 200 A. A quench was initiated by heating up the test coil. The results in Fig. 9 show that a Bi-2223 HTS magnet can be protected by a conventional protection circuit. Another point to note is that the coil can be protected with a decay time constant of four seconds even if it generates over 100 W. A Bi-2223 HTS coil can be operated stably and reliably.

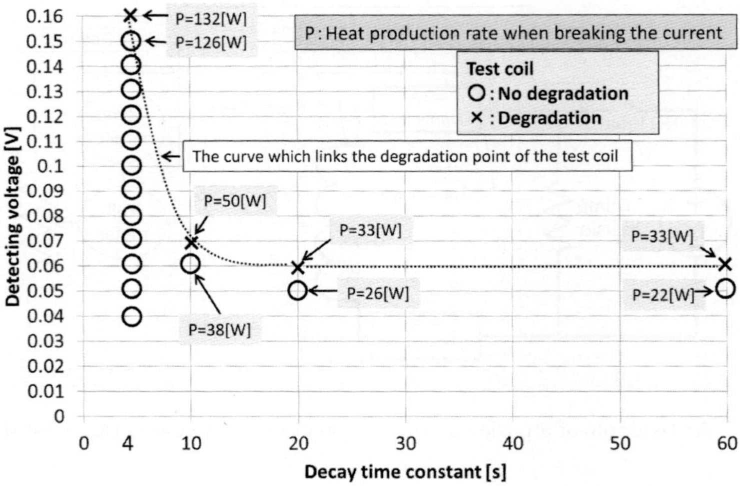

Fig. 9. Relation between decay time constant and detecting voltage for quench protection.

References

1. E. Ueno, T. Kato, and K. Hayashi, Race-track coils for a 3 MW HTS ship motor, *Physica C: Supercond.* **504**, 111–114 (2014).
2. E. S. Otabel, V. S. Vyatkin, M. Kiuchi, T. Matsushita, B. Ni, T. Kato, T. Nishimura, and R. Uetsuki, Fabrication of 1T Bi-2223 superconducting magnet with 92 mm bore diameter at 77 K, *AIP Conf. Proc.* **1573**, 754 (2014).
3. G. Nishijima, H. Kitaguchi, Y. Tshuchiya, T. Nishimura, and T. Kato, Transport critical current measurement apparatus using liquid nitrogen cooled high-T_c superconducting magnet with variable temperature insert, *Rev. Sci. Instrum.* **84**, 015113 (2013).
4. S. Hanai *et al.*, Upgraded cryogen-free 20T superconducting magnet, *IEEE Trans. Appl. Supercond.* **24**(3), 4301204 (2014).
5. K. Hatanaka *et al.*, HTS Sector Magnet for the RCNP New Injector, *Proc. The* 8th *Annual Meeting of Particle Accelerator Society of Japan*, 117–119 (2011).
6. Y. Arai, H. Seino, K. Yoshizawa, and K. Nagashima, Development of superconducting magnetic bearing with superconducting coil and bulk superconductor for flywheel energy storage system, *Physica C: Supercond.* **494**, 250–254 (2013).
7. Y. Terao, Newly designed 3T MRI magnet wound with Bi-2223 tape conductors, *IEEE Trans. Appl. Supercond.* **23**, 710–713 (2010).

8. J. H. Bae, S. H. Kim, H. J. Kim, M. H. Sohn, K. C. Seong, and H. M. Kim, Design, fabrication and evaluation of a conduction cooled HTS magnet for SMES, *Physica C: Supercond.* **469**(15–20), 1794–1798 (2009).
9. S. Kusada *et al.*, The project overview of the HTS magnet for superconducting Maglev, *IEEE Trans. Appl. Supercond.* **17**(2), 2111–2116 (2007).
10. M. Runde *et al.*, Commercial induction heaters with high-temperature superconductor coils, *IEEE Trans. Appl. Supercond.* **21**(3), 1379–383 (2011).
11. K. Tasaki, Testing of the world's largest HTS experimental magnet with Ag-sheathed Bi-2223 tapes for Si single crystal growth applications, *IEEE Trans. Appl. Supercond.* **11**(1), 2260–2263 (2001).
12. http://www.sei.co.jp/super/magnet_coil/system.html#example.
13. T. Yamaguchi, E. Ueno, T. Kato, and K. Hayashi, Quench Protection of DI-BSCCO Coil, submitted to *Physics Procedia* (2015).

Chapter 3.17

Dipole Magnet for Beam Line Switching

Jun Yoshida
Sumitomo Heavy Industries, Ltd.,
Technology Research Center,
Yokosuka, Kanagawa, Japan 237-8555

We are developing a Bi-2223 HTS dipole magnet for beam line switching for use in the cyclotron facility of RCNP, Osaka University. Exit beam lines are periodically switched by increasing and decreasing of the magnetic field between 0 T and 1.6 T with a switching time of 10 sec. A Bi-2223 coil assembly was designed with the electromagnetic force support and the suppression of temperature rise by AC loss and eddy current loss. In this chapter, we introduce this magnet as a practical example of conduction-cooled Bi-2223-HTS magnet for accelerator application.

1. Introduction

HTS wire such as Bi-2223 is expected to be applied to a compact high field magnet using a high critical field or an AC/pulse medium field magnet using a large temperature margin from the viewpoint of an accelerator application.[1–3] As a step to the latter accelerator application, we have fabricated a Bi-2223-HTS magnet for beam line switching at the Research Center of Nuclear Physics (RCNP), Osaka University in collaboration with Osaka University and Sumitomo Electric Industries, Ltd.

Table 1 summarizes specifications of this magnet. There are two entrance channels from ring cyclotron and AVF (azimuthal varying

Table 1. Main specifications of HTS-SW magnet.

Parameters	Specifications
Entrance Angle of Beam Lines	$+2^\circ$ and $+12^\circ$
Exit Angle of Beam Lines	$+2^\circ$, $+12^\circ$ and -12°
Wire	DI-Bi-2223 type HT-CA[4] (from Sumitomo Electric Industries)
Maximum Magnetic Field	1.6 T
Switching Time	10 sec
Maximum Current	200 A
Number of Turns	512 turns
Cryocooler	SRDK-408S2[5] (from Sumitomo Heavy Industries)
Cooling	Conduction Cooling
Peak Perpendicular Field	0.9 T
Coil Temperature	$\sim$10 K

field) cyclotron at upstream side of Bi-2223-HTS switching (HTS-SW) magnet. Three exit channels at downstream side can be rapidly switched by increasing and decreasing of magnetic field. Therefore, it enables beams to be delivered through experimental halls by time sharing. Maximum magnetic field is 1.6 T. The fastest switching time is 10 sec between 1.6 T and 0 T. Bi-2223 wire with reinforcing copper alloy and insulation tape are wound into some double pancake coils, which are conduction-cooled by cryocoolers. Critical current corresponding to peak perpendicular field of 0.9 T is estimated to be about 400 A at 10 K from wire typical performance,[4] which is twice as large as the maximum current.

2. Magnet Structure

Figure 1 shows the appearance of HTS-SW magnet. The magnet consists of a laminated yoke and two cryostats. The structure is similar to that of ordinary H-shaped normal conducting magnet with substitution of two cryostats for copper coil. The upper and the lower cryostat are exactly same, and are connected by four columns for support of electromagnet force.

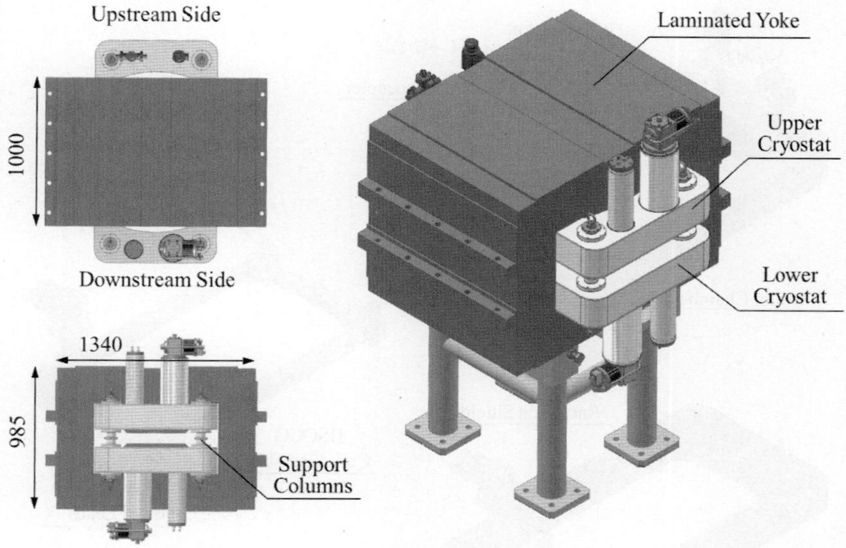

Fig. 1. Appearance of HTS-SW magnet.

Figure 2 shows appearance and inside of the cryostat. Each cryostat contains a radiation shield and a Bi-2223 coil assembly, and is equipped with a cryocooler outside of yoke. A commonly used 4 K GM cryocooler for low temperature superconducting magnet is more expensive than a 10 K GM cryocooler due to magnetic cooling storage materials. We chose a two-stage 10 K GM cryocooler because HTS-SW magnet do not require cooling capacity of 4 K. The cooling capacity of 1st stage and 2nd stage are respectively 40 W at 45 K and 6.3 W at 10 K in the used cryocooler.[5] The radiation shield and the Bi-2223 coil assembly are respectively thermally linked with 1st stage and 2nd stage of the cryocooler, whose temperature is about 40 K and 10 K. A current feeder module is located near the cryocooler to reduce temperature difference. HTS current leads by Bi-2223 wire are located between the radiation shield and the Bi-2223 coil assembly. Upper and lower Bi-2223 coil are connected in series with one power supply, and generate dipole magnetic field. There are four FRP columns between the coil assembly corners and vacuum chamber for unbalanced electromagnetic force support.

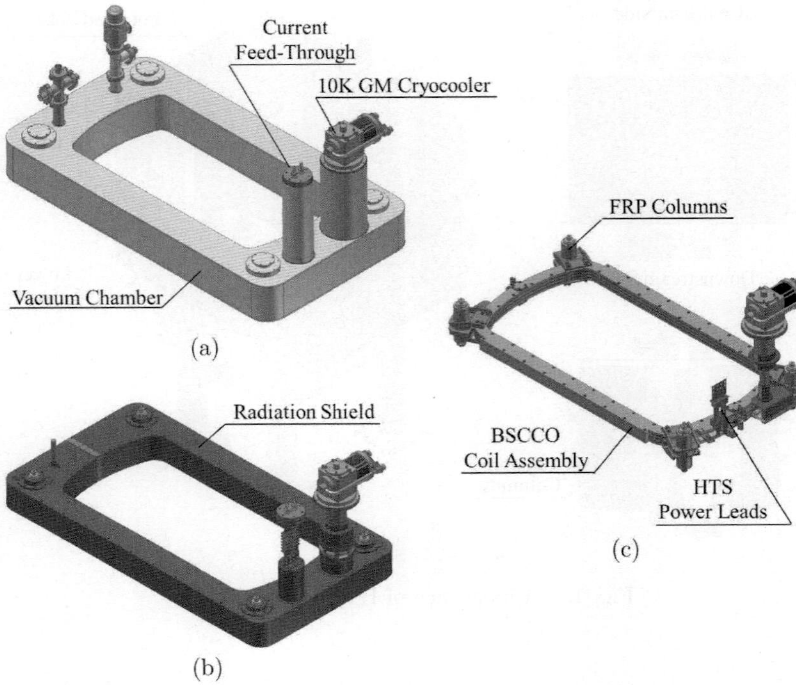

Fig. 2. Cryostat structure of HTS-SW magnet. (a) Appearance of the cryostat. (b) Inside of the vacuum chamber. (c) Inside of the radiation shield.

3. Bi-2223 Coil Assembly

3.1. *Technical issue and strategy*

There are two major technical issues for design of the Bi-2223 coil assembly. The first issue is the support for electromagnetic force. Figure 3 shows electromagnetic force pattern and magnitude obtained by electromagnetic field finite element analysis. Electromagnetic force works as an expansive force on the long-side of rectangle-shaped Bi-2223 double pancake coil. In contrast, electromagnetic force works as a contractive force at the narrow-side due to attraction between Bi-2223 coils and a magnetic pole. Interaction between upper and lower Bi-2223 coils, F_z is repulsive, but that in usual split superconducting coils is attractive. This is because magnetic saturation of yoke is not so high and attraction between coil and yoke is

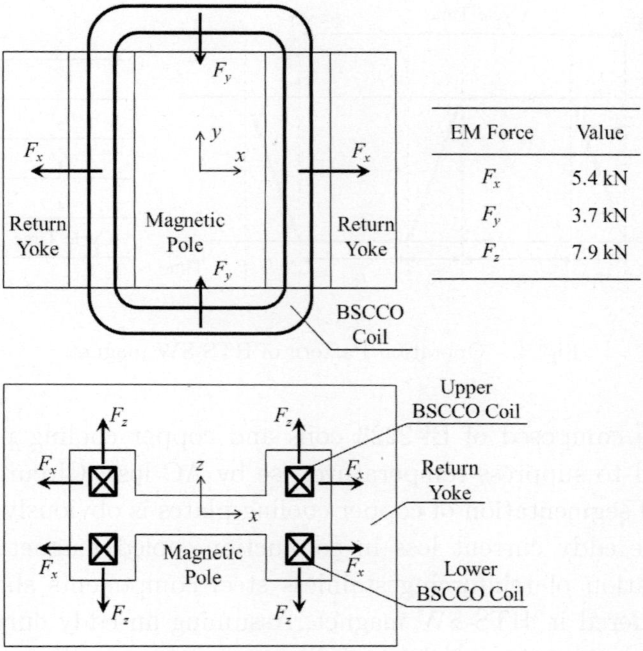

EM Force	Value
F_x	5.4 kN
F_y	3.7 kN
F_z	7.9 kN

Fig. 3. Electromagnetic force pattern of HTS-SW magnet. Right table shows the magnitude of force obtained by electromagnetic field finite element analysis.

dominant in HTS-SW magnet. Although Bi-2223 wire is reinforced by copper alloy, a reinforcing structure is required to suppress deformation and stress by above electromagnetic force. Used Bi-2223 wire has similar thermal contraction to austenitic stainless steel. Therefore, stacked Bi-2223 double pancake coils are reinforced by surrounding austenitic stainless steel components to avoid thermal stress. The reinforcing structure was designed so that the coil assembly deformation can be below 1 mm.

The second issue is the loss generated during periodic increasing and decreasing of magnetic field as shown in Fig. 3. Both rise time t_1 and fall time t_3 are 10 sec. Flat top time t_2 is 100 sec, and interval from completion of fall to start of next rise, t_4 is 300 sec. AC loss of Bi-2223 wire and eddy current loss of metal components are significantly generated in this operation pattern. AC loss was estimated as 200 J during t_1 and t_3 in the whole coil assembly. Sandwich

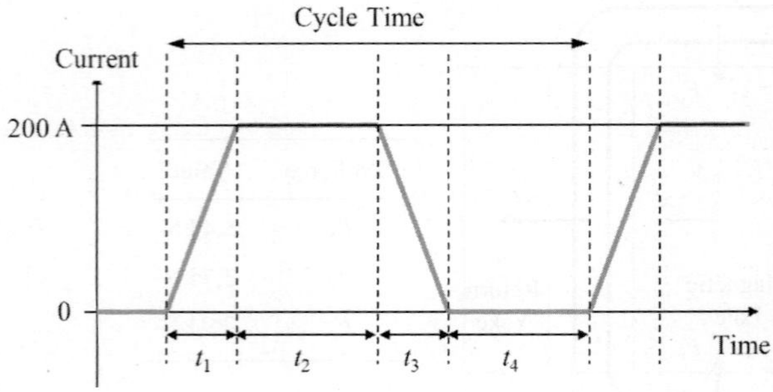

t_1	10 sec
t_2	100 sec
t_3	10 sec
t_4	300 sec
Cycle Time	420 sec

Fig. 4. Operation Pattern of HTS-SW magnet.

structure composed of Bi-2223 coils and copper cooling plates are employed to suppress temperature rise by AC loss. Circumferential electrical segmentation of copper cooling plates is obviously required to reduce eddy current loss in conduction-cooled magnet, but the segmentation of reinforcing stainless steel components should also be considered in HTS-SW magnet. Assuming an eddy current loop around a magnetic pole by metallic components, uniform metallic components cross section and uniform field rate, eddy current loss during t_1 (or t_3) is roughly represented by following equation:

$$Q_{\mathrm{eddy}} = \frac{S_{\mathrm{metal}}}{\rho \ell} \left(\frac{B_{\mathrm{max}} S_{\mathrm{pole}}}{t_1} \right)^2, \tag{1}$$

where ρ, ℓ and S_{metal} are respectively electrical resistivity, loop length, cross-section area of metallic components. In case of reinforcing components by austenitic stainless steel, ρ and ℓ are 4.0×10^{-7} Ωm and 3.4 m, respectively. B_{max} and S_{pole} are maximum magnetic field and magnetic pole area, and their value are respectively 1.6 T and 0.42 m^2. Assuming that S_{metal} is 2.4×10^{-3} m^2, the eddy current loss of reinforcing components may reach 8.1 W. This value is comparable to that of segmented copper cooling plates calculated by finite element analysis, and is not so small compared with time-averaged AC loss of 20 W. Furthermore, eddy current loss can become significant than AC loss in magnet with faster field rate because eddy current loss is proportional to the inverse square of field rise (or fall) time. We reduced effective cross-section area S_{metal} to 0.3×10^{-3} m^2

by the segmentation in HTS-SW magnet as study for a magnet with faster field rate. In the results, eddy current loss of stainless steel components is suppressed to about 1 W.

3.2. *Fabrication and unit test of coil assembly*

Coil assembly structure should eliminate above technical issues, and be built without degradation of superconducting properties. Figure 5 shows a 3D CAD drawing and a cross section view of the Bi-2223 coil assembly. Each coil assembly mainly consists of Bi-2223 double pancake coils, winding frames, cooling plates, reinforcing plates, and

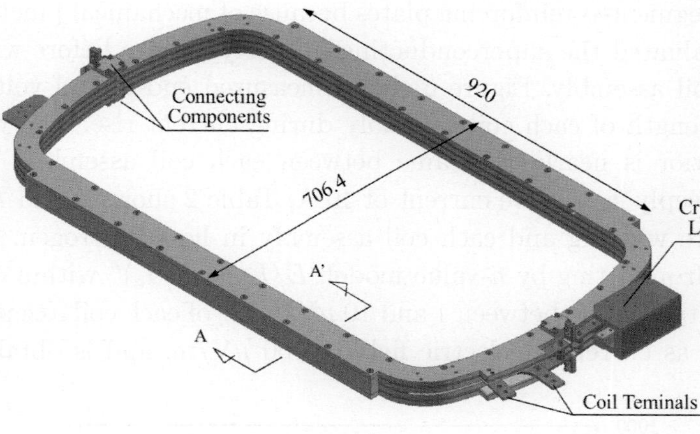

(a)

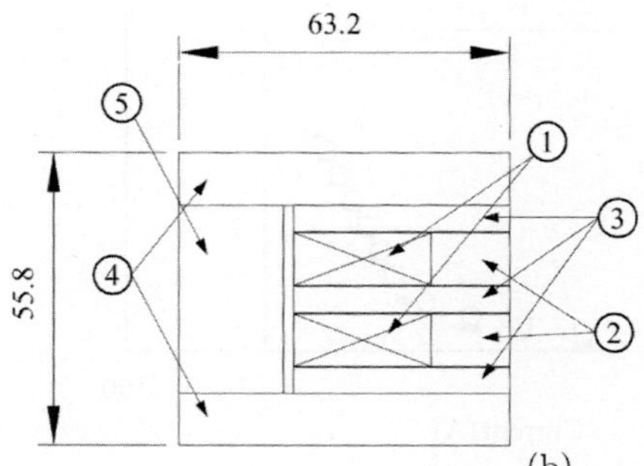

No.	Description	Material
1	BSCCO DPC	
2	Winding Frame	Stainless Steel
3	Cooling Plate	Copper
4	Reinforcing Plate	Stainless Steel
5	Reinforcing Bar	Stainless Steel

(b)

Fig. 5. The Bi-2223 coil assembly structure. (a) Appearance of the coil assembly. (b) A-A′ cross-sectional view and major component list.

reinforcing bars. Bi-2223 wire was wound around the winding frame, and the Bi-2223 double pancake coil was formed. The winding frame also plays a role in the support for compressive force F_y in Fig. 3. The Bi-2223 coil with the winding frames was sandwiched by glam fibre reinforced polymer (GFRP) insulation sheet, and was epoxy impregnated together. Integrated coils with winding frames, cooling plates and reinforcing plates were stacked. Then, reinforcing bars were inserted at only long side of the coil assembly as shown in Fig. 5(b). Finally, the coil assembly was epoxy impregnated again to fill some gaps between components. Cooling plates and reinforcing plates are segmented at midpoint of the coil assembly narrow side to reduce above eddy current loss. There are connecting components between segmented reinforcing plates because of mechanical junction.

We evaluated the superconducting property of wire before winding the coil assembly. Figure 6 shows measured end-to-end voltage per wire length of each coil assembly during current rise. The voltage behavior is nearly the same between each coil assembly. The voltage steeply rise above current of 40 A. Table 2 shows n and I_c of wire before winding and each coil assembly in liquid nitrogen. n is obtained from fitting by n-value model, $E/E_c = (I/I_c)^n$ within end-to-end voltage range between 1 and 10 μV/m. I_c of each coil assembly is defined as current at electric field of 100 μV/m, and is obtained

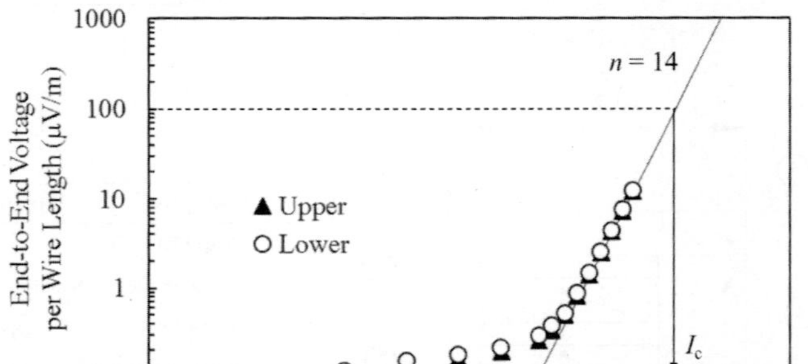

Fig. 6. Unit applying current test results of upper and lower coil assembly in liquid nitrogen. Solid lines represents fitting curves by n-value model within electric field range between 1 and 10 μV/m.

Table 2. Comparison of n and I_c between the wound Bi-2223 wire and the coil assembly in liquid nitrogen.

Test sample	n at end-to-end	I_c at end-to-end
Wire for Upper DPC 1	17	186 A
Wire for Upper DPC 2	17	195 A
Wire for Lower DPC 1	17	188 A
Wire for Lower DPC 2	17	187 A
Upper coil assembly	14	66 A (Extrapolation)
Lower coil assembly	14	66 A (Extrapolation)

by extrapolation using n-value model n and I_c of coil assemblies are lower than that of wire, but this is due to self field. There were no degradation in wire performance during assembly process.

4. Performance Test

We have not yet performed periodic increasing and decreasing of magnetic field as shown in Fig. 3 because of some troubles except for the coil assembly. However, we shall report about present performance test results in this section. Initial cooling was performed, and the lowest temperature of the coil assembly was 7.1 K. Initial cooling time was 54 hours. We have not yet achieved ramping up of magnetic field to 1.6 T with 10 sec due to incompletion of the power supply adjustment at this moment. However, we achieved excitation of 1.6 T with increasing and decreasing time of 80 sec. Figure 7 shows measured coil temperatures during increasing and decreasing of current. When current decreases from 200 A to 0 A, current rate was not constant owing to imperfection of power supply control. Measured coil assembly temperatures, T_1 and T_2 are respectively temperature at near side and far side to the cryocooler. Both T_1 and T_2 increased due to AC loss and Joule loss after start of current increase, and showed peaks at the end of current increase. Even peak temperature of T_2 was 7.5 K, which is sufficiently low for constant magnetic field generation. The measurement of the peak temperature at periodic operation is an important future task.

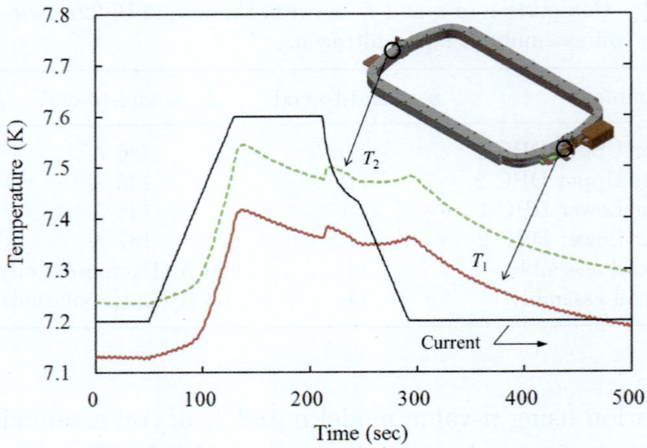

Fig. 7. Coil temperatures during increasing and decreasing of current between 200 A and 0 A. Increasing and decreasing time are 80 sec, but current rate is not constant owing to imperfection of power supply control.

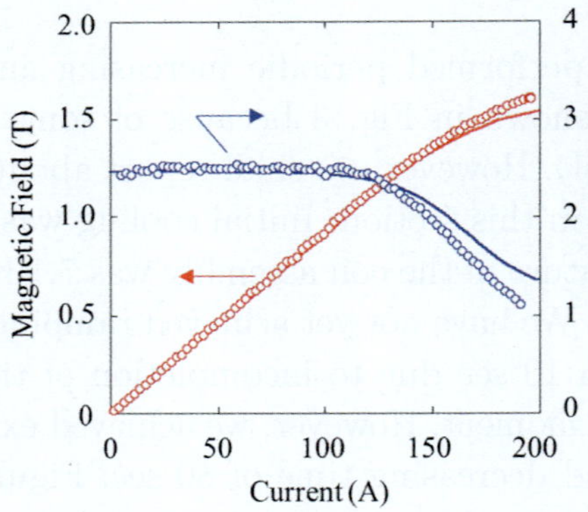

Fig. 8. Current dependence of the magnetic field on median plane and the inductance. Symbols and solid lines are respectively measured data and calculated curves by finite element analysis.

The magnetic field generation of 1.6 T also was confirmed using a Hall sensor. Figure 8 shows measured current dependence of the magnetic field on the median plane and the inductance. The inductance data were calculated from measured, induced-voltage between coil terminals. Current dependence of the magnetic field and the inductance displays good agreement with calculated curves from

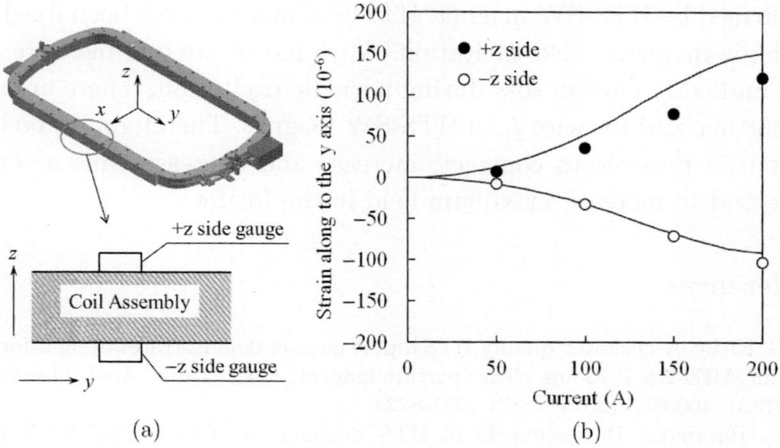

Fig. 9. Strain measurements of the coil assembly. (a) shows bond location of strain gauges. (b) shows current dependence of strain along to the y axis. Symbols and solid lines respectively represent measured data and calculated curves from finite element analysis.

finite element analysis, and maximum magnetic field was 1.6 T. There is slight difference between measured data and calculated curves, but we believe that the magnetic properties of laminated yoke cause it.

Strains of the coil assembly were also investigated during the magnetic field generation for verification of mechanical reinforcing structure. Figure 9 shows measured strains of the coil assembly. The long side of the coil assembly was expected to be deformed as a bent beam by electromagnetic force F_x and F_z shown in Fig. 3. Therefore, strains along to the y axis were measured on two surfaces of reinforcing components as shown in the left of Fig. 9. Strains of $+z$ side and $-z$ side are opposite sign. It indicates bending deformation as expected. Current dependences of strain are also nearly the same as calculated curves. Therefore, stress and deformation of Bi-2223 wire are considered to be successfully suppressed by the reinforcing structure.

5. Summary

We introduce mainly a characteristic Bi-2223 coil assembly and its technical issue in this chapter. Magnetic field generation of 1.6 T was confirmed by HTS-SW magnet. HTS-SW magnet have been fixed for periodic magnetic field excitation. After fixure, we will measure AC loss and eddy current loss during periodic excitation. There are sufficient margins for wire I_c in HTS-SW magnet. Therefore, we believe that it is possible to contract, increase and decrease time or cycle time and to increase maximum field in the future.

References

1. G. Kirby, Accelerator-quality HTS dipole magnet demonstrator designs for the EuCARD-2 5-T 40-mm clear aperture magnet, *IEEE Trans. Appl. Supercond.* **25**(3), 4000805 (2014) ISSN 1051-8223.
2. K. Hatanaka, Developments of HTS magnets, in *Proc. IPAC 2012*, New Orleans, Louisiana, US, (May 2012) p. 1095.
3. R. Gupta and W. Sampson, Medium and low field HTS magnets for particle accelerators and beam lines, *IEEE Trans. Appl. Supercond.* **19**(3), (2009) 1095–1099.
4. http://global-sei.com/super/hts_e/type_ht.html.
5. http://www.shicryogenics.com/products/specialty-cryocoolers/rdk-408s2-10k-cryocooler-series/.

Chapter 3.18

Ship Propulsion Motor Employing Bi-2223 and MgB_2 Superconductors

Swarn Kalsi

Kalsi Green Power Systems

46 Renfield Dr, Princeton, NJ 08540, USA

Compact and light weight direct-drive machines in large rating are desired as ship propulsion motors, and as generators for off-shore wind farm applications. A key goal for such machines is to be shipped to the site as fully assembled units. In order to achieve this goal, it is essential to construct both rotor and stator windings also using high-temperature superconducting (HTS) materials. Two commercially available HTS conductors are Bi-2223 ($Bi_2Sr_2Ca_2Cu_3O$) HTS with a critical temperature of about 110 K, and Magnesium Diboride (MgB_2) with a critical temperature of about 40 K. The MgB_2, available in small diameter wires, is suitable for manufacturing stator coils operating in high AC magnetic field environment. This chapter presents a concept design for a 40 MW, 120-RPM ship propulsion motor employing Bi-2223 for field winding and MgB_2 for stator winding. Ambient temperature magnetic iron is employed on the rotor and the stator. The field winding consists of race track shaped Bi-2223 coils operating at 35 K. The stator winding, made up of MgB_2 race track coils, operates at 20 K. Available off-the-shelf cryo-coolers are used for cooling all coils. The concept 40 MW motor is expected to be about 3 m in diameter, 2.3 m in axial length, and weigh around 80,000 kg. The design approach discussed here could also be used for designing large rating generators for wind farm applications.

1. Introduction

Many ship builders prefer employing propulsion motors directly coupled to the propeller. Likewise, wind power developers (users) and manufacturers around the world are seeking generators directly coupled to turbine for large off-shore wind power generators. To date, ship propulsion system has employed synchronous motors with permanent magnet (PM) field poles or induction motors. Likewise, most of wind turbine generators are land-based and are manufactured using a variety of technologies, like synchronous generators with wound and PM poles, and doubly-fed induction generators for wind farms. Both induction and synchronous machines have reached their upper limit in size and weight for both ship and wind farm applications. Ship owners are looking for compact and light weight motors that could fit into a tight space. On the other hand, in order to catch more wind and avoid noise pollution concerns of the land-based generators, wind power developers are moving the wind generators out into the shallow seas. Currently, the machines employing PM tend to be most compact and light weight. Most users prefer factory built machines, which could be transported to site using the existing roadways or other public modes of transportation. To meet this criterion, only a maximum rating of 20 MW is possible with the PM technology for ship propulsion applications and 5 MW for wind farm application. These applications limit generator's outside diameter and axial length to less than 5 m and 4 m, respectively. Currently, the only technology that could satisfy these requirements is based on using superconductors for the field winding. A 40 MW, 120-RPM superconducting machine is expected to be more compact and light weight by more than a factor of two to three when compared with a 20 MW machine utilizing PM poles. A large 36.5 MW, 120-RPM motor utilizing high-temperature superconductor (HTS) poles has already been successfully built and tested,[1,2] — this machine is about $1/3^{rd}$ the size of the 10 MW, 10-RPM generator. On the basis of experience with this motor, construction of a 10 MW superconducting generator looks quite feasible. The reluctance to construction of such large units is mostly due to high cost (attributed to high cost of superconductor) and reliability of the refrigeration system needed for keeping the superconductor at desired cryogenic temperature. The reliability issue is very important for both ship and offshore wind power application, as it is expensive to replace or maintain such units. Available HTS materials, Bi-2223 and MgB_2 appear to be economically attractive for applications in electrical rotating machines.

2. Ship Propulsion Motor Requirements

The objective of this study is to develop a conceptual design for a 40 MW, 120-RPM ship propulsion motor using Bi-2223 winding on rotor and MgB_2 winding on stator. The motor is expected to have outside diameter less than 4 m and axial length less than 3 m. Table 1 lists specifications for such a motor.

Table 2 summarizes selected design assumptions and approach. The motor operates using a 6-phase power at 5 kV/phase voltage. The rotor and stator employ B-2223 and MgB_2 windings, respectively. Cryo-coolers, available off-the-shelf, are used for cooling the rotor and stator windings; single-stage cryo-coolers for the field winding, and two-stage cryo-cooler for the stator winding. Gaseous helium (GHe) is employed as a link medium between the cryo-coolers and the windings. Separate cooling systems for the rotor and stator windings enable operation of each, independent of the other for achieving improved reliability and low service downtime.

Table 1. Design specifications for the 40 MW, 120-RPM ship propulsion motor.

Parameter	Value
Rated output, MW	40
Rated speed, RPM	120
Rated power factor	0.99 lag
Number of phases	3
Number of poles	12
Maximum diameter, m	<4
Maximum axial length, m	<3

Table 2. Propulsion motor design assumptions and approach.

Parameter	Value
Number of poles	12
Rotor assumptions	
Field winding type	Bi-2223
Conductor configuration	0.4×4.6 mm
N-value	14
Winding temperature, K	35
Coil shape	Race track
Winding cooling	GHe
Refrigerators	Single-stage cryo-coolers
Iron yoke	Solid at ambient temp.
Cryostat	Independent
Electromagnetic (EM) shield	Copper at ambient temperature
Stator Assumptions	
Armature winding type	MgB_2
Conductor configuration	7-strand cable
N-value	20
Winding temperature, K	20
Coil shape	Race track
Winding cooling	GHe
Refrigerators	Multiple 2-stage cryo-coolers
Iron yoke	Laminated magnetic steel at ambient temp.
Cryostat	Independent
Armature voltage/phase, kV	5.3
Armature number of phases	6

The Bi-2223 conductor has been commercially available from a number of sources since 1995. This study employs a DI-BSCCO-HT-CA (High Current Density and Tough with Copper Stabilizer)[3] wire with an N value of 14 from Sumitomo Electric Industries (SEI) of Japan. Characteristics of this wire are shown in Fig. 1 and Table 3. In the figure, the top figure shows I_c when field (B) is perpendicular to the broad face of the conductor, and the bottom figure shows I_c when field is parallel to the broad face of conductor.

The Magnesium Diboride (MgB_2) superconductor is also available from several sources in round wire shapes, which makes it insensitive to orientation of magnetic field. It is available in a range of

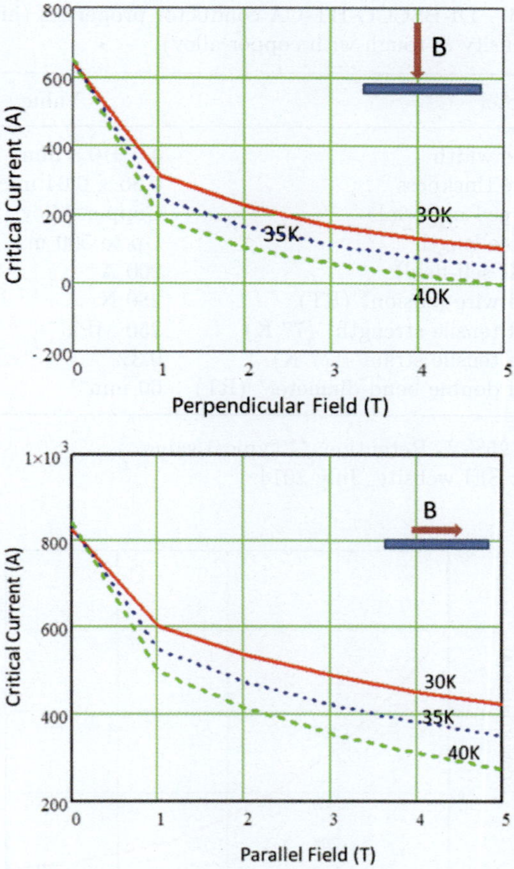

Fig. 1. Critical current (I_c) for DI-BSCCO conductor at three temperatures.
Source: SEI website, July 2014.

diameters from 0.3 mm to 2.0 mm, and has a superconductor fraction from 7–25%.

This design study uses wire[4] available from HyperTech Research of Columbus, OH with critical current of 465 A (self-field, 20 K) in a 0.83 mm diameter wire, 18% superconductor fraction and an N value of more than 20. The performance of this wire is expected to improve by a factor of two or more by 2015. The critical current of MgB_2 wire as a function of field and temperature is shown in Fig. 2 for the year 2015 for a 0.83 mm diameter wire (a conservative

Table 3. DI-BSCCO-HT-CA conductor properties (high current density & tough with copper alloy).

Parameter	Value
Average width	4.5 ± 0.3 mm
Average thickness	0.36 ± 0.04 mm
Reinforced material	Copper Alloy (50 μm)
Available length	Up to 500 m
I_c(77 K, self-field)	200 A
Critical wire tension* (RT)	280 N
Critical tensile strength* (77 K)	250 MPa**
Critical tensile strain* (77 K)	0.3%**
Critical double bend diameter* (RT)	60 mm**

Note: *95% I_c Retention; **Typical value.
Source: SEI website, July 2014.

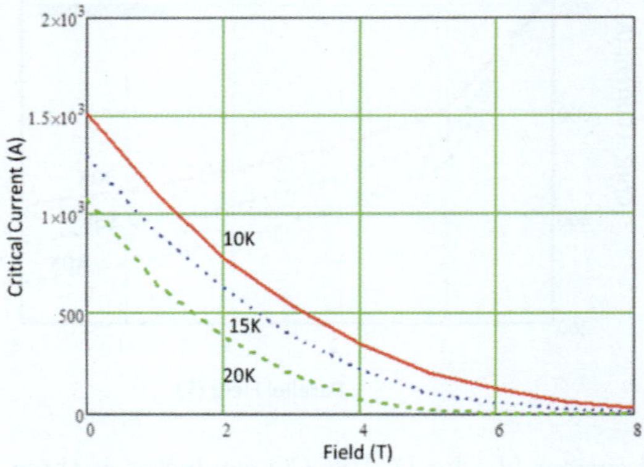

Fig. 2. Critical current (I_c) of MgB_2 wire (0.83 mm diameter with 18% superconductor fraction) at different fields and temperatures (10 K, 15 K, and 20 K), as expected in 2015.

Source: Hyper Tech Research.

estimate per reference).[4,5] Expected properties of the MgB_2 wire and cables as employed in this study are summarized in Table 4. The construction of multifilament wire is typically Niobium (Nb) barrier around monofilament MgB_2, Cu or Cu/Ni matrix around filaments and Cu in center of multifilament strands. The outside of the multifilament wire is Monel. Each strand includes ~10–12% copper.

Table 4. MgB_2 wire characteristics expected in 2015.[5]

Parameter	Value
MgB_2 wire source	HyperTech, USA
Strand diameter, mm	0.83
Superconductor fraction	20%
Copper fraction	10–12%
N value	>20
Performance of wire expected in 2015	1080 A (self-field, 20K)
Stator winding — 7-strand twisted cable	
Number of filament in a strand	36
Cable operating current at 20 K	1360 A
Critical current (self-field, 20 K)	6000 A
Self-field with the critical current	0.47 T
Expected cost of 0.83 mm wire in 2015	\$3/m

The Table 4 includes critical currents for a 7-strand cable at 20 K in self-field (specified in the table). The cables are made by twisting strands along the length using a suitable twist pitch. HyperTech is developing these wires for rotating machine coils experiencing 1–2T field at 400 Hz.

Since the magnetic field varies throughout a coil cross section, the operating current is selected on the basis of a desired total integrated conduction loss in a coil. A conduction loss of 1 W per coil is assumed for the rotor coils, which carry DC only. A conduction loss of 2 W per coil is assumed for the stator coils — AC losses in the 7-strand stator coils are accounted for by increasing the calculated conduction loss to 5 W per coil. This (5 W per coil) loss at operating frequency of 12 Hz is considered reasonable for this prototype motor. Nevertheless, it would be necessary to recalculate this loss component once more detailed data becomes available for the MgB_2 cable.

3. Ship Propulsion Motor Configuration

The motor (40 MW, 120-RPM) shown in Fig. 3 is synchronous type with DC excitation poles on the rotor and AC armature winding on the stator. The cross section of the motor shows locations of armature and field windings along with their cryostat boundaries. Ambient

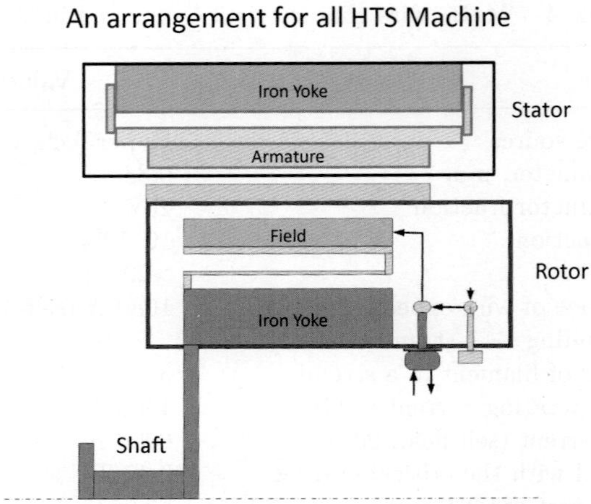

Fig. 3. Overall motor configuration.

temperature iron yokes on stator and rotor are also shown. Cryo-coolers are directly mounted on the rotor cryostat.

Both rotor and stator assemblies are mechanically independent of each other for providing most flexibility in design and for ease of operation and service. The 12-pole field excitation winding employing DI-BSCCO superconducting coil is cooled to 35 K with GHe, which in turn is cooled with single-stage cryo-coolers. The field winding poles are supplied from a brushless exciter. The stator winding employs single-layer armature concept[6] consisting of race-track shaped coils using MgB_2 superconductor cable. These coils are cooled to 20 K with GHe, which in turn is cooled with 2-stage cryo-coolers. Both rotor and stator assemblies are independently cooled. The key dimensions of various components in the machine cross section are listed in Table 5. Motor case includes 50 mm thick cylinder outside the stator iron yoke and 50 mm thick side covers.

4. Rotor Design

The field excitation winding on the rotor consists of 12 poles. An arrangement of components on the rotor is shown configurationally in Fig. 4. All field coil assemblies are supported on a stainless pole

Table 5. Physical, radial, and axial dimensions of motor components.

Parameter	Value
Rotor	
Inside radius of iron yoke, mm	1043
Outside radius of iron yoke, mm	1150
Inside radius of field winding, mm	1200
Outside radius of field winding, mm	1230
Inside radius of EM shield, mm	1280
Outside radius of EM shield, mm	1290
Stator	
Air-gap length, mm	20
Inside radius of cryostat, mm	1310
Inside radius of armature winding, mm	1360
Outside radius of armature winding, mm	1370
Inside radius of iron yoke, mm	1421
Outside radius of iron yoke, mm	1528
Axial active length of machine, mm	1950
Outside diameter of stub shaft, mm	586

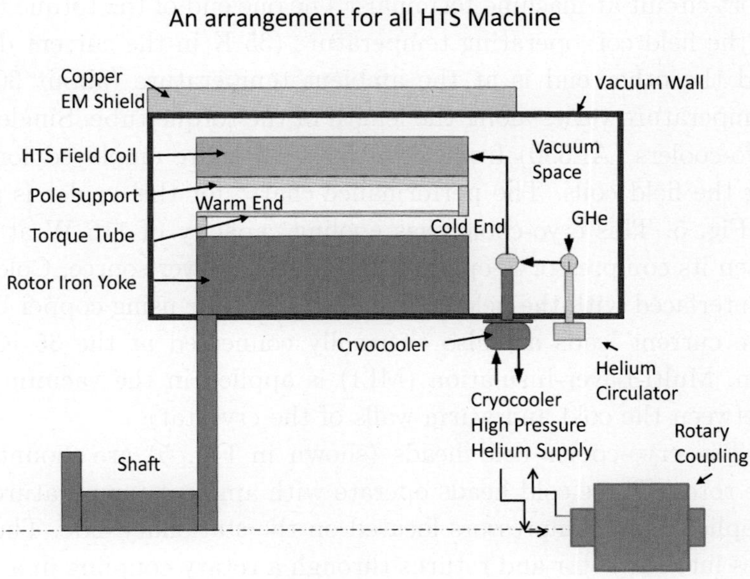

Fig. 4. An arrangement of components on the rotor.

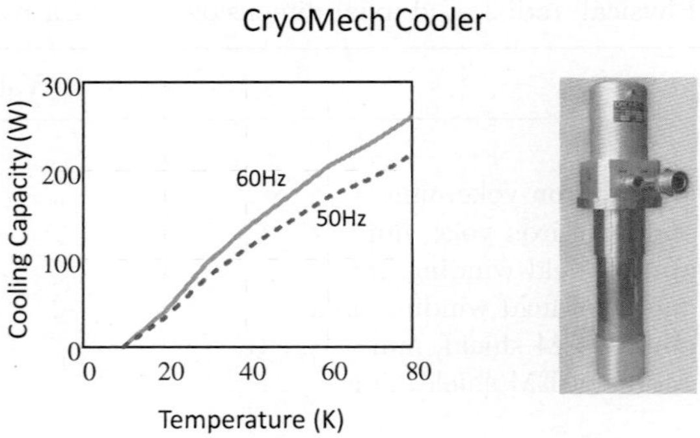

Fig. 5. Performance of a cryo-cooler AL330 (CryoMech, USA).

support cylinder. This cylinder is connected to torque tube on one end, as shown in the figure. The other end of the torque tube is connected to the ambient temperature iron yoke. The torque tube is designed for transferring full torque experienced during a line–line short-circuit at machine terminals. The one end of the torque tube is at the field coil operating temperature (35 K in the current design) and the other end is at the ambient temperature (about 300 K). Temperature varies along the length of the torque tube. Single-stage cryo-coolers (AL330) from CryoMech, USA are employed for cooling the field coils. The performance chart[7] for this cooler is shown in Fig. 5. This cryo-cooler has cooling capacity of 116 W at 35 K, when its compressor is operated from 60Hz power source. Cold head is interfaced with the field coils and torque tube using copper braids. The current leads are also thermally connected at the 35 K location. Multi-Layer-Insulation (MLI) is applied in the vacuum space (between the cold and warm walls of the cryostat).

The cryo-cooler cold heads (shown in Fig. 5) are mounted on the rotor. These cold heads operate with ambient temperature GHe supplied from compressors located on the stationary side. The GHe goes into the rotor and returns through a rotary coupling in a closed loop. Rotary couplings for transferring GHe back and forth have been demonstrated.[8] An EM shield in form of a copper cylinder (10 mm thick) is applied on the outside surface of the rotor cryostat.

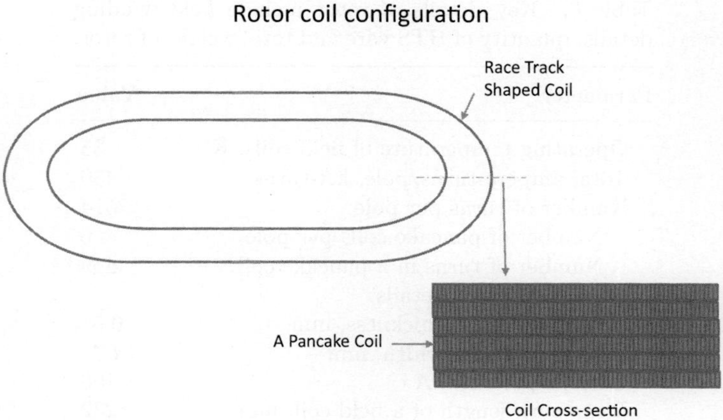

Fig. 6. A race track shaped field excitation coil. Each coil consists of 6 individual pancake coils built from insulated DI-BSCCO wire (0.5 mm × 4.7 mm).

This EM shield at ambient temperature intercepts any high frequency harmonic fields created by the armature winding. These high frequency fields dissipate most of their energies in form of eddy current heating in the EM shield, which being at ambient-temperature are easily removed. This reduces losses in the superconducting field winding on the rotor. The ambient temperature EM shield also acts as a damper winding and helps to damp low frequency oscillations of the rotor.

Each pole consists of a race track shaped coil shown in Fig. 6. Coils are constructed using DI-BSCCO wire. The finished wire is 0.5 mm × 4.7 mm including turn insulation. Each coil has 2514 turns in 6 pancakes. Additional details of the rotor winding are summarized in Table 6.

5. Stator Design

The stator employs a single-layer armature winding as shown in Fig. 7 and is constructed using an MgB_2 cable. This figure shows coils for a 16-pole machine; 12-pole winding will be similar. Individual coils are rectangular (or race-track) in shape, similar to those shown in Fig. 8. This single layer winding has two sets of identical coils.[9] One

Table 6. Key details of rotor include field winding details, quantity of HTS wire and total weight of rotor.

Parameter	Value
Operating temperature of field coils, K	35
Total ampere-turns/pole, kA-turns	450
Number of turns per pole	2514
Number of pancake coils per pole	6
Number of turns in a pancake coil	419
DI-BSCCO wire details	
Insulated wire thickness, mm	0.5
Insulated wire width, mm	4.7
Current per turn, A	160
Mean-turn length of a field coil, mm	5232
DI-BSCCO wire used in all poles, km	158
Loss assumed per coil, W	1
Overall Dimensions and Weight	
Outside diameter, mm	2580
Overall axial length, mm	2306
Weight, kg	41908

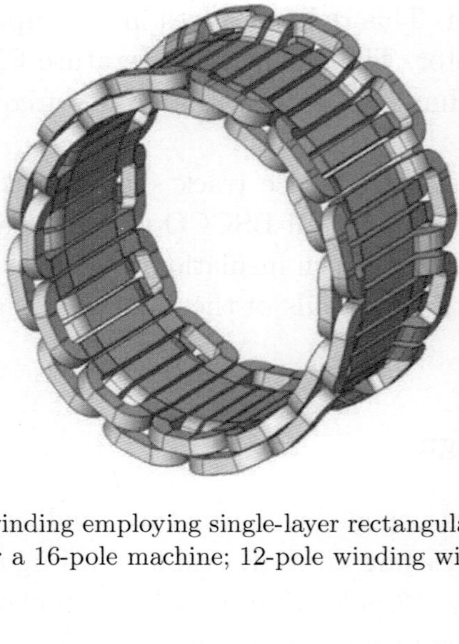

Fig. 7. Armature winding employing single-layer rectangular shape coils. Winding shown here is for a 16-pole machine; 12-pole winding will be similar.

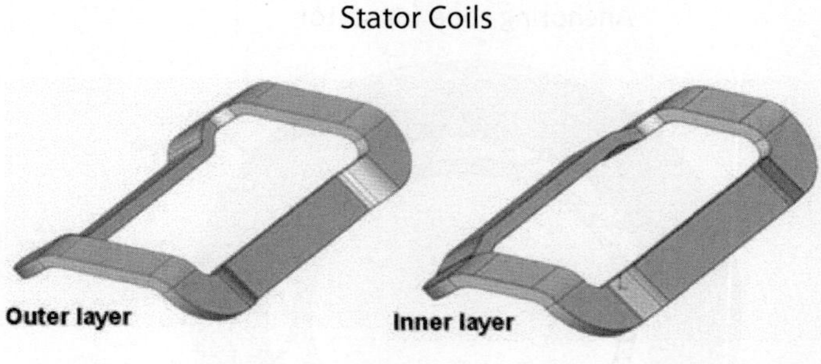

Fig. 8. Armature coil configuration for a single layer winding, employing two types of coils.

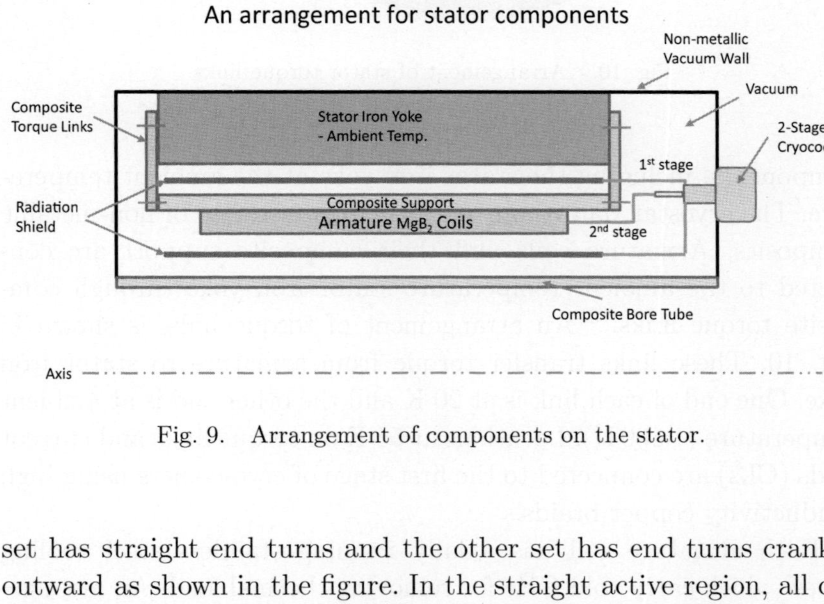

Fig. 9. Arrangement of components on the stator.

set has straight end turns and the other set has end turns cranked outward as shown in the figure. In the straight active region, all coil sides lay at identical radial locations, but the end windings are in two radial locations. Each coil is constructed using a 7-strand MgB_2 cable. The winding is constructed to form a 3-phase system. Each phase has equal number of straight and cranked coils connected in series. The armature winding has two 3-phase circuits. The MgB_2 cable used in these coils consists of 7 strands, each 0.83 mm.

An arrangement of components on the stator is shown configurationally in Fig. 9. The cryostat vacuum wall encloses the armature

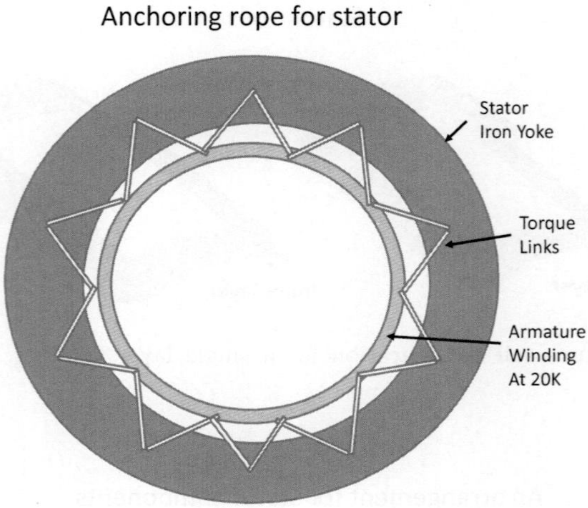

Fig. 10. Arrangement of stator torque links.

components, including the stator iron yoke at the ambient temperature. The cryostat wall facing the air-gap is made up of non-metallic composite. Armature coils and their composite support are connected to the ambient temperature stator iron yoke through composite torque links.[10] An arrangement of torque links is shown in Fig. 10. These links transfer torque from armature to stator iron yoke. One end of each link is at 20 K and the other end is at ambient temperature (~300 K). Locations at 55 K on torque links and current leads (CLs) are connected to the first stage of cryo-coolers using high conductivity copper braids.

Each armature coil has axially running stainless steel cooling tubes, which carry cold GHe for removing thermal load. An arrangement of these tubes is shown in Fig. 11. Stator coils employ 7-strand MgB_2 cable. Each coil is sandwiched between cooling tubes, which carry cold GHe (at 20 K) for cooling. All cooling tubes are electrically insulated from each other and from the MgB_2 coil. Cold GHe is circulated in a closed loop, which consists of the cooling tubes in the stator coils and cooling coils mounted on the 2nd stage of cryo-coolers. A pump is employed for circulating the GHe. Details of the stator winding are summarized in Table 7.

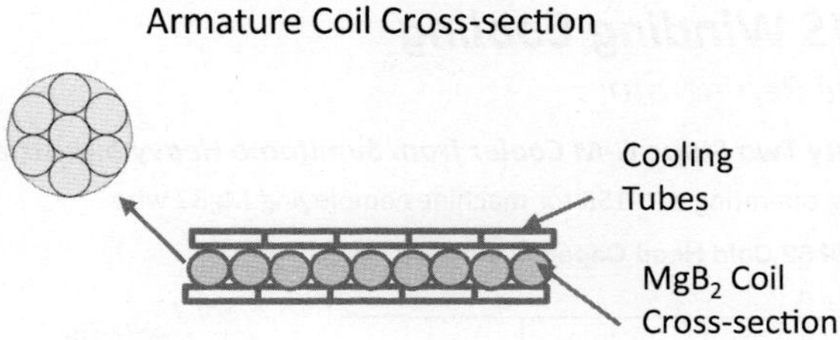

Fig. 11. Arrangement of stator coil cross section and cooling tubes.

Table 7. Key details of stator including armature winding details, quantity of MgB_2 wire and total weight of stator.

Parameter	Value
Number of stator slots	144
Winding type	Single layer
Mean turn length of in a coil, mm	5646
Tooth width, mm	10
Coil width, mm	49
Coil height, mm	10
Number turns in a coil	11
Average field in the armature coil, T	1.25
Current per turn, A	1300
MgB_2 cable used in armature winding, km	4.5
Number of strands in a cable	7
Strand diameter, mm	0.83
Assumed loss per coil, W	5
Overall Dimensions and Weight	
Overall diameter, mm	3156
Overall axial length, mm	2306
Weight, kg	38738

Two stage cryo-coolers (RDK-408S from Sumitomo Heavy Industries) are used for cooling the stator components. The performance chart for this cooler is shown in Fig. 12, when operating from a 60 Hz power source. Each cryo-cooler is operated with 1st stage at 55 K and 2nd stage at 20 K. The cooling capabilities at 1st and 2nd stages are 55 W and 17 W, respectively, for the selected operating temperatures. The 2nd stage maintains the stator coils at 20 K and the 1st stage operating at 55 K cools the thermal radiation shield

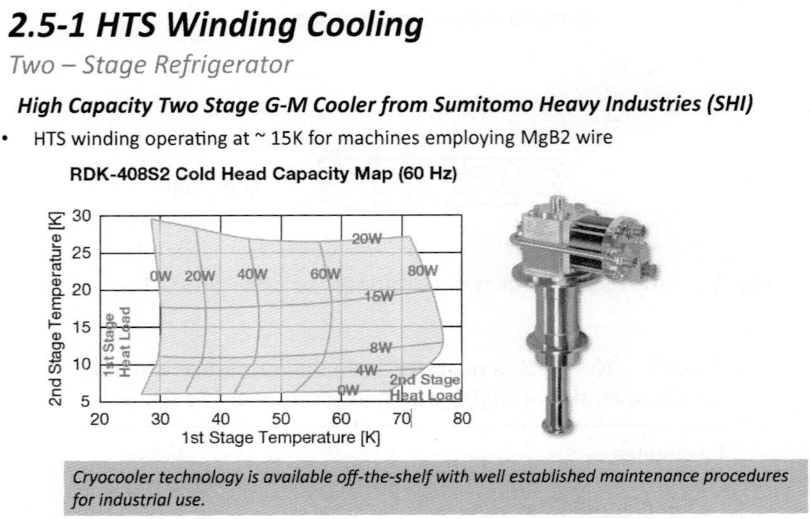

Fig. 12. Sumitomo Heavy industries RDK-408S2 cryo-cooler.

(absorbs energy radiated from the warm walls of the cryostat). The CLs are also thermally connected at the 55 K location. MLI is applied between the ambient wall of the cryostat and the radiation shield.

6. Design Analysis

6.1. *Motor sizing calculations*

The motor rotor has 12 poles, each excited with a race track coil. The cross section of a field coil is shown in Fig. 6. All coils are enclosed in a cryostat as shown in Fig. 4. A solid iron yoke is included on the rotor. Each field coil provides an excitation of 450 kA-turns. The cross section of the motor for a 1/2 pole-pitch is shown in Fig. 13 along with flux line contours in all regions. Field winding, stator coils, and EM shield regions are highlighted. An EM shield is included on the outside surface of the cryostat. Figure 14 shows magnetic field contours (calculated with a 2-D finite-element program) with the field coil excitation of 450 kA-turns. Local field is utilized for calculating conduction loss in the DI-BSCCO wire. The rotor EM shield and

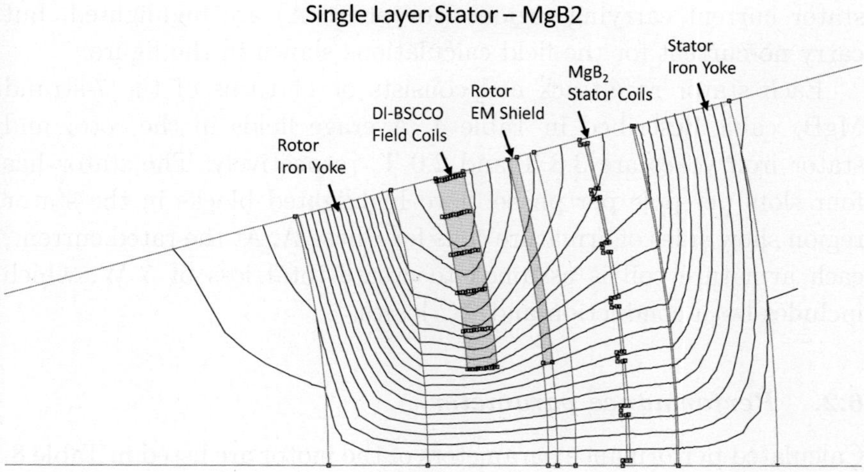

Fig. 13. Motor cross section for 1/2 pole-pitch.

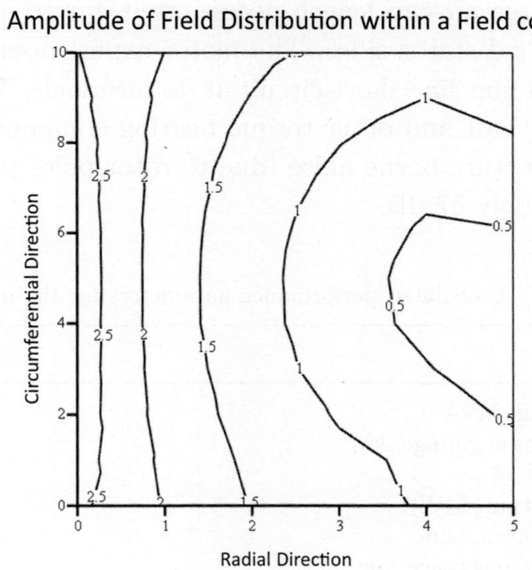

Fig. 14. Magnetic field contours within cross section of a field coil (contours are marked in tesla).

stator current carrying regions (for phase A) are highlighted, but carry no current for the field calculations shown in the figure.

Each stator race-track coil consists of 11 turns of the 7-strand MgB_2 cable described in Table 4. Average fields in the rotor and stator iron yokes are 3.8 T and 2.0 T, respectively. The stator has four slots per pole per phase. Two highlighted blocks in the stator region show sides of armature coils for Phase A. At the rated current, each armature coil is assumed to create total loss of 5 W, which includes both conduction and AC losses.

6.2. *Performance parameter*

Calculated performance parameters of the motor are listed in Table 8. Output power is 40.1 MW at 5.3 kV (phase voltage). The motor has low synchronous reactance (0.74 pu), which is a characteristic of such large pole count machines. Sub-transient reactance is high (0.53 pu) because of a large air-gap length necessitated due to cryostat walls on both rotor and stator sides. The motor experiences 3.3 pu peak torque during a line–line short-circuit at its terminals. This torque is used for sizing shaft and other torque bearing components on rotor and stator. Structure borne noise (due to rotor poles pulling on the stator yoke) is only 57 dB.

Table 8. Calculated performance parameters for the motor.

Parameter	Value
Motor rating, MVA	41.3
Terminal phase voltage, kV	5.3
Line current, A	1300
Machine output, MW	40.1
Base impedance, ohm	8.15
Synchronous reactance, pu	0.74
Transient reactance, pu	0.62
Sub-transient reactance, pu	0.53
Peak torque during a line-line short-circuit at machine, pu	3.3
Structure borne noise at rated speed dB	57

6.3. *Efficiency calculations*

The motor experiences energy loss (conduction loss) in the superconducting coils due to inherent losses in the superconducting wire. Additionally, stator coils also experience losses due to AC field from the rotor and AC current in the coils. Even the low frequency (12 Hz) AC field and current could cause significant losses in the stator coils. These losses are included in thermal analysis by increasing the DC conduction loss (2W/coil) in the stator coils to 5W/coil. Other thermal loads include thermal conduction along components mechanically supporting the superconducting windings, CLs supplying current to the field coils, and thermal radiation from the ambient cryostat wall to the cold region. Thermal loads are removed with 2-stage cryo-coolers discussed earlier. The colder end of each CL is connected to the 50 K stage of the cryo-coolers. Typically a 100 A lead could be designed to limit conduction loss to 8W per lead, i.e. thermal conduction along each lead is calculated using scaling 8W/100A. Table 9 summarizes thermal loads and number of cryo-coolers needed for cooling rotor and stator. Losses for rotor and stator are separately listed. Rotor employs Cryomech AL330 coolers and stator uses Sumitomo Heavy Industries RDK-408S2 cryo-coolers. Efficiency at the rated load is >99%, which is high for such large motors. At lower speeds, efficiency drops due to constant loss component associated with power consumption by the cryo-cooler (248 kW). Figure 15 shows efficiency of the motor as a function of propeller speed.

6.4. *Weight calculations*

Table 10 summarizes weights of various components. This represents a reasonable estimate for total motor weight. More detailed mechanical analysis would be necessary for better weight estimate. Total weight of this motor is 80646 kg, including weight of cryo-cooler compressors.

Table 9. Summary of losses and efficiency at rated load.

Parameter	Rotor	Stator
Winding conduction loss, W	18	360
Current lead loss, W	13	160
Radiation through MLI, W	147	124
Conduction along torque transfer components	—	—
Rotor at 35 K, W	13	—
Stator		
First stage at 55 K, W	—	33
Second stage at 20 K, W	—	11
Safety margin for calculated losses	50%	50%
Total load on rotor cryocoolers (including safety margin), W	285	—
Number of cryocoolers	3	—
Input power to compressors, kW	23	—
Total load on stator cryocoolers		
Load on the 1st stage (including safety margin), W	—	17
Load on the 2nd stage (including safety margin), W	—	540
Number of cryocoolers needed		30
Input power to cooler compressors, kW	—	225
Iron yoke losses, W	—	60
Total losses at rated load, kW	252	
Efficiency at rated load	99.4%	

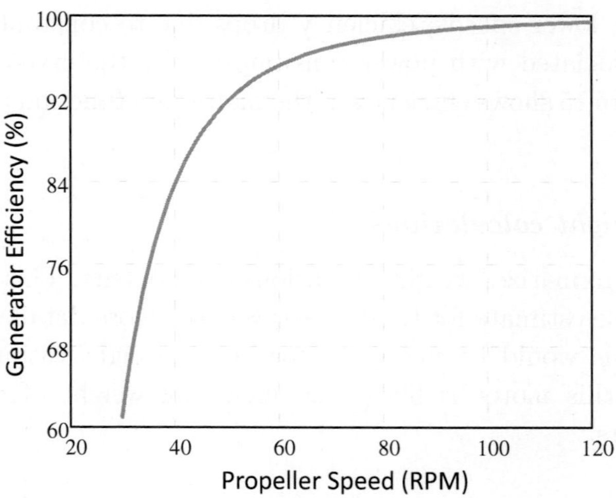

Fig. 15. Motor efficiency as a function of propeller speed. High efficiency (>99%) at rated speed.

Table 10. Summary of motor component weights.

Component	Weight, kg
Rotor	
Stub shaft	4969
Torque Tube	555
Field winding support	2874
Field winding	6603
Cryostat	3656
EM shield	503
End covers	690
Brushless exciter	223
Iron yoke	13454
Sum of rotor component	33526
Total rotor with 25% safety margin	42770
Stator	
Iron yoke	16698
Armature winding	135
Case	8603
Cryocoolers (both rotor and stator)	4017
Miscellaneous	1537
Sum of stator components	30990
Total stator with 25% safety margin	38738
Total motor weight	80646

7. Possible Applications

Highly efficient, compact and light weight superconducting motor described in this chapter could be employed for propelling Navy ships (surface and submarine) as well as commercial ships. A few possible applications are discussed below.

United States' DDG 1000 Navy surface combatant employs highly survivable Integrated Power System (IPS). Key design features[11] that make the DDG 1000 IPS architecture unique include the ability to provide power to propulsion, ship's service, and combat system loads from common gas turbine prime movers. DDG 1000's power allocation flexibility allows for potentially significant energy savings and is well-suited to enable future high energy weapons and sensors. The

40 MW propulsion motor discussed here could be employed on such ships.

These motors are also highly suitable for large aircraft carriers, which require high power propulsion motors. These motors could be ideal choices for the 'rumored' Chinese catamaran hull aircraft carriers.[12]

ABB has developed Azipod[13] that is widely employed for propulsion of large cruise ships. An Azipod unit consists of a fixed pitch propeller mounted on a steerable gondola ("pod") which also contains the electric motor driving the propeller, where the propeller is connected directly to the motor shaft. Compact and light weight motor discussed in this chapter could be an obvious choice for powering the Azipod units.

Largest cargo ship size is expected to keep growing with widening of the Panama Canal. Container ships, capable of carrying >22,000 twenty-foot equivalent unit (teu) containers, are expected in near future. Such ships require about 100 MW of propulsion power while cruising at 25 knots. Two 40 MW class of motors discussed in this chapter could be employed for propelling such ships.

8. Conclusions and Outlook

Large superconducting motors up to a rating of 36.5 MW have been successfully built and tested for a variety of applications around the world. For example, the 40 MW motor design presented here, employing DI-BSCCO (Bi-2223) conductor for the field coils and MgB_2 conductor for the stator, could be the most compact and light weight option available for propelling a variety of military and commercial ships. These motors are also highly suitable for podded ship propulsion systems currently employed on most cruise ships. Very high efficiency (>99%) of such superconducting motors could result in substantial reduction in fuel cost. Moreover, savings in space and weight, realized by employing compact and light weight superconducting motors, could be utilized for meeting other needs on a ship.

References

1. B. Gamble, G. Snitchler, and T. MacDonald, Full power test of a 36.5 MW HTS propulsion motor, *IEEE Trans. Appl. Supercond.* **21**(3), 1083 (2011).
2. S. S. Kalsi, *Applications of High Temperature Superconductors to Electric Power Equipment.* Hoboken, NJ: Wiley–IEEE Press (2011). ISBN: 978-0-470-16768-7.
3. Sumitomo Electric Industries website, July 23 (2014).
4. D. Doll, M. Rindfleisch, M. Tomsic, J. Yue, C. J. Thong and X. Peng, Development of MgB_2 superconductor wire and coils for practical applications at Hyper Tech Research, Presented at *the 16th US-Japan Workshop on Advanced Superconductors*, Dayton, OH: University of Dayton Research Institute, July 9–12 (2013).
5. Email messages from Michael Tomsic of Hyper Tech, July 7 (2014).
6. S. S. Kalsi *et al.*, *Superconducting Rotating Machines*, U.S. Patent 6489701, December 3 (2002).
7. CryoMech website, July (2014).
8. G. Snitchler, B. Gamble, C. King, and P. Winn, 10 MW class superconductor wind turbine generators, *IEEE Trans. Appl. Supercond.* **21**(3), 1089 (2011).
9. S. S. Kalsi, Superconducting wind turbine generator employing MgB_2 windings both on rotor and stator, *IEEE Trans. Appl. Supercond.* **24**(1), 5201907 (2014).
10. C. Lewis, Direct drive superconducting wind generators, In *Wind Power Generation and Wind Turbine Design.* UK: WIT Press (2010). ISBN: 978-84564-205-1.
11. US Navy Facts File, Aug. 7 (2014). (http://www.navy.mil/navydata/fact_display.asp?cid=4200&tid=900&ct=4).
12. ChinaDailyMail (*via* mil.huanqiu.com), Oct. 7 (2013).
13. ABB Azipod gearless propulsors, (http://new.abb.com/marine/systems-and-solutions/electric-propulsion/azipod), August 7 (2014).

Chapter 3.19

Axial-Gap Superconducting Synchronous Motors Cooled by Liquid Nitrogen

Tomoya Oota and Atsuko Fukaya

Corporate Business Development Division, IHI Corporation
1-1 Toyosu 3-chome, Koto-ku, Tokyo 135-8710 Japan

IHI corporation succeeded the trial manufacture of the world's first 12.5 kW motor with superconducting armature windings cooled by liquid nitrogen (LN_2) in January 2005. With the aim of the practical application of larger-scale motors, various technical problems peculiar to the axial-gap motors with superconducting armatures have been solved, for example, cryostats, AC losses, and so forth. Then, we completed load tests on the motors with the rated power of 400 kW, which are at present the world's largest in LN_2-cooled superconducting motors. The features of those motors and technical breakthroughs are described in this chapter. Problems for their practical applications are also discussed.

1. Introduction

Diesel engines have been commonly used for ship propulsion. Recently, some vessels are equipped with propulsion systems using electric motors to drive propellers instead of diesel main engines.

We introduce two examples of vessels equipped with electric propulsion systems. The first example is Oasis of the Seas, operated by Royal Caribbean International.[1] Oasis of the Seas is a cruise passenger ship equipped with three 20 MW Azipod propulsion systems.[2]

Azipod is the brand name for azimuth thrusters manufactured by ABB. The electric motors connected to the Azipod system are powered by gas turbine generators. The second example is Shinei Maru, constructed by Niijima Bussan Corporation.[3] Shinei Maru is a freighter for cargo and oil transport, has diesel–electric propulsion systems, and is Japan's first domestic vessel equipped with contra-rotating propellers. The inner and outer propellers are rotated separately with different induction motors powered by diesel generators.

Conventional diesel main engines, which make loud noises and vibrations (particularly when operated at low or medium speeds), are not the best engines for passenger ships or cargo-passenger ones, since such ships are preferred to be quiet. In contrast, noises are reduced in electric propulsion systems, in which propellers are rotated by electric motors powered by high-speed diesel generators or gas turbines.

Diesel engines are designed to run with high fuel efficiency during ordinary navigation, while the efficiency is low during low-load operation. On the other hand, in electric propulsion systems with multiple diesel generators, the number of operated diesel generators is controlled depending on the load in order to reduce fuel consumption. In addition, if their electric connection is established, the generators can be arranged anywhere in the engine room, which is an advantage for design of engine room.

Although electric propulsion systems have advantages as mentioned above, the losses in the energy conversion process from generating electrical power to rotation of propellers by motors lowers the system efficiency. To solve this problem, it is necessary to improve the efficiency of individual devices, such as power generators, inverters, and motors. In the point of efficiency, propellers should be directly driven by low-speed high-torque motors without mechanical reduction gears. This configuration is also suitable in order to reduce noises from gears.

Superconducting motors is expected to be one of the promising device for such electric propulsion systems from the points of noise, efficiency, size, and weight. Therefore, IHI carried out the development of motors with high-temperature superconducting

(HTS) windings. In this chapter, we introduce the features of HTS motors developed by IHI. The problems are also described.

2. Approach to Liquid Nitrogen (LN_2)-Cooled Superconducting Motors

A radial-gap type of motor with HTS fields has been extensively studied as a typical type of HTS motors. This type of motor consists of a rotor with helium-cooled HTS field coils and a stator with copper armature coils.[4] IHI, however, took different approaches in order to realize HTS motors with high torque and high efficiency, even at low speeds. We chose the axial-gap type of motors with HTS armatures from the following viewpoints.

2.1. *Cooling medium for HTS coils*

Critical current (I_c) of superconducting wire increases with decreasing temperatures. Helium (boiling temperature 4.2 K) is often used to cool superconducting windings. However, the amount of helium production has been recently decreasing, and the price has rapidly risen. In addition, the coefficient of performance (COP) of refrigerators decreases with decreasing operating temperature. It is necessary not only to reduce the heat load during the motor operation, but also to improve the thermal insulation performance for the cryogenic components.

In contrast, nitrogen is the main component of the air, and can be easily produced with low cost. The COP of refrigerators is rather high at LN_2 temperatures (64–77 K under ambient pressure) compared with liquid helium. Moreover, LN_2 has the large heat capacity and the large heat transfer coefficient. These features are preferable for cooling HTS coils. However, the operating temperatures obtained by using LN_2 are rather higher than helium. Thus, the I_c of HTS coils is not high enough, especially in high magnetic fields. If we choose LN_2 as cooling medium to cool HTS coils, a coreless motor is not feasible. Thus, we decided to use iron cores in the armatures.

Fig. 1. Prototype of a contra-rotating propeller system with two superconducting motors manufactured by IHI.

2.2. *Conception of the HTS motors for propulsion systems*

In order to realize the practical application of HTS motors for ship propulsion systems, high efficiency of contra-rotating propeller system is required as a whole, because of the electric conversion losses.

The prototype of the contra-rotating propeller system with two HTS motors is shown in Fig. 1. One motor drives the front propeller *via* an outer hollow shaft with no reduction gear, and the other drives the rear propeller *via* an inner solid shaft also with no reduction gear. Such structure cannot be applied to the radial-gap type of HTS motors with rotating HTS-fields, in which the cooling medium is introduced to the rotator through a shaft with a so-called rotary joint. In contrast with such radial-gap motors, IHI installed HTS armature coils in the stator and eliminated a rotary joint for reliability.

Slip rings are generally used in order to supply the electric power to rotors. We, however, selected the type of rotating fields with permanent-magnet to eliminate the slip ring for power supply for durability.[a]

[a]In a series of development of HTS motors, IHI had performed the trial manufacture of an inductor-type fully superconducting motor in which HTS coils were used in both armatures and fields.[5] We, however, found that the size of fields with permanent-magnets was smaller than such type of motors. In addition, the cost of

2.3. *Structure of the axial-gap motors with a HTS armature*

Most of industrial motors are the radial-gap type. The radial-gap motor consists of a cylindrical rotor and a cylindrical stator. Another type of industrial motors is the axial-gap type. The axial-gap motor consists of disk-shape rotor(s) and stator(s). The motor shape can be flat, and they are installed to flat space, for example, optical disk drives and recently-developed in-wheel motors for electric vehicles.

The schematic structure of the HTS axial-gap motor manufactured by IHI is shown in Fig. 2. As mentioned above, the contra-rotating propeller system requires two motors driving the front and rear propellers. The axial-gap motors with the short dimension in the axial direction are suitable for such applications. They are also suitable for simplifying the shape of the HTS armature coils and that

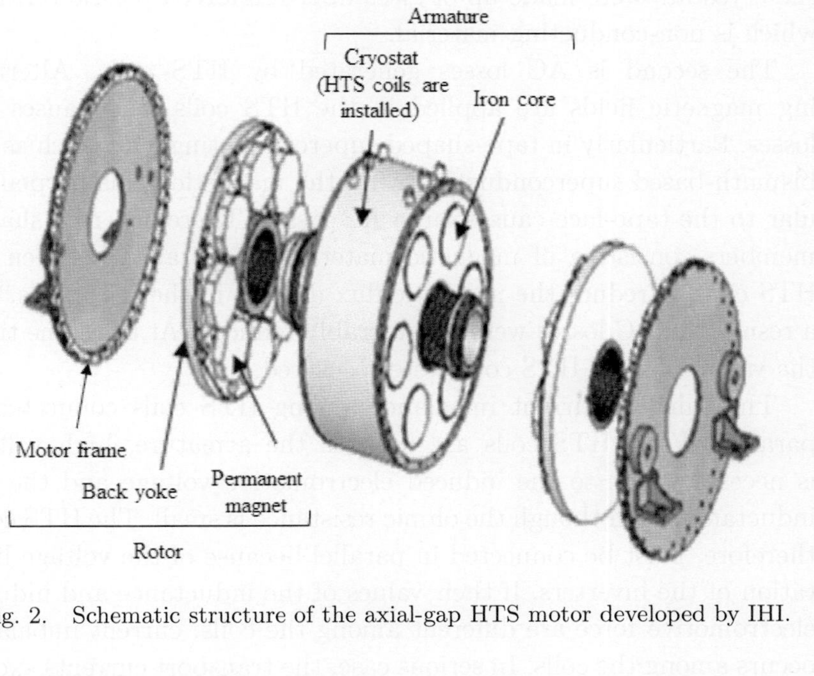

Fig. 2. Schematic structure of the axial-gap HTS motor developed by IHI.

---

permanent-magnets is lower than the sum of cryostat and HTS coils. Therefore, the motor type with permanent-magnet rotors was selected.

of the iron cores. Two rotators with permanent magnets were placed at both ends to stabilize the attracting force between the permanent magnets and the iron core in the axial direction.

2.4. *Technical problems of axial-gap motors with HTS armature*

Several technical problems peculiar to the axial-gap motors with HTS armature were found during designing, manufacturing, and testing. In this section, we briefly describe the problems and the means that we had.

The first is the cryostat for installation of the HTS coils. The cryostat must have high thermal insulation performance. Since the cryostat is one of the members of the armature in this type of the motor, it is directly affected by alternating magnetic fields. Therefore, the cryostats were made up of glass-fiber reinforced plastic (GFRP) which is non-conducting material.

The second is AC losses generated by HTS coils. Alternating magnetic fields are applied in the HTS coils, and causes AC losses. Particularly in tape-shaped superconducting wire, such as the bismuth-based superconducting wire, the magnetic fields perpendicular to the tape-face causes large AC losses. Therefore, ring-shaped members consisting of magnetic material are installed between the HTS coils to reduce the magnetic-flux density in the HTS coils.[6] As a result, the AC losses were considerably reduced. At the same time, the values of I_c of HTS coils were recovered.

The third is current imbalance among HTS coils connected in parallel. When HTS coils are used as the armature, high voltage is necessary due to the induced electromotive voltage and the coil inductance, even though the ohmic resistance is small. The HTS coils, therefore, must be connected in parallel because of the voltage limitation of the inverters. If their values of the inductance and induced electromotive force are different among the coils, current imbalance occurs among the coils. In serious case, the transport currents exceed I_c of HTS coils. To prevent such current imbalance, we manufactured new device named current distributor, which consisted of reactors.[6]

Problems also exist in the assembly of the motor, especially, concerning permanent-magnets. In the HTS motors manufactured by IHI, the neodymium magnets with high magnetic force and the iron cores are installed in the fields and armature, respectively. The magnetic force strongly attracts the iron cores and the permanent-magnets, even when the armature is not magnetically excited. The assembly of the motor under such strong magnetic attracting force is a dangerous work. Therefore, we first assembled the rotators without permanent-magnets and the stators with iron cores, and then installed the magnets to the rotors.

3. Results of the Load Test

After the technical problems were solved, the design of HTS motors with the rated power of 400 kW was performed using the simulations of the JMAG for electromagnetic field analysis. Then, the trial manufactures and the load tests were carried out. The specifications of the latest HTS motor manufactured by IHI are shown in Table 1. The appearance and the schematic structure are also shown in Fig. 3.

View of the HTS motor and other equipment during setup for the load test is shown in Fig. 4. The inverter supplied electric power to the HTS motor *via* the current distributor. The gear was installed between the motor and the dynamometer. The LN_2 refrigerator and the tank were connected to the cryostat of the HTS motor, and the LN_2 pump transported LN_2 to the cryostat. In the load test, we

Table 1. Specifications for the latest HTS motor cooled by LN_2, which was manufactured by IHI.

Type	Three-phase permanent-magnet synchronous motor with HTS armature
Rated power	400 kW
Rated speed	250 rpm
Rated voltage	440 V
Efficiency	98.0% (measured value, motor only)
Superconducting wire	BSCCO, type H[7]
Operating temperature of HTS coils	66–70 K

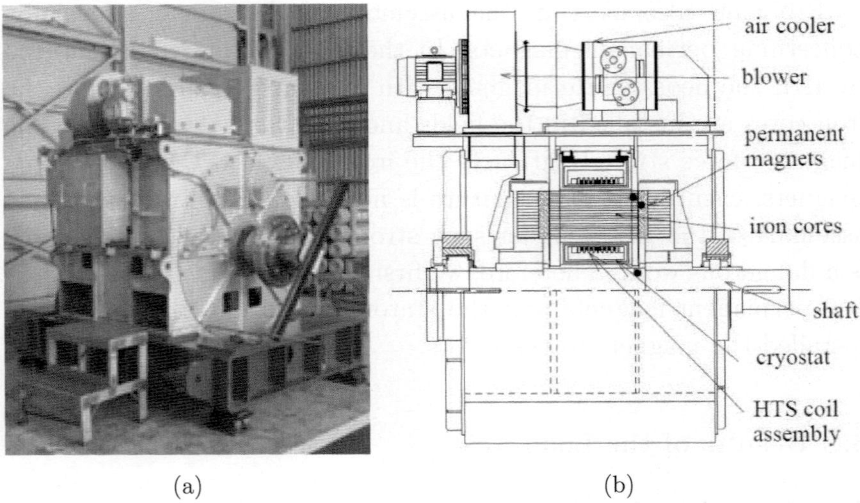

Fig. 3. 400 kW LN_2-cooled superconducting motor. (a) Appearance of the HTS motor. (b) Cross section of the HTS motor. The closed circles show the positions where the temperatures in Fig. 5(b) were measured.

Fig. 4. View of the HTS motor, LN2 cooling system, inverter, and other equipment during set-up for the load test.

measured the motor performance; the electro motive force, the currents, the temperature rises, the losses, and so forth. These results showed good agreement with the designed values.

The results in the continuous load test are shown in Fig. 5. The time when the motor started to rotate is shown as time $t = 0$. The vertical dotted lines show $t = 0$, the time when the load reached

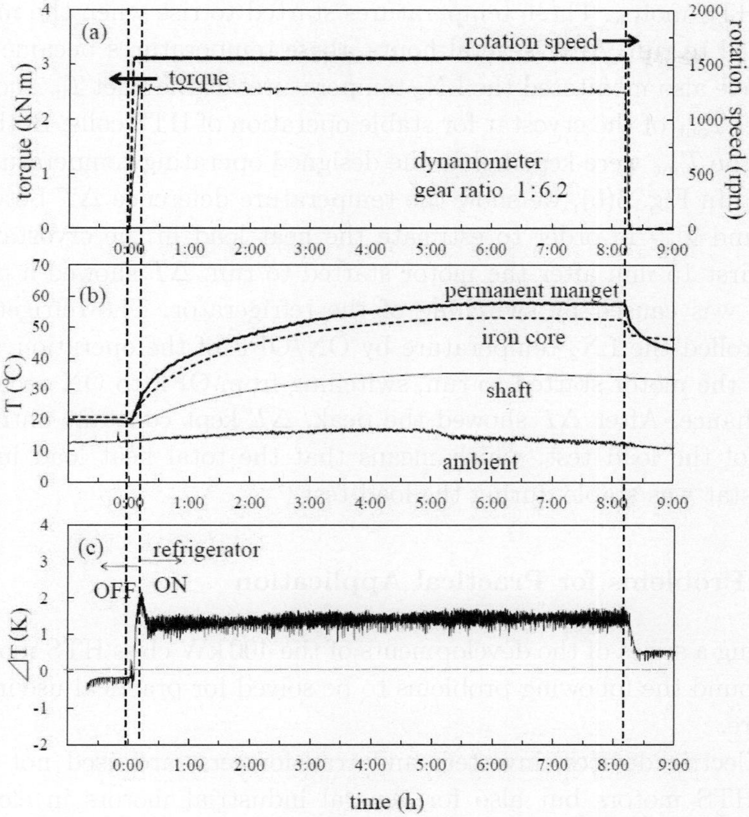

Fig. 5. The results of the continuous load test. (a) Time variation of the torque and the rotation speed of the dynamometer. (b) Temperatures at representative positions of the iron-cores and permanent-magnets, and the shaft. Ambient temperature is also shown for reference. (c) Temperature deference ΔT between the inlet and the outlet of the cryostat.

400 kW and the time when the load started to decrease at the end of the test. The torque and the rotation speed of the dynamometer are shown in Fig. 5(a). After the load reached 400 kW, both torque and the rotation speed were stable until the end of the load test. In this test, the motor output was estimated to be 410 kW, and the measured value of the armature current was 940 Arms. We show the temperatures at the iron-core, the permanent-magnet, and the shaft in Fig. 5(b). The closed circles in Fig. 3(b) indicate the measured positions in

the HTS motor. These temperatures started to rise when the motor started to run. After several hours, these temperatures became stable. We also monitored the LN_2 temperatures at the inlet T_{in} and the outlet T_{out} of the cryostat for stable operation of HTS coils. Both T_{in} and the T_{out} were kept below the designed operating temperature of 70 K. In Fig. 5(c), we show the temperature deference ΔT between T_{in} and T_{out} in order to estimate the heat load in the cryostat. In the first 15 min after the motor started to run, ΔT showed a peak. This was caused by switching of the refrigerator. The refrigerator controlled the LN_2 temperature by ON/OFF of the operation. Just after the motor started to run, switching from OFF to ON occurred by chance. After ΔT showed the peak, ΔT kept constant until the end of the load test, which means that the total heat load in the cryostat was stable during the load test.

4. Problems for Practical Application

During a series of the developments of the 400 kW class HTS motors, we found the following problems to be solved for practical use in the future.

Electric devices, inverters and transformers, are used not only for HTS motors but also for general industrial motors in electric propulsion systems. These electric devices are generally large and heavy. It is necessary to reduce the sizes and weight of these devices as well as those of HTS motors.

Neodymium magnets are often selected as permanent-magnets for industrial motors, since its coercive force is large. The neodymium magnets, however, are expensive due to both material cost and the mechanical processing. The permanent-magnets are formed into small blocks to reduce eddy currents, and these small blocks are adhered together after the magnetization. Moreover, in large-scale applications, the handling of the permanent-magnets will be more difficult because of the stronger magnetic force, since larger and heavier magnets are required.

The cost is a problem also for GFRP cryostat. The structure of the cryostat is complicated because of a double container structure for the vacuum thermal insulation. No vacuum leak is permitted, though the internal tank is exposed to the strong thermal stress in the cryogenic temperatures. We selected an autoclave method for the manufacture of the inner container. Various element tests had been carried out, and then the cryostat without vacuum leak was manufactured. It was confirmed that the heat leakage of whole of the cryostat agreed with the designed value.

A large part of the heat load is caused by AC losses in the cryogenic parts of HTS motor during running with the rated power of 400 kW. The AC losses are generated not only in the HTS coils, but also in other materials. The AC losses include the eddy current losses generated by conducting materials and the magnetization losses generated by magnetic materials. As mentioned above, we installed the members with magnetic material between the HTS coils in order to control magnetic flux density in HTS coils. Since the magnetic material causes the magnetization losses, they must be designed to minimize the total loss of the HTS coils and the magnetic materials. If AC loss of HTS material itself is small, the amount of the magnetic materials installed to the HTS coil assemblies can be reduced. HTS wires with small AC losses are desired.

Finally, we should point out problems of the cryogenic cooling system. The system consists of a refrigerator, a LN_2 tank, a LN_2 pump, pipes, and joints, in addition to the cryostat. Since COP of refrigerators is not high enough at cryogenic temperatures, even LN_2 temperatures, we must reduce the heat load including the heat leakage. The improvements of the efficiency of the refrigerator and the pump are also required.

5. Conclusions

IHI designed and performed the trial manufacture of the HTS motors cooled by LN_2 and the LN_2 cooling systems with the aim of its practical use. The continuous load tests with the rated power of 400 kW were also completed. These HTS motors have different structures from the typical radial-gap HTS motors operated at much lower temperatures than LN_2. We have demonstrated that HTS motors cooled by LN_2 are feasible in the point of efficiency. It, however, seems to be difficult to reduce the size and the weight compared with typical HTS motors with HTS fields. This disadvantage, however, must be accepted as the result from taking the choice of LN_2 as the cooling medium. When HTS wires with higher in-field performance at LN_2 temperatures will be developed in the future, feasible solutions for the typical type of HTS motors with HTS fields cooled by LN_2 will be hopefully obtained.

References

1. For information about *Oasis of the Seas*, http://www.oasisoftheseas.com.
2. For information about Azipod propulsion systems, http://new.abb.com/marine.
3. H. Yamada, H. Miyabe, and A. Saeki, *IHI Eng. Rev.* **44**, 12 (2011).
4. J. F. Gieras, *Przeglad Elektrotechnizny* (*Elec. Rev.*) **85**, 1 (2009).
5. T. Takeda, H. Togawaand, and T. Oota, *IHI Eng. Rev.* **39**, 89 (2006).
6. H. Sugimoto, T. Morishita, T. Tsuda, T. Takeda, H. Togawa, T. Oota, K. Ohmatsu, and S. Yoshida, *J. Phys. Conf. Ser.* **97**, 012203 (2008).
7. N. Ayai and S. Kobayashi, *et al.*, *Physica C* **468**, 1747 (2008).

Chapter 3.20

Radial-Gap Motor for Ship Propulsion

Toshiyuki Yanamoto and Minoru Yokoyama

Corporate Technology Division,
Kawasaki Heavy Industries, Ltd.
P.O. Box 673-8666, 1-1
Kawasaki-cho, Akashi-city, Japan

The KHI team has developed radial gap high-temperature superconducting (HTS) motors of three sizes, 1 MW-class, 3 MW, and 20 MW, to be used for electric propulsion systems for ships. The volumetric torque density of the assembled 3 MW HTS motor was recorded at 40 kNm/m^3 in the load test; the world's highest in the class.

1. Introduction

Recently, in the field of maritime transport, the International Maritime Organization (IMO), an agency of the United Nations, adopted international environmental regulations in which CO_2 emissions from ships must be reduced from the past decade average in stages up to 30% by 2025. The IMO also adopted regulations to reduce NO_X and SO_X emissions from ships in 1997. These regulations were applied in 2005 and changed in 2010 so as to gradually reduce the emissions further.

At present, most ships have diesel engines as the power unit for propulsion, however, of late, due to the above environmental regulations, different propulsion systems have been used in the field of maritime transport. For example, Liquefied Natural Gas (LNG) carriers with Dual Fuel Diesel Electric (DFDE) propulsion have been introduced recently. In Japan, the Minister of Land, Infrastructure, Transport and Tourism (MLIT) and the Japan Railway Construction, Transport and Technology Agency (JRTT) have been promoting the super ECO ship which has an electric propulsion system installed. These ships have been operated in domestic waters. On the other hand, electric propulsion systems have been adopted for offshore vessels, specialty vessels such as icebreakers, research vessels, and some passenger vessels in the world since the early 19th century, because of easy control with higher power. For example, Japan's first electric propulsion vessel was built in 1920.

Because of the situation mentioned above, it is expected that ships with electric propulsion systems will increase in number globally in the future, and a smaller and more powerful electric propulsion system with higher energy efficiency achieved by using a high temperature superconducting (HTS) motor will be required.

The KHI team has been developing HTS motors for such an electric propulsion system. A radial-type field-winding synchronous motor (a radial gap motor) was selected for this purpose. AMSC,[1] Siemens,[2] and others have also developed radial-gap HTS motors respectively. First, the KHI team developed a 1 MW-class motor[3] with the aim of developing the basic technology structure and of carrying out a load test at 450 kW during 2007–2010. In addition, the team developed a 3 MW motor[4] applicable to coastal vessels in 2010–2013, succeeding in a 3 MW load test and achieving the world's highest torque density. These developments were carried out with the assistance of the New Energy and Industrial Technology Development Organization (NEDO). The team also developed key hardware technologies for a 20 MW HTS motor destined for future ocean-going vessels with assistance of the Japan Science and Technology Agency (JST) in 2009–2013.

The members of this team included Kawasaki Heavy Industries, Ltd., Sumitomo Electric Industries, Ltd., the Tokyo University of Marine Science and Technology, Yokohama National University,

Sophia University, Niigata University, and the National Maritime Research Institute.

2. Development of Radial-Gap HTS Motor

A 1 MW-class motor and a 3 MW motor were designed and assembled, and load examinations were carried out for both. A 20 MW motor was also designed and key hardware was developed for it. The type of motor selected for each was a radial-type field-winding synchronous motor (a radial-gap motor). The motor power equation for the radial-gap motor is shown in Eq. (1).

$$P = K \cdot B_g \cdot AC \cdot D^2 \cdot L \cdot \omega, \tag{1}$$

where P is the motor power, K is constant, B_g is the magnetic field strength of the radial gap [T], AC is the electric loading [A/m], D is the radius of the rotor, L is the length of the motor, and ω is the angular velocity [1/sec].

The magnetic field strength B_g of an HTS coil installed in a rotor is much stronger than that of a permanent magnet. This is because the HTS coil made by winding thin DI-BISCCO® tape has a high winding number, to which an electric current of over 100 A is supplied. So, Eq. (1) demonstrates that for the same value of the angular velocity and electric loading, the HTS motor is much smaller in size than a conventional permanent magnet motor. In addition, the angular velocity ω can be matched to that of a rotating propeller directly without use of gears. Also the motor efficiency is very good because the electric power consumed in the rotor is almost zero.

Details of each of the three HTS motors developed are provided in the following sub-sections.

(a) 1 MW-class motor

The 1MW-class radial-gap motor was developed in 2007–2010 with the aim of establishing a basic method for designing HTS motors. This motor was designed so as to enable downsizing, whilst achieving high efficiency. Figure 1 shows the schematic structure, Fig. 2, the external view, and Table 1, the specifications of the motor.

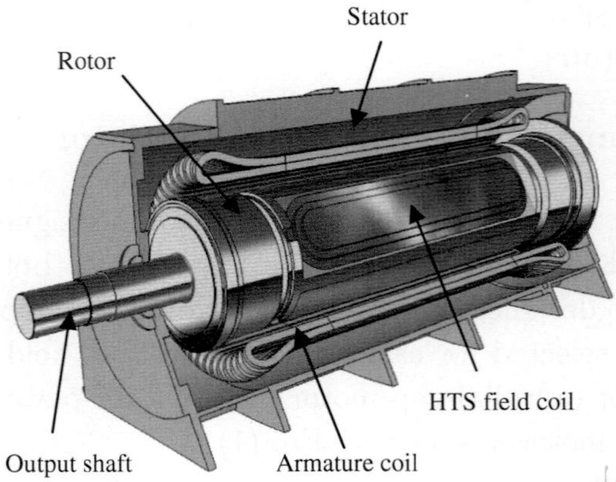

Fig. 1. Schematic structure of 1 MW-class superconducting motor.

Fig. 2. 1 MW-class superconducting motor.

The rotor has four poles. Six HTS field coils with a race track type structure are installed and piled up in each pole of the rotor without an iron core so as to avoid magnetic field saturation. The coils are cooled down to 30 K by low-temperature helium gas that is provided from an external cooling system. A 200 A electric current is supplied to the coils, generating a magnetic field of maximum 5 T which is strong enough to enable downsizing.

Table 1. Specifications of the 1 MW-class superconducting motor.

Rated power	1 MW
Rated speed	190 rpm
Number of pole	4
Number of slot	48
Rated voltage of armature coil	1400 V
Rated current of field coil	200 A
Field coil windings	DI-BSCCO
Field coil cooling	Helium gas
Field coil operating temperature	30 K
Number of field coil unit layer	6
Armature coil cooling	Air-cooled
Outer diameter of back yoke	900 mm
Bearing distance	2789
Volumetric torque density	28 kNm/m^3

To improve the motor efficiency, the eddy current loss and the cooling loss are reduced. The armature coil forming the stator is composed of an ordinary copper winding without any iron core so as not to generate an eddy current, because it is exposed to a high magnetic field from the HTS field coil. This makes it possible to effectively convert a large electromagnetic force by the HTS field coil unit into large torque with high efficiency.

In order to reduce the cooling loss in the HTS field coils and improve the reliability in keeping the superconducting state, the incoming heat from outside of the rotor caused by radiation and conduction is minimized by forming a multiple heat insulation structure in the rotor of the cryostat. A circulating low-temperature helium gas cooling system is adopted. The incoming heat is transferred outside of the rotor by low-temperature helium gas, which is re-cooled by an external refrigerator.

The helium gas is fed by a helium feed-exhaust device between an external refrigerator and the rotor which is rotating at 190 rpm. The flow rate of the helium gas here is 4 g/s with a pressure of 0.4 MPa at temperatures from 25 K to 30 K. Helium gas leakages by the helium feed-exhaust device is below 5% and the resulting heat loss is below 5 W. In the assembly of the 1 MW-class motor, two out of six superconducting coils were installed at each pole. The motor's output power reached 450 kW at 190 rpm in the load test. It is estimated from the results of these tests that the motor's output would reach 1 MW with an outstanding efficiency of 98% if all HTS coils were installed.

The motor was operated at an output of 450 kW continuously for 24 h, and the temperature rise in the low-temperature section inside the rotor was confirmed to be within 1 K of the allowable value.

(b) 3 MW motor

Development of the radial-gap 3 MW motor for the propulsion system of coastal-vessels started in 2010.

This motor is of the same construction as that of the 1 MW-class motor based on a radial-type field winding synchronous motor with race track-type HTS field coils in the rotor. Four coils are piled up and installed in the coil box without use of an iron core to form each pole. The rotor has six of these poles.

Figure 3 shows a photograph of the HTS field coil made by using two tape materials; thin H tape which has no reinforcement, and HT tape which is reinforced by laminating type H tape with stainless steel tape in order to achieve a smaller bending radius. Table 2 shows the comparison of the two tape materials used. The HT tape was used in the core of the coils where the smaller bending radius is required while the type H tape was used on the outer periphery with a larger minimum bending radius to shorten the total width of the coil. The width of coils made in this way is approximately 20% less than coils in the 1 MW-class motor made only of the HT tape.

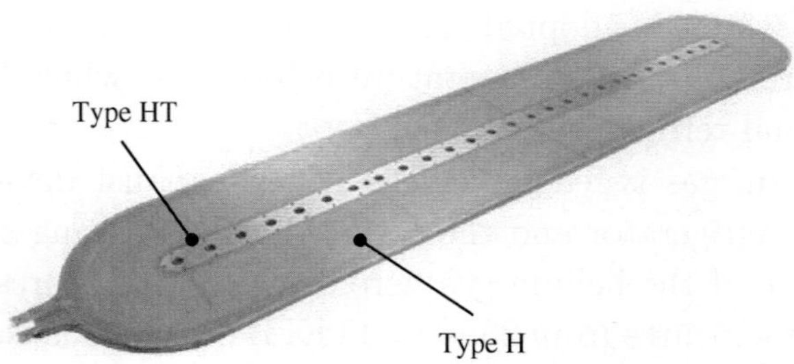

Fig. 3. Superconducting field coils for the 3 MW motor.

Table 2. Comparison of BSCCO tapes.

Type	Cross-section view	Thickness (mm)	Minimum bending radius (mm)
H		0.23	70
HT		0.30	50

Fig. 4. 3 MW superconducting motor.

The armature coils forming the stator are also composed of an ordinary copper winding without an iron core. The HTS field coil unit is cooled by helium gas supplied from the external refrigerator to the rotor *via* the helium gas feed and exhaust device. The volumetric torque density of the 3 MW motor is 40 kNm/m^3, approximately 1.4 times that of the 1 MW-class motor. The torque density of the world's highest in the class was achieved.

Figure 4 shows an external view of the motor and Table 3 shows the specifications. The assembled 3 MW motor was connected to a hydraulic dynamometer and a load test was conducted. The results of the load test are shown in Table 4. The rated output of 3 MW was confirmed.

(c) 20 MW motor design

Design of a 20 MW HTS motor for ocean-going vessels started in 2009. It is a radial-type field winding synchronous motor of the same

Table 3. Specifications of the 3 MW superconducting motor.

Rated power	3 MW
Rated speed	160 rpm
Number of pole	6
Number of slot	72
Rated voltage ofarmature coil	1200 V
Rated current of field coil	200 A
Field coil windings	DI-BSCCO
Field coil cooling	Helium gas
Field coil operating temperature	30 K
Number of field coil unit layer	4
Armature coil cooling	Air-cooled
Outer diameter of back yoke	1400 mm
Bearing distance	2845
Volumetric torque density	40 kNm/m^3

type as the 1 MW-class and the 3 MW motor, with the armature coil unit of no iron-core. Figure 5 shows the schematic structure and Table 5 shows the specifications.

The field coil without iron-core is fabricated by winding high temperature superconducting tape and installed in the coil boxes on the periphery of rotor inside.

A thermal siphon system is adopted and developed as a key part of the hardware to cool the coil boxes. The motor and the cooling device are integrated to reduce the size of the motor unit. Neon gas is used as the refrigerant. On the stator side, a condenser is connected to the refrigerator to produce a liquid phase refrigerant which is introduced through the rotary joint into the evaporator at the center of the rotor.

The refrigerant gas evaporated by the evaporator returns to the refrigerant condenser. The coil box and the evaporator are connected by a heat conducting bus bar, the HTS field coils are cooled down to 37 K by thermal conduction with the evaporator. The downsizing of the motor is achieved by the integration of the refrigerator and the rotor.

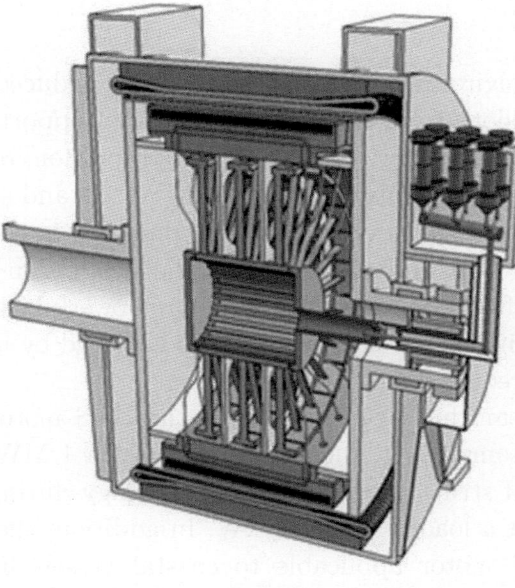

Fig. 5. Schematic structure of 20 MW motor.

Table 5. Specifications of the 20 MW superconducting motor basic design.

Rated power	20 MW
Rated speed	90 rpm
Number of pole	24
Number of slot	216
Rated voltage of armature coil	5200 V
Rated current of field coil	200 A
Field coil windings	300A-class@77 K
Field coil cooling	He + Ne gas
Field coil operating temperature	37 K
Number of field coil unit layer	4
Outer diameter of back yoke	4000 mm
Bearing distance	3000 mm
Volumetric torque density	56 kNm/m^3

3. Summary

International environmental regulations specify reduced CO_2, NO_X, and SO_X emissions from ships. This creates an opportunity for the introduction of LNG carriers with DFDE propulsion, or super ECO ships with electric propulsion for coasting vessels and so on.

Because of the situation mentioned above, it is estimated that the number of ships globally with electric propulsion systems will increase in the future, and smaller and more powerful electric propulsion systems with higher energy efficiency achieved by using an HTS motor will be required.

The KHI team has been developing the HTS motor for electric propulsion systems. First, the team developed a 1 MW-class motor with the aim of structuring the basic technology during 2007–2010, and carried out a load test at 450 kW. In addition, the team developed a 3 MW motor applicable to coastal vessels in 2010–2013, succeeding in a 3 MW load test and achieving the world's highest torque density. The team also developed key hardware technologies for a 20 MW HTS motor destined for future ocean-going vessels in 2009–2013.

References

1. B. Gamble, G. Snitchler, and T. MacDonald, Full power test of a 36.5 MW HTS propulsion motor, *IEEE Trans. Appl. Supercond.* **21**(3), 1083–1088 (2011).
2. W. Nick, J. Grundmann, and J. Frauenhofer, Test results from Siemens low-speed, high-torque HTS machine and description of further steps towards commercialisation of HTS machines, *Physica C* **482**, 105–110 (2012).
3. K. Umemoto, K. Aizawa, M. Yokoyama, K. Yoshikawa, Y. Kimura, M. Izumi, K. Ohashi, M. Numano, K. Okumura, M. Yamaguchi, Y. Gocho, and E. Kosuge, Development of 1 MW-class HTS motor for podded ship propulsion system, *J. Phys.: Conf. Series* **234**, 032060 (2010).
4. http://www.khi.co.jp/news/detail/20101101_1.html.

Chapter 3.21

Induction/Synchronous Motor for Automobile

Taketsune Nakamura

Department of Electrical Engineering,
Graduate School of Engineering, Kyoto University
P.O. Box 615-8510, 1 Kyoto-Daigaku-Katsura,
Nishikyo-Ku, Kyoto, Japan

Rotating principle and fundamental performance of high temperature superconducting induction/synchronous motor employing Bi-2223 tapes are reviewed. Development status of 20 kW class prototype motors is also introduced, and then the future perspective of the motors to the next generation automobile is discussed.

1. Introduction

In a background of technological advance in high quality and long length Bi-2223 tapes, various types of industrial and power applications for the effective utilization of the tapes have actively been underway worldwide. Industrial application is one of the most important targets for realizing superconductivity innovation with the use of such tapes. For instance, more than 50% of the electric power has been consumed by means of operation of the electric motors in Japan, and then the realization of the highly efficient motors is crucial for the environmentally-friendly as well as low carbon emission society. Especially, recent trends of electric drive of the automobile will be accelerated, and then high efficiency and high torque density drive motors have been demanded.

We have proposed and carried out fundamental study of High Temperature Superconducting Induction/Synchronous Motor (hereafter, designated as HTS-ISM),[1–10] and then tried to apply to the automobile,[11–18] electric train[19–21] and liquid hydrogen circulation pump.[22–25] In this section, rotating principle and fundamental performance of the HTS-ISM are firstly presented. And then, R&D activities of the HTS-ISM to the next generation automobile are reviewed and discussed.

2. Principle of AC Rotating Machine

Mechanical output of the motor, $P_{\rm m}$, is formally expressed as follows.

$$P_{\rm m} = D^2 L \cdot AC \cdot B \cdot N, \tag{1}$$

where D and L are, respectively, the diameter and the length of the rotor. AC, B and N are Ampère turns (electric loading), airgap magnetic flux density (magnetic loading), and rotating speed, respectively. Allocation of the electric loading and the magnetic one is important for designing the rotating machine. There exist mainly two types of conventional (normal conducting) AC drive motor, i.e., synchronous (permanent magnet: PM) motor and induction motor. The PM motor is the major drive motor in Japan, and many induction motors are utilized in USA and EU.

Figure 1 illustrates schematic diagrams of the rotating principle of AC motors.[15] It is mandatory to generate the rotating magnetic field for the generation of the torque, and this field is simply expressed by means of the rotating magnets in this figure. It should be noted, however, that such rotating field is realized with the use of the stationary three phase windings in a practical case. For the PM motor (Fig. 1(a)), the rotor consists of the permanent magnets, and the torque is generated by the attractive force in between different magnet poles. In this type of motors, the rotating speed of the rotor, N, always equals to that of the outer magnets, N_s, at the steady state, and such motor is called as "synchronous motor". On the other hand, the induction motor possesses the shortened windings in the rotor, and the induced current flows in such windings by interlinking

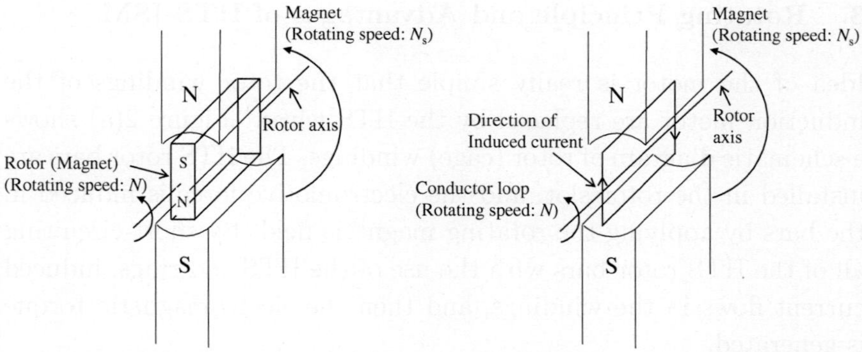

(a) Synchronous motor (N=Ns) (b) Induction motor (0<N<Ns)

Fig. 1. Rotating principle of AC motor.[15]

the rotating magnetic fluxes. And then, the torque is generated by interacting such current with the magnetic fluxes. It is important to note that the rotating speed of the rotor, N, has to be slower than N_s in order to maintain the induced current. Such speed's difference is called slip, s, and is defined as follows.

$$s = (N_s - N)/N_s. \tag{2}$$

From this equation, we can see that $s = 0$ for the synchronous speed and 1 for the stationary state. The synchronous motor rotates at $s = 0$ and the induction motor at $0 < s < 1$ for the steady state.

As I mentioned, the slip is mandatory to induce the current in the rotor windings of the induction motor. Such current, however, also generates Joule-type dissipation, and could lower the power conversion efficiency of the motor. The synchronous motor has the permanent magnet in the rotor, and then it is generally said that the efficiency of such motor is superior to that of the induction motor.

The advantages of the induction motor, on the other hand, are its simple structure, easy maintenance, etc. Therefore, the possibility of the induction motor could spread out for the major introduction of the drive motor, and our research group has tried to realize this breakthrough with the aid of the superconductivity technology. In the following section, the rotating principle and advantages of our motor are explained and discussed.

3. Rotating Principle and Advantages of HTS-ISM

Idea of the motor is really simple that the rotor windings of the induction motor are replaced by the HTS wires.[1] Figure 2(a) shows a schematic diagram of rotor (cage) windings. The HTS rotor bars are installed in the rotor slot, and the electromotive force is induced in the bars by applying the rotating magnetic field. By short-circuiting all of the HTS rotor bars with the use of the HTS end rings, induced current flows in the windings, and then the electromagnetic torque is generated.

Figure 2(b) shows a schematic of voltage (V) versus current (I) curve in rotor bars (I_c: critical current), and the corresponding magnetic states are schematically illustrated in Fig. 3. When the HTS-ISM is in a starting or overload condition, large current ($I > I_c$) flows in the windings and the rotor will be at the slip rotation

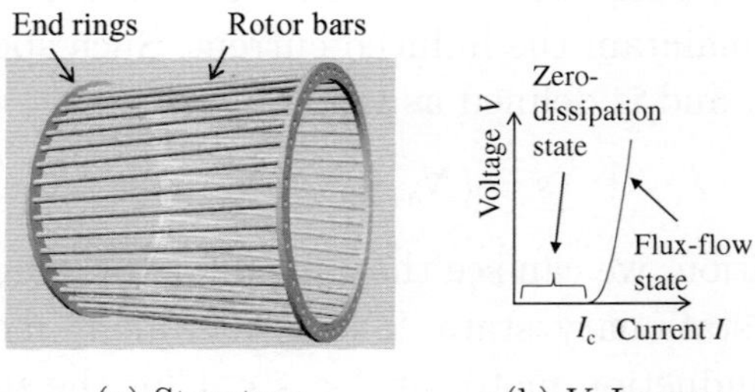

(a) Structure (b) V–I curve

Fig. 2. Schematic diagram of rotor (cage) windings and its $V - I$ curve (I_c: critical current).

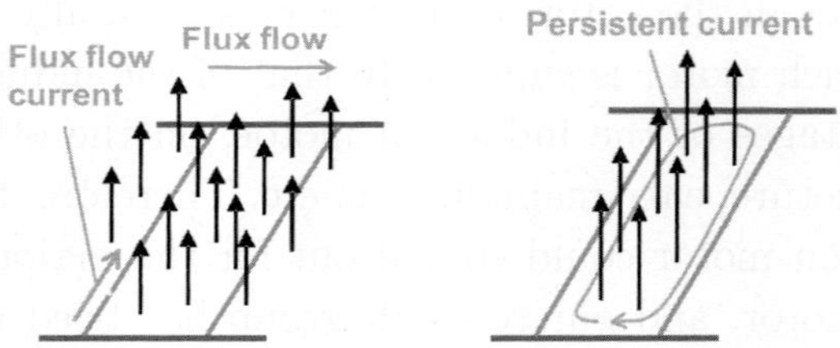

(a) Flux-flow state (slip mode) (b) Trapped flux state (synchronous mode)

Fig. 3. Schematic diagram of magnetic state in one loop of cage windings.

mode, because the rotor bars are in the dissipative flux-flow state (Fig. 3(a)). On the other hand, the magnetic fluxes will be trapped in the windings at a steady rotation mode ($I < I_c$), and then the highly efficient synchronous rotation is possible to be realized (Fig. 3(b)). The advantages of the HTS-ISM are summarized as follows:

(1) Co-existence of synchronous and asynchronous (slip) torque[1–10]
The HTS-ISM possesses both of synchronous and asynchronous (slip) torque, and the corresponding electromagnetic state of the rotor bars is, respectively, zero resistivity and flux-flow (dissipative) state. Synchronous rotation mode is used for the steady state operation, and the slip one is for the starting as well as the overload (short term) condition.

(2) High efficiency and high toque density[11]
The concept of the HTS-ISM is the maximization of the electric loading (Ampère turns), and the high efficiency for the variable speed control and high torque density can be realized without high magnetic field.

(3) Stable rotation against variable speed control[26]
The control method of the HTS-ISM effectively utilizes not only zero resistivity but also the flux-flow state, and the motor is stable against the variable speed control thanks to such dissipative flux-flow state.

(4) Possibility of non-superconducting state operation[10]
By applying the rotating magnetic field in the HTS rotor bars, an electromotive force is firstly induced in the windings, and then the induced current flows in such bars. If the HTS rotor bars are in the non-superconducting state, therefore, the current is limited to small value. In this induction phenomenon, the HTS-ISM can rotate as the normal conducting induction motor, and then the motor could be rotated at non-superconducting state for a while.

4. Development of 20 kW Class HTS-ISMs

Based on the fundamental study of the HTS-ISM, we have tried to develop two types of 20 kW class prototype motor in a framework of

Japan's national projects. Final target of the project is to realize the low carbon emission automobile. In order to overcome the conventional (normal conducting) drive motor, it is mandatory to realize not only the high efficiency against the variable speed control but also the high torque density at low speed. That is, if we attain the high torque density motor, the transmission gears, of which is heavy in weight and generates much losses, can be removed. And then, the overall energy conversion efficiency of the system will be increased even if we consider the power consumption of the cryocooler. In these project, the maximum rotation speed of the motor is targeted to 1800 rpm.

4.1. *First generation*

First generation motor has been developed for confirmation of the integrity of our design method. Figure 4 shows the photographs of fabricated Bi-2223 cage rotor. 11 pieces of Bi-2223 tapes (I_c = 70 A@77 K) are stacked together, and then are attached to an oxygen-free copper as one rotor bar (Fig. 4(a)). The Bi-2223 rotor bars, 44 in number, are then installed in the rotor slot, and are shorted by means of different type of Bi-2223 tapes (I_c = 190 A@77 K). Figure 4(b) shows a fabricated Bi-2223 rotor.[13]

We have fabricated two kinds of stator (3-phase, 8-pole, turn number: 30), i.e., Cu stator and Bi-2223 one, as shown in Fig. 5.[13] Figure 6 shows a photograph of the first generation motor.[13]

As a tested result of Bi-2223 rotor/Cu stator combination motor, we could successfully realize the full load power (20 kW) at 77 K

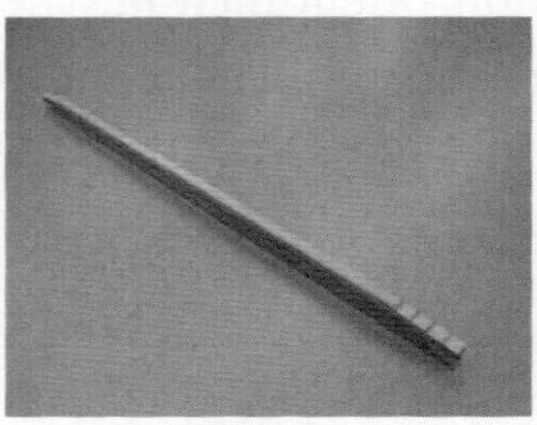

(a) Bi-2223 rotor bar

(b) Completed rotor

Fig. 4. Photographs of Bi-2223 rotor.[13]

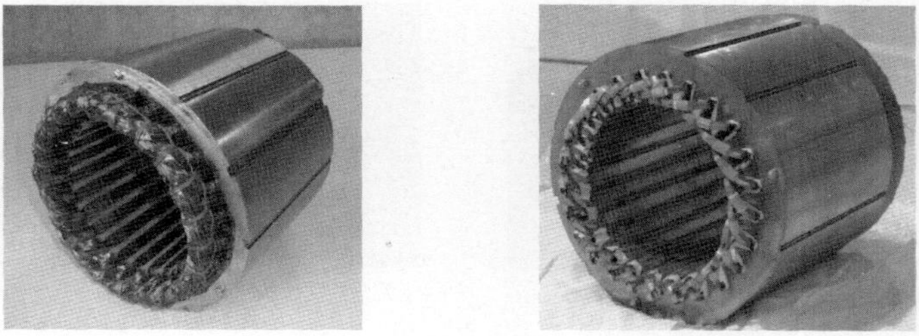

(a) Cu stator (b) Bi-2223 stator

Fig. 5. Photographs of fabricated stators.[13]

Fig. 6. Photograph of first generation motor.[13]

(atmospheric liquid nitrogen).[17] Furthermore, we combined Bi-2223 stator with Bi-2223 rotor, i.e., fully superconducting motor, tried to test at no-load condition. As a result, we succeeded in the continuous operation at 1200 rpm[13] (after that, we have also succeeded in 1800 rpm rotation test).

4.2. *Second generation*

Based on the successful development of the first generation motor, we have further moved on to the development of small scale, same power

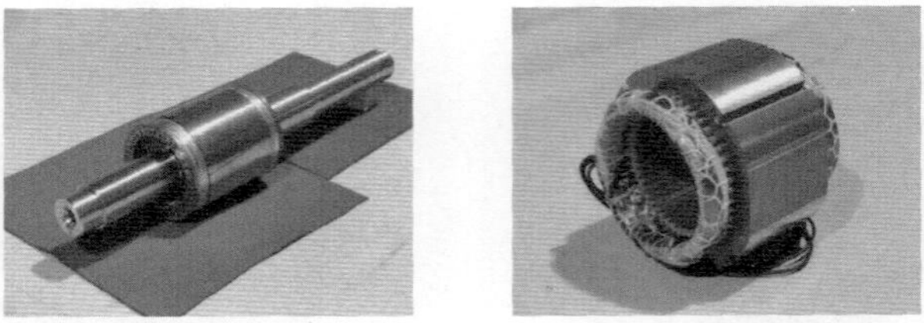

(a) Bi-2223 rotor (b) Cu stator

Fig. 7. Photographs of second generation motor.[17]

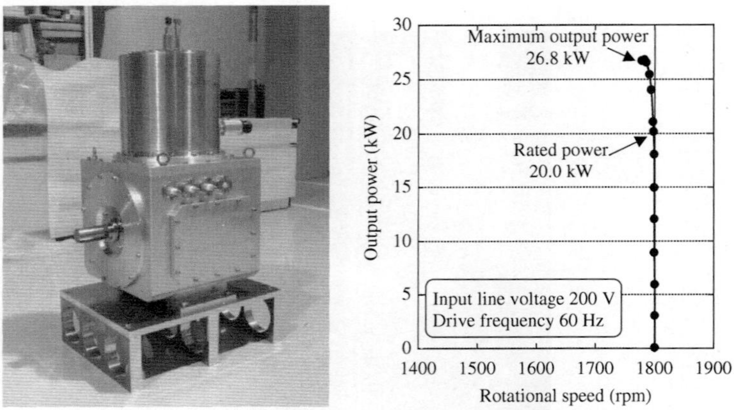

(a) Photograph (b) Load test result(77K)

Fig. 8. Photograph and load test result of second generation motor.[17]

motor. We have tried to reduce 70% in magnetic volume at this stage. Figure 7 shows the photographs of fabricated Bi-2223 rotor and Cu stator, and Fig. 8(a) shows the completed motor. Figure 8(b) shows the typical experimental result at 77 K. As can be clearly seen in the figure, the fabricated motor realizes the synchronous power at 20 kW as designed. This indicates the integrity of our design method. Further application of the load transits the rotation mode from the synchronous to the slip, and the maximum slip power reaches 26.8 kW. From this result, the fabricated motor possesses a tolerance against the overload, and this operation mode can be used as the short time rating.

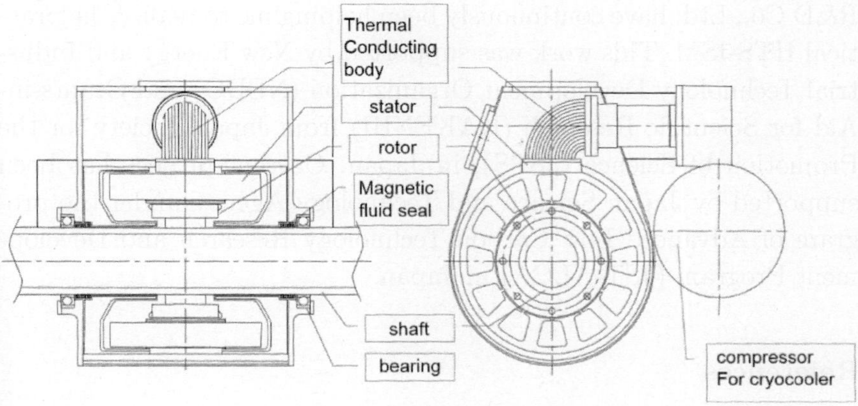

Fig. 9. Future image of HTS-ISM drive system.

5. Future Plan

Our group has now been developing a 50 kW class fully superconducting model motor. We have also developed optimal control code of the HTS-ISM, low thermal invasion and small sized cryostat, and highly efficient Stirling-type cryocooler. By combining the above technology, we will target a 100 kW class practical sized motor in near future. Figure 9 illustrates the future image of the HTS-ISM drive system.

6. Conclusion

This section introduces the fundamental characteristics and current status of HTS induction/synchronous motor for the next generation automobile. Two types of 20 kW class prototype motor have been designed, fabricated and successfully tested. Future prospective is discussed for the practical application of the HTS-ISM.

Acknowledgments

I would like to thank all of my co-workers for their support of our project. Especially, AISIN SEIKI Co., Ltd. and IMRA MATERIAL

R&D Co., Ltd. have continuously been helping me to realize the practical HTS-ISM. This work was supported by New Energy and Industrial Technology Development Organization (NEDO) and Grants-in-Aid for Scientific Research (KAKENHI) from Japan Society for the Promotion of Science (JSPS), in Japan. Current project has been supported by Japan Science and Technology Agency under the program of Advanced Low Carbon Technology Research and Development Program (JST-ALCA) in Japan.

References

1. G. Morita, T. Nakamura, and I. Muta, *Supercond. Sci. Technol.* **19**, 473 (2006).
2. T. Nakamura, H. Miyake, Y. Ogama, G. Morita, I. Muta, and T. Hoshino, *IEEE Trans. Appl. Supercond.* **16**(2), 1469–1472 (2006).
3. T. Nakamura, Y. Ogama, and H. Miyake, *IEEE Trans. Appl. Supercond.* **17**(2), 1615 (2007).
4. T. Nakamura, Y. Ogama, H. Miyake, K. Nagao, and T. Nishimura, *Supercond. Sci. Technol.* **20**, 911 (2007).
5. K. Nagao, T. Nakamura, T. Nishimura, Y. Ogama, N. Kashima, S. Nagaya, K. Suzuki, T. Izumi, and Y. Shiohara, *Supercond. Sci. Technol.* **21**, 015022 (2008).
6. T. Nakamura, K. Nagao, T. Nishimura, Y. Ogama, M. Kawamoto, T. Okazaki, N. Ayai, and H. Oyama, *Supercond. Sci. Technol.* **21**, 085006 (2008).
7. T. Nakamura, T. Nishimura, K. Nagao, K. Matsumura, and Y. Ogama, In *Proc. 18th Int. Conf. Elec. Mach.* 2008 *(ICEM'08)*, 1278 (2008).
8. T. Nakamura, K. Nagao, T. Nishimura, and K. Matsumura, *Supercond. Sci. Technol.* **22**, 045022 (2009).
9. T. Nakamura, Y. Ogama, K. Matsumura, T. Nishimura, K. Nagao, N. Kashima, and S. Nagaya, *J. Cryogenics Supercond. Soc. Jpn.* **44**(3), 112 (2009) (in Japanese).
10. H. Shimura, T. Nakamura, H. Kitano, T. Nishimura, N. Amemiya, and Y. Itoh, *IEEE Trans. Appl. Supercond.* **23**(3), 5201705 (2013).
11. T. Nakamura, K. Matsumura, T. Nishimura, K. Nagao, Y. Yamada, N. Amemiya, Y. Itoh, T. Terazawa, and K. Osamura, *Supercond. Sci. Technol.* **24**, 015014 (2011).
12. M. Miyamoto, T. Matsuo, and T. Nakamura, *Przeglad Elektrotechniczny* **87**(9b), 111 (2011).
13. D. Sekiguchi, T. Nakamura, S. Misawa, Y. Kawanishi, H. Kitano, M. Fukui, T. Matsuo, M. Miyamoto, N. Amemiya, Y. Itoh, M. Yoshikawa, T. Terazawa, K. Osamura, Y. Ohashi, and N. Okumura, *IEEE Trans. Appl. Supercond.* **22**(3), 5200904 (2012).

14. K. Kitano, T. Nakamura, D. Sekiguchi, N. Amemiya, Y. Itoh, M. Yoshikawa, N. Okumura, and Y. Ohashi, *IEEE Trans. Appl. Supercond.* **23**(3), 5202505 (2013).
15. T. Nakamura, *OYO BUTURI (The Japan Society of Applied Physics)*, **82**(7), 579 (2013) (in Japanese).
16. T. Nishimura, T. Nakamura, Q. Li, N. Amemiya, and Y. Itoh, *IEEE Trans. Appl. Supercond.* **24**(3), 5200504 (2014).
17. T. Nakamura, Y. Itoh, M. Yoshikawa, T. Nishimura, T. Ogasa, N. Amemiya, Y. Ohashi, S. Fukui, and M. Furuse, *IEEE Trans. Appl. Supercond.* **25**(3), 5202304 (2015).
18. M. Furuse, M. Yoshikawa, Y. Itoh, S. Fukui, and T. Nakamura, *IEEE Trans. Appl. Supercond.* **25**(3), 5200504 (2015).
19. T. Konishi, T. Nakamura, T. Nishimura, and N. Amemiya, *IEEE Trans. Appl. Supercond.* **21**(3), 1123 (2011).
20. T. Konishi, T. Nakamura, T. Nishimura, and N. Amemiya, *IEEJ Trans. EIS* **131**(6), 493 (2011) (in Japanese).
21. T. Konishi, T. Nakamura, and N. Amemiya, *IEEJ Trans. EIS* **132**(5), 398 (2012) (in Japanese).
22. K. Kajikawa and T. Nakamura, *IEEE Trans. Appl. Supercond.* **19**(3), 1669 (2009).
23. K. Kajikawa, R. Osaka, T. Nakamura, M. Sugano, and T. Wakuda, *J. Supercond. Novel Magnetism* **24**, 987–991 (2011).
24. T. Nakamura, Y. Yamada, H. Nishio, K. Kajikawa, M. Sugano, N. Amemiya, T. Wakuda, M. Takahashi, and M. Okada, *Supercond. Sci. Technol.* **25**, 014004 (2012).
25. T. Nakamura, H. Nishio, N. Amemiya, K. Kajikawa, and T. Wakuda, *IEEE Trans. Appl. Supercond.* **22**(3), 5200404 (2012).
26. T. Nakamura, K. Kitano, and N. Amemiya, to be published.

Chapter 3.22

Development of High-Temperature Superconducting DC Motor for Automobiles

Hitoshi Oyama and Tsuyoshi Shinzato
Sumitomo Electric Industries, Ltd.
Automotive Technology R&D Laboratories
1-1-3, Shimaya, Konohana-ku Osaka, Japan

In recent years, electrification of automobiles is in progress. Following the advent of passenger electric vehicles, large size commercial vehicles with electric drive are also being developed. One of the problems in the development of large electric vehicles is the heavy weight which leads to short driving distances. Energy saving by the use of high-efficiency motors will be a solution. The authors have developed a prototype electric vehicle equipped with a high-temperature superconducting (HTS) motor and a refrigerator. The test results showed that the motor has torque of 136 Nm and an output of 30 kW, and the prototype vehicle obtains the maximum speed of 80 km/h.

1. Introduction

In recent years, electrification of automobiles is in progress. As for passenger vehicles, hybrid electric vehicles have been popular, and some mass-production models of pure electric vehicles have appeared for consumer use. As for large size commercial vehicles, a number of experimental programs that use electric buses as public transportation have been done in some cities in Japan since 2010. In the near future, electric buses are to be utilized in cities that are environmental conscious and require new transportation systems based on the smart-city concept.

The electric bus market is expected to grow.

One of the problems in the electrification of large size vehicles is the heavy weight of a vehicle and the load or passengers that the vehicle carries. In many cases, the space given for batteries is limited not to suppress the load space, thus the capacity of batteries eliminates driving ranges. Reduction in the energy consumption of batteries by the use of high-efficiency motors will be a good solution. In general, the permanent magnet synchronous motor (PMSM) is most efficient, and thus, passenger electric vehicles in the market commonly use the PMSM. Large size vehicles have also started using PMSMs or induction motors (IMs) on an experimental basis. The PMSMs and IMs have employed advanced technologies and it is not easy to increase the motor efficiency any further.

As an innovative material for electric power applications, research is in progress on high-temperature superconducting (HTS) wire. HTS wire, whose electrical resistance is zero at liquid nitrogen (LN_2) temperature, has shown dramatic improvement in performance and has been experimentally applied to electric power cable, ship propulsion, and other applications.[1–3]

In 2007 and 2008, the authors developed motors using Bi-2223 superconducting wire and mounted it on electric vehicles to investigate the potentials and challenges of applying superconducting wire to automobiles.[4] Conventional superconducting wires were simply cooled by LN_2, however, the current prototype has an on-vehicle refrigerator for investigation of a cooling system.

2. Environmental-Friendly Vehicle and Application of Superconducting Motor

To cope with the depletion of fossil fuels and global warming caused by CO_2 emissions and to realize sustainable motorization, significant improvements in energy efficiency and shift to petroleum alternative fuels are required. There are mainly three types of environmental friendly vehicles under development by automakers and research institutes: Hybrid electric vehicles (HEVs), electric vehicles (EVs), and fuel cell vehicles (FCVs). HEVs have been mass-produced since 1997, spreading in the market much faster than EVs and FCVs. As HEVs use both a combination of a gasoline engine and an electric motor, the existing infrastructure for gasoline-driven vehicles can be utilized. EVs and FCVs, on the other hand, produce zero emissions during operation, making them the ultimate eco-cars. What these three types of environmental-friendly vehicles have in common is the use of electricity to deliver power from a motor. Realizing a high efficiency, high performance motor is the key to the development of electric vehicle systems. Thanks to recent improvements in the performance of HTS wire and in the adiabatic cooling technology, the feasibility of more efficient electric vehicle systems by applying superconducting motors is being actively investigated.

Figure 1 shows the features of a superconducting motor coil. A superconducting coil provides a high magnetic flux density, and therefore, delivers much higher torque than ordinary motors. Furthermore, a superconducting motor can be used without copper loss, and an air-core superconducting motor may be developed in the future to reduce iron loss and increase motor efficiency. Automobile motors need to respond to a wide range of operations from low speed driving to high speed driving and from low torque output for constant-speed cruising to high torque output for acceleration. While ordinary motors exhibit considerable copper loss and poor efficiency during high torque output, superconducting motors provide high

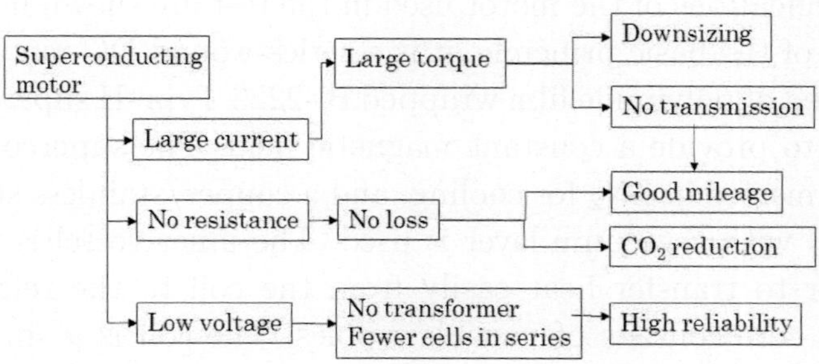

Fig. 1. Features of superconducting motor.

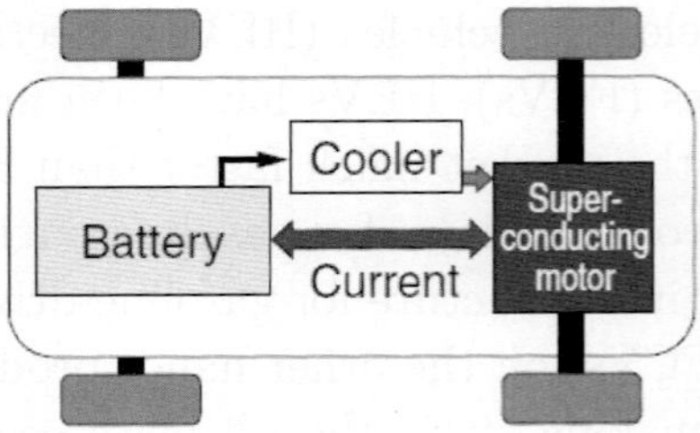

Fig. 2. System configuration with on vehicle refrigerator.

efficiency over a wide range. The high torque feature leads to the development of a system in which the motor directly drives a shaft without variable-speed gears, and thus, results in the reduction of transmission loss caused by the gear system. The major challenge in applying a superconducting motor to an automobile is that the superconducting coil must be kept at a low temperature, below the critical temperature of the superconducting material. Cryogenic coolants, such as LN_2, are suitable for keeping materials in a superconducting condition, allowing simple motor designs with vessels. But in a long operation without the refilling of coolant, an on-vehicle refrigerating unit is required to keep the superconducting motor at a low temperature while the vehicle is running.

3. Prototype Superconductive Motor

3.1. *Principle and specifications of the motor*

The specifications of the motor used in the test are shown in Table 1. In terms of the basic principle, it is a series-wound DC motor with a coil formed of polyimide film wrapped Bi-2223 Type-H superconducting wire to provide a constant magnetic field. The superconducting coil is immersed in LN_2 for cooling, and a copper/stainless steel cooling vessel with a vacuum layer is used. The inner vessel is made up of copper to transfer heat easily from the coil to the refrigerator. The iron core consists of four claw poles. The coil is a simple pancake coil formed of 186 turns of superconducting tape. The claw pole design was adopted to make the coil larger in winding radius and

Table 1. Specifications.

	Wire type	Type H (polyimide film insulated)
Wire	Dimensions	4.2 × 0.22 mm
	Critical current (Ic)	140 A
	Maximum tensile strength (77K)	150 MPa
	Minimum bend radius (room temperature)	70 mm
Coil	Shape	Inner dia.: 216 mm, Outer dis.: 241 mm, Width: 35 mm
	Turns	186 turns/coil
	Motor type	Series-wound DC (field superconductor)
	Coil refrigerator method	Liquid nitrogen immersion, with on-vehicle refrigerator
Motor	Maximum voltage	144 V
	Maximum current	500 A
	Dimensions	3.7 (dia.) × 400 (axis len.)
	Weight	About 110 kg

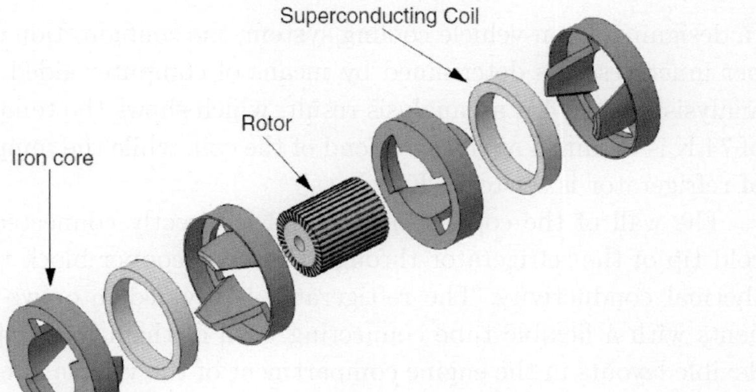

Fig. 3. Components of superconducting motor

thus reduce the number of coils, and to simplify the design of the cooling apparatus and other elements. Because the motor has a long shaft, the coil is divided into two parts as shown in Fig. 3. The same armature rotor as one used in a commercial DC motor is used.

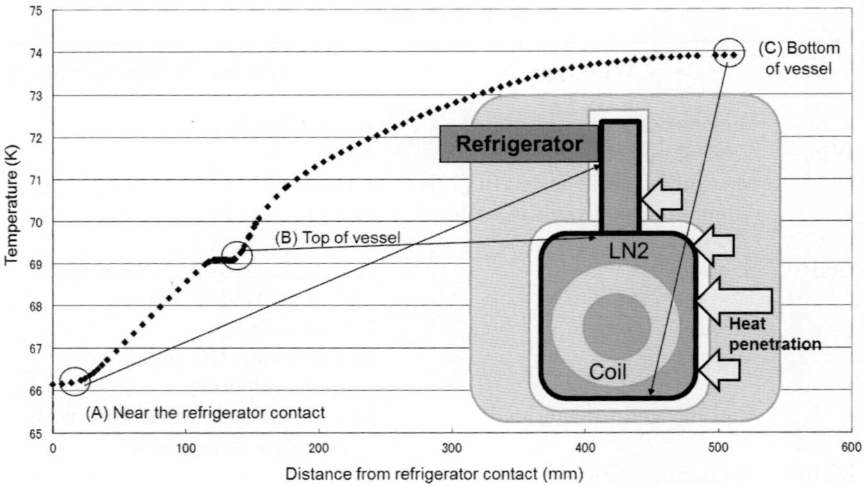

Fig. 4. Temperature of copper inner vessel.

3.2. *Design*

In designing an on-vehicle cooling system, the configuration of a copper inner vessel is determined by means of computer-aided thermal analysis. Figure 4 is an analysis result, which shows the temperature of 74 K is obtained at the lower end of the coil, while the temperature of refrigerator is set to 65 K.

The wall of the copper inner vessel is directly connected to the cold tip of the refrigerator through the thick copper block to attain thermal conductivity. The refrigerator is divided into two components with a flexible tube connecting both of them, which provides flexible layouts in the engine compartment of the vehicle. The height of the refrigerator is designed as short as possible to avoid the interference to the hood. The inertance tube of the refrigerator is directed upward and bent downward to set the large compliance tank in the lower place.

The superconducting coil is immersed in LN_2 in the thermal insulation vessel. LN_2 reservoirs are placed on the top of the motor. In operation, LN_2 is provided first for pre-cooling, and then the refrigerator starts to cool down LN_2. When the temperature of LN_2 is cooled below the boiling point, the evaporation of LN_2 stops. Therefore, LN_2 does not need to be refilled during the refrigerator operation.

4. Prototype Vehicle with Superconducting Motor

The authors developed two prototype superconducting electric vehicles in 2007 and 2008. In the present study, a superconducting motor with a refrigerator is mounted on a converted vehicle (a Toyota Crown Athlete) to verify its feasibility in a vehicle environment. The drive system of the vehicle is diagrammed in Fig. 5. The power source of the motor is twelve 12-V lead-acid batteries connected in series (144 V). The rotation angle on the accelerator pedal is converted to electric resistance with a potentiometer, and a commercial current controller is used for supplying the corresponding amounts of current to the motor. Torque from the motor is conveyed to the wheels as drive force by the original vehicle's FR transmission. Batteries are charged by the contact-type method, using a cord running from an

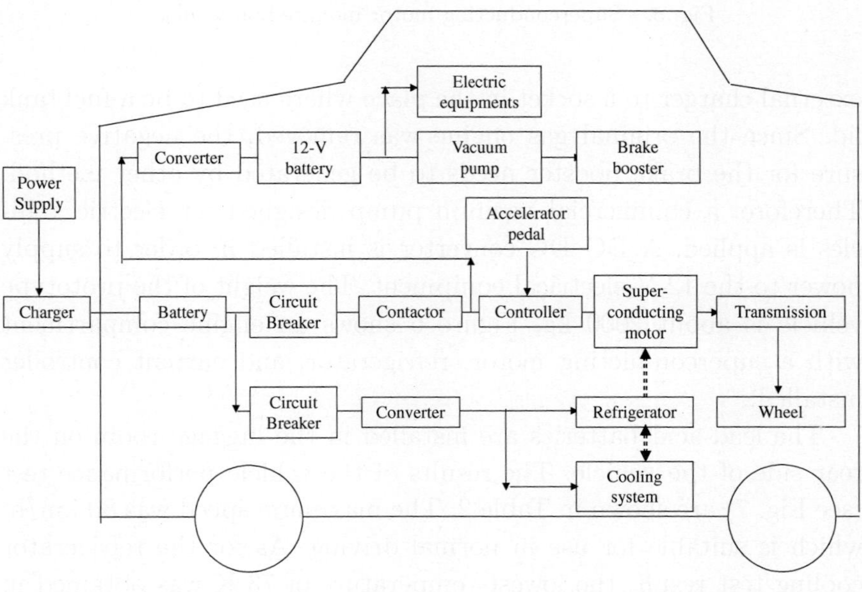

Fig. 5. Configuration of superconducting vehicle.

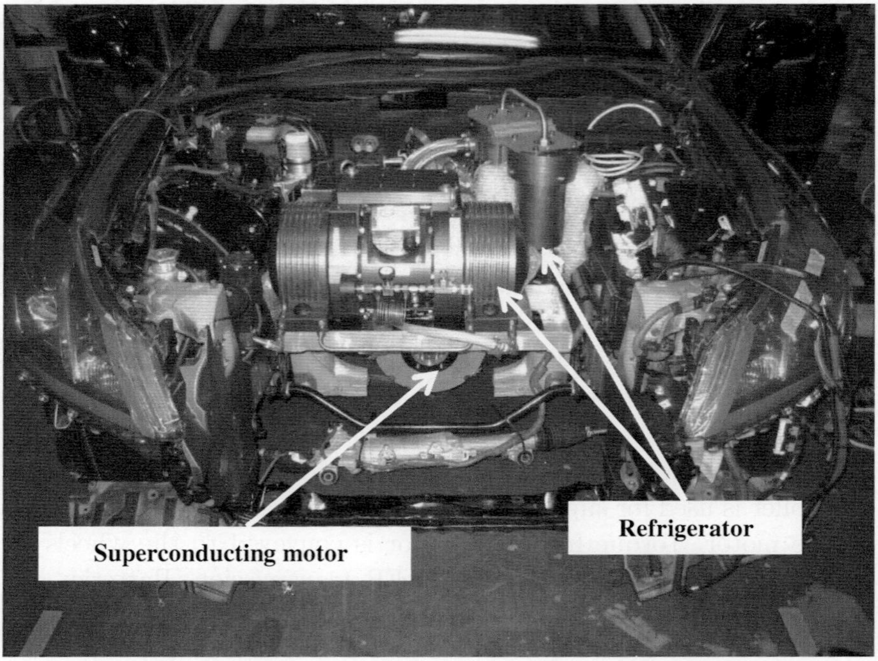

Fig. 6. Superconducting motor mounted on vehicle.

external charger to a socket in the place where used to be a fuel tank lid. Since the original gas engine was removed, the negative pressure for the brake booster needs to be generated by other method. Therefore, a commercial vacuum pump designed for electric vehicles is applied. A DC–DC converter is installed in order to supply power to the 12-V electrical equipment. The weight of the prototype vehicle is about 2000 kg. Figure 6 shows an engine compartment with a superconducting motor, refrigerator, and current controller installed.

The lead-acid batteries are installed in the luggage room on the rear side of the vehicle. The results of the vehicle performance test (see Fig. 7) are shown in Table 2. The maximum speed was 80 km/h, which is suitable for use in normal driving. As for the refrigerator cooling test result, the lowest temperature of 73 K was obtained in the superconducting coil, even though in a transition condition.

Fig. 7. Test drive.

Table 2. Vehicle performance.

	Items	Performance
Vehicle performance	Maximum torque (1,540 rpm)	136 Nm
	Maximum output (2,200 rpm)	30 kW
	Maximum speed (4th gear)	Abbout 80 km/hr

5. Conclusion

The authors have developed a prototype passenger EV which consists of a motor using HTS wire and devices to supply power to the motor. In the driving test held from December 2011–May 2012, the vehicle operated smoothly without problems. As for the on-vehicle cooling system, components, such as a refrigerator, power supplier, water cooling pump and radiator, operated properly to keep the temperature under 77 K, the boiling point of LN_2. The authors will continue compiling test-driving records to verify the reliability of the superconducting motor. Furthermore, the authors will move forward with research on further advanced superconducting motor systems, with the aim of developing superconductor-applied products with excellent commercial advantages. A superconducting motor system for electric buses is one example. Since the operation rate of the bus motor system is higher than that of the passenger vehicle's motor system, energy required for cooling the motor in the out-of-service state is relatively small, and thus the total energy efficiency of the system increases. Also, in a heavy vehicle that requires a high output, the output of the cooling mechanism has less impact on the increase of the motor efficiency and the decrease of transmission loss. Furthermore, as heavy vehicles require high output for acceleration and deceleration regeneration, a superconducting motor would be suitable for buses and other mass-transit vehicles that are subjected to frequent stop-and-go operations.

References

1. T. Okazaki, *SEI Tech. Rev.*, (168), (March 2006).
2. H. Sugimoto *et al.*, *IEEE Trans. Appl. Supercond.* **17**(2), 1637 (2007).
3. H. Yumura, *SEI Tech. Rev.*, (170), 69 (January 2007).
4. H. Oyama, *SEI Tech. Rev.*, (67), 22 (October 2008).
5. T. Shinzato, *SEI Tech. Rev.*, (75), 62 (October 2012).